Springer Series in Measurement Science and Technology

Volume 2

For further volumes:
http://www.springer.com/series/13337

The Springer Series in Measurement Science and Technology comprehensively covers the science and technology of measurement, addressing all aspects of the subject from the fundamental physical principles through to the state-of-the-art in applied and industrial metrology. Volumes published in the series cover theoretical developments, experimental techniques and measurement best practice, devices and technology, data analysis, uncertainty, and standards, with application to physics, chemistry, materials science, engineering and the life sciences.

The series includes textbooks for advanced students and research monographs for established researchers needing to stay up to date with the latest developments in the field.

Markys G. Cain
Editor

Characterisation of Ferroelectric Bulk Materials and Thin Films

Springer

Piezo INSTITUTE
The European Institute of Piezoelectric Materials and Devices

Editor
Markys G. Cain
National Physical Laboratory
Teddington, Middlesex
UK

ISSN 2198-7807 ISSN 2198-7815 (electronic)
ISBN 978-94-017-7730-8 ISBN 978-1-4020-9311-1 (eBook)
DOI 10.1007/978-1-4020-9311-1
Springer Dordrecht Heidelberg New York London

Printed on acid-free paper

Springer is part of Springer Science+Business Media (www.springer.com)

I would like to acknowledge all those who helped in writing this book including all the book's co-authors and the Functional Materials research team at the UK's National Physical Laboratory. I am also indebted to the UK's National Measurement Office, the European Commission and UK industry that supported much of the long-term research and metrology described in this volume. Finally, I would like to acknowledge my publisher who has been incredibly patient whilst I complete the book, and to my family for their understanding and unconditional support!

Preface

Piezoelectric materials (and pyroelectrics and ferroelectrics) are used by industry in a very large range of applications. These include ultrasonic detectors, cleaners, imaging systems and sonar devices; ink-jet printer heads, diesel and gasoline fuel injectors for automobiles, trucks and vans; electronic memory devices such as Ferroelectric RAM; micro and nano positioning actuators, valves, motors and translators; RF filters, resonators and VCOs; among many other applications. There are many excellent books and reviews detailing the significance of this class of material—which can be polymer or ceramic; in bulk and thin film form; as 1D nano rods or wires to nanotubes and may also be processed as quantum dots. The properties of the materials have been shown, in some cases, to be dependent on their scale and dimension (as with many nanoscale material in fact) and perhaps, more importantly, the properties are highly nonlinear as regards to excitation voltage, mechanical stress and temperature. These factors make the development of measurement good practice and ultimately setting that down into documentary standards rather difficult.

In this book we are publishing, for the first time, a selection of NPL's Materials Measurement Good Practice guides and practically written reviews for the measurement of a large number of the important multifunctional parameters for a range of materials including ferroelectric bulk and thin films, piezoelectric actuators and sensors, pyroelectric devices and electrostrictive materials. We first set out reminding ourselves the nature of polarisation in bulk materials and how this property is practically measured. We highlight some of the measurement pitfalls and common sources of error when you carry out the ubiquitous PE-loop measurement! Next, we turn our attention to one of the standard measurement methods (available as an international standard in fact) for the evaluation of low field properties of piezoelectric ceramics, based on resonance analysis. A practical worked example takes the reader through the precise details of this test method and assists in the calculation of the materials properties. Another 'standard' test method, based on the original ideas set out by Don Berlincourt back in the 1950s is presented in our next chapter and measurement good practice has been developed alongside one of the manufacturers of the test equipment. The results of an inter laboratory round robin are included in this chapter to show how the test method has been improved as regards to accuracy of result. The pyroelectric properties of crystals (where a temperature change induces a polarisation in the material) are incredibly important for low light level thermal imaging cameras, motion detectors, people detectors and so on, and one of the leading experts in this field describes the details of how the tensor is measured accurately on a number of materials types in chapter. The use of interferometry to traceably measure the actuation displacement or strain of piezoelectric materials is next described using the double beam interferometric method. There are many additional details associated with the assessment of the material properties of thin films which are not dealt with in this chapter, but are the subject of current intense research effort worldwide. The thermal properties of piezoelectrics often dominate high power use—such as in high power sonar or ultrasonic welding, for example. The particular issues as regards assessing the evaluation of the materials polarisation or strain characteristics at high temperatures is a complicated measurement problem which is explored in some detail in this chapter. An extension to the problems associated with self-heating (when a piezo transducer is electrically 'over-driven') in such high power applications is the subject of the next chapter, and here we propose various ways in which the thermal properties of the transducer may be modelled using fairly simple methods. One of the more recent additions to the family of measurement methods for piezoelectrics (especially piezo thin films) is Piezoresponse Force Microscopy (PFM). Now, there are many excellent reviews published on this technique and indeed, the method is constantly evolving and new operational modes are being discovered every year. So, in this chapter we force ourselves to focus on the measurement apparatus and the principle mode of operation. Issues such as contact electromechanics and surface quality of the piezo film are evaluated and methods for quantifying the (to date) qualitative method are proposed. The mechanical properties of piezo thin films are of great importance in piezo-MEMS technology for example, and we devote a chapter to describe the operational principles behind an industry-standard mechanical indentation method for evaluating the elastic properties of piezo ceramics. The technique, on its own, is not sufficient to quantify all the elastic properties but, used with complementary methods such as SAW/ultrasonic propagation methods, it is a very useful tool that also allows for in situ electrical excitation of the piezo material. Finally, we spend some time discussing the measurement of dielectric breakdown in bulk piezoelectric ceramics based on the standards already developed for bulk electronic substrate dielectric materials. A final chapter on current standards (with links online) completes this volume.

The book is intended to occupy space in the research or technical lab where worked examples take the reader through some of the more tricky experimental methods.

Teddington, August 2013 Markys G. Cain

Contents

Electrical Measurement of Ferroelectric Properties

Mark Stewart, Markys G. Cain and Paul Weaver

1 Electrical Measurement of Ferroelectric Properties

Ferroelectric materials are defined by the existence of a finite polarisation at zero electric field, the direction of which can be switched by the application of an external field. Measurement of the electrical properties is therefore an important tool in identifying ferroelectricity and in characterising the ferroelectric properties of a material. In discussing the electrostatics of materials it is common to assume that a material becomes polarised only by the application of an electric field and that the polarisation returns to zero on removal of the field. This is not the case for ferroelectrics. This is illustrated by brief review of the electrostatics of a simple capacitor, followed by the application to the measurement of switchable polarisation in a ferroelectric and methods for the measurement thereof.

2 Electrostatics of Polar Materials

The following equations describe the macroscopic electrostatics:
Gauss's Law (integral form):

$$Q = \oint_S \mathbf{D} \cdot d\mathbf{S} \tag{1}$$

Gauss's Law (differential form):

$$\nabla \cdot \mathbf{D} = \rho \tag{2}$$

M. Stewart (✉) · M. G. Cain · P. Weaver
National Physical Laboratory, Hampton Road, Teddington, Middlesex TW11 0LW, UK
e-mail: mark.stewart@npl.co.uk

M. G. Cain (ed.), *Characterisation of Ferroelectric Bulk Materials and Thin Films*, Springer Series in Measurement Science and Technology 2,
DOI: 10.1007/978-1-4020-9311-1_1,

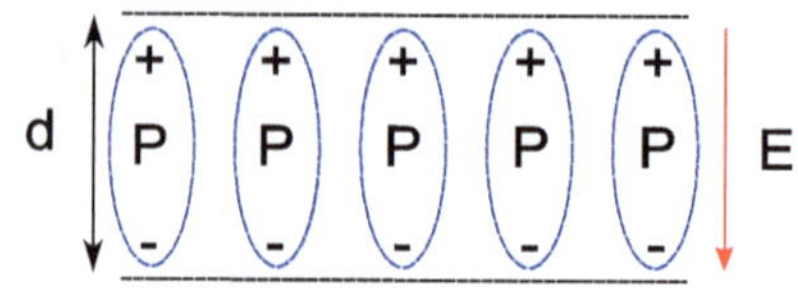

Fig. 1 Polar material—open circuit

Constitutive relation:

$$\mathbf{D} = \epsilon_0 \mathbf{E} + \mathbf{P} \tag{3}$$

If the system is isolated from external charge sources (open circuit) there are no free charges present. Figure 1 describes the fields associated with a polar material in isolation. It is assumed that the material has a permanent polarisation, P_0, in the absence of any applied field and, for simplicity, that this polarisation is not changed by the application of a field. A similar analysis to that performed above can be used to calculate the fields inside and outside of the capacitor. As there are no free charges present anywhere **D** is zero everywhere. Applying the constitutive relation (Eq. 3) to the material between the plates gives (omitting the vector notation because of the 1D nature of the problem):

$$D = 0 = \epsilon_0 E + P_0$$

$$E = -\frac{P_0}{\epsilon_0} \tag{4}$$

This equation describes a real macroscopic electric field within the material resulting in a potential difference between the plates:

$$\Delta V = -\int E \mathrm{d}z \tag{5}$$

$$\Delta V = \frac{P_0 d}{\epsilon_0} \tag{6}$$

This electric field opposes the polarisation direction. Consideration of some typical figures shows that this electric field can be quite large e.g. Barium Titanate has a spontaneous polarisation of 0.25 C m^{-2}, giving a figure of 28×10^9 V m^{-1}. This large field can drive movement of charges within and from outside the capacitor to reduce the field over time. The field can also drive depolarisation and domain formation. It is therefore commonly referred to as the depolarisation field.

If the polar capacitor described above is short-circuited by connecting the top and bottom electrodes together, charges will flow until the potential difference across the material becomes zero. This situation is described in Fig. 2.

The free charges on the electrodes create a displacement field $D_2 = \sigma$ and a consequent electric field in the material which opposes that from the polarisation. When the electrodes are short circuited there can be no potential difference between

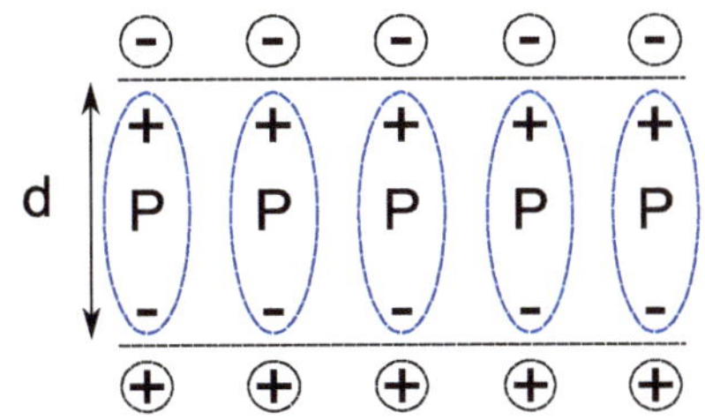

Fig. 2 Polar material—short circuit

the electrodes and therefore no net electric field (the field from the polarisation and that from the free charges on the electrodes cancel exactly) i.e. $E = 0$. Applying the constitutive relation (Eq. 3) now gives:

$$D_2 = \sigma = 0 + P$$

$$\sigma = P$$

3 Polarisation Loops

The above analysis represents a limiting case where the polarisation of the material is unchanged by external electric fields. In reality ferroelectrics are characterised by their ability to change their polarisation in response to an external field and to switch between two or more stable directions of the polarisation vector. Measurement of the variation of polarisation with electric field can therefore tell us much about the ferroelectric properties. The variation of polarisation with applied electric field is typically non-linear and hysteretic.

A typical ferroelectric P-E loop is shown in Fig. 3. The coercive field, E_c is the electric field at which the polarisation reaches zero. The polarisation at zero electric field is termed the remnant polarization, P_R. This is important for small field piezoelectric applications as a large and stable remnant polarisation will provide the maximum piezoelectric coefficient for a given material. P_{sat} is the saturation polarisation which is achieved at high electric fields when all the domains are aligned as closely as possible to the electric field axis. Ps (saturation) is still dependent on electric field as some polarisation is developed by further distortion of the crystal with increasing electric field even though the domain movement is saturated. P_S is termed the spontaneous polarisation. For a single crystal with the electric field applied along the polar axis this would be equal to the spontaneous polarisation of the crystal. In a ceramic it will be less than this due to the random orientation of the grains [2].

It should also be noted that the P-E loop and the parameters, particularly the coercive field, can vary significantly with frequency. All the ferroelectric properties are typically temperature and stress dependent (the spontaneous ferroelectric ordering

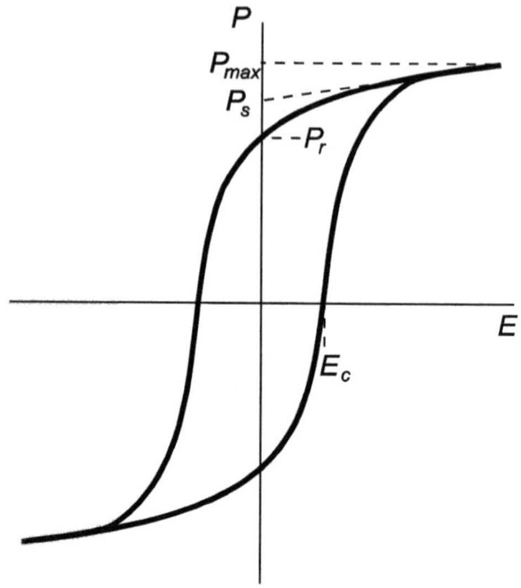

Fig. 3 P-E hysteresis loop parameters for a ferroelectric material [1]

vanishes above the Curie temperature). Microstructure, composition, history, defect structure, and substrate can all have an effect of the P-E loop [3]. This makes this basic electrical measurement a powerful tool in the investigation of the physics of ferroelectrics and the performance of ferroelectric devices. All crystalline ferroelectric materials are also piezoelectric, and electrically induced strain is very closely related to the polarisation, through the electrostriction coefficient [4]. P-E loop measurement can therefore provide important information on piezoelectric properties in situations where a mechanical strain measurement may not be possible or practical. P-E loops are obtained from measurements of charge supplied to a sample and the voltage across it (see Sect. 4). Charge–voltage schematics for some other common circuit elements are shown in Fig. 4. In some circumstances, such as a thin loop ferroelectric with low remanent polarisation, or a material with high leakage current, these may represent limiting cases. For an ideal linear capacitor the charge is proportional to the electric field voltage so the loop (Fig. 4a) is a straight line whose gradient is equal the capacitance of the material and is proportional to the permittivity. In this ideal case the charge, Q, is proportional to the displacement field D:

$$D = \frac{Q}{A} \tag{7}$$

where A is the electrode area. Only for high permittivity materials is $P \approx D$. It is common in the literature to refer to these loops as “P-E loops” but it as well to remember these limitations in the interpretation of these experiments.

For an ideal resistor the current is proportional to the voltage (Ohm’s Law). The amount of charge that flows is therefore dependent on time, and for a cyclical voltage there is a phase difference between the charge and the electric field. Because of this,

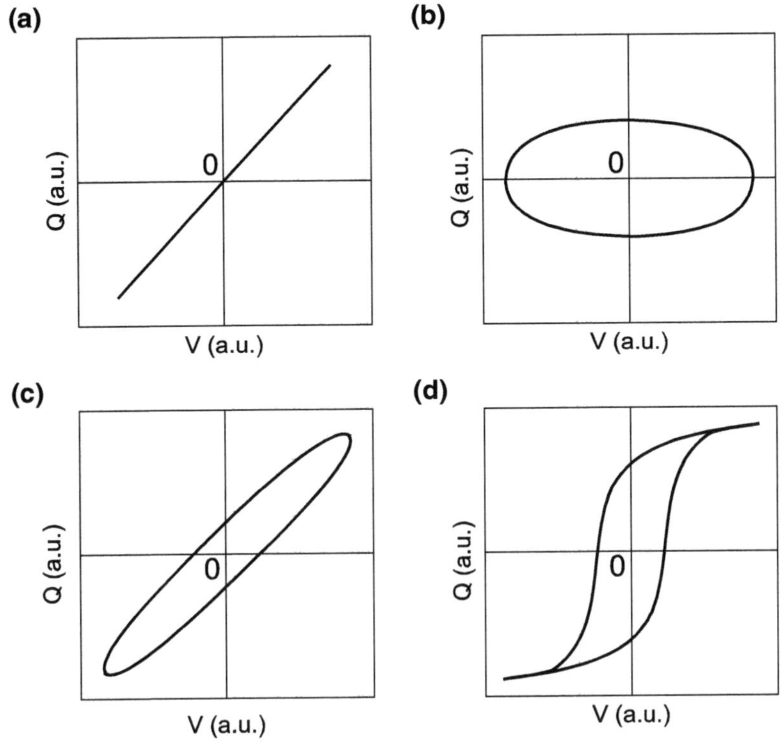

Fig. 4 Schematic charge voltage response (in arbitrary units) of **a** linear capacitor, **b** resistor, **c** lossy capacitor and **d** ferroelectric

the loop is a circle with the centre at the origin (Fig. 4b). In this case the measured charge relates to the current rather than the polarisation. The loop in Fig. 4c is a lossy capacitor, where the area within the loop is proportional to the loss tangent of the device, and the slope proportional to the capacitance. Loss can be due to dielectric hysteresis or leakage current or both. Figure 4d shows the loop for a true ferroelectric. These loops are usually centred around zero and for both the lossy capacitor and the ferroelectric they cross the y axis at a non-zero value. In the ferroelectric case this crossing point provided a measure of the remanent polarisation. For the lossy capacitor the non-zero crossing point does not indicate any remanence. Caution must be exercised in interpreting the crossing point on the charge (polarisation) axis as ferroelectric remanence, particularly where there may be leakage currents or the ferroelectric behaviour is not clearly established. There are many examples where lossy dielectric loops have been incorrectly presented as evidence of ferroelectric behaviour [1, 5].

The degree of non-linearity in the polarisation response is often an indicator of non-linearity in the piezoelectric response. The polarisation is related to the strain through the electrostriction coefficient [4]. At low fields the P-E loop will resemble that of Fig. 4c for a lossy capacitor, but the loops will begin to open out at much higher drive fields as saturation is neared (Fig. 4d). For achieving a controlled displacement with a piezoelectric actuator it may be better to confine the fields to values where the behaviour is less hysteretic. As small currents are simpler to measure than small displacements the P-E loop can can be used in a feedback loop to drive a piezoelectric actuator with a non-linear field to give a linear displacement [6].

The P-E loop can provide information on the capacitance and loss of a device at high fields and at different frequencies—information that is needed for tuning the drive electronics for piezoelectric applications and determining the self-heating effect of components. In some applications, such as thin film ferroelectric memories the hysteresis of the material is put to good use, and measurement of the P-E loop helps define the drive parameters and can be used to investigate the long and short term performance.

There is an implication from definitions of parameters such as P_s, and E_c that in hysteresis measurements the field must always be driven to saturation. However, we have seen that valuable information can be gained from measurements at fields well below this. Conversely, measurements need not stop at saturation, and the field can be increased until breakdown occurs. This gives the opportunity to study breakdown behaviour, and it may even be possible to determine a pre-breakdown characteristic.

4 Polarisation Measurement

The polarisation in a material cannot be directly measured by electrical means. However, we can measure the charge flowing through an external circuit. This forms the basis of the techniques for the measurement of polarisation in a material. In an ideal experiment all the measured current flows onto the electrodes of the ferroelectric capacitor and there is no charge transport through the material and no charge sources or sinks in the measurement circuit i.e.

$$\mathbf{D} = \int i dt \tag{8}$$

In this case the external movement of charge corresponds to a change in D in the ferroelectric. If we further assume that we can measure the voltage across the electrodes, and that the material is homogeneous throughout then we can also infer the electric field in the material. We can therefore calculate the polarisation from Eq. 3:

$$\mathbf{P} = \mathbf{D} - \epsilon_0 \mathbf{E} \tag{9}$$

For many ferroelectric materials $\mathbf{D} \gg \epsilon_0 \mathbf{E}$ so D and P are used interchangeably. In many situations this does not matter, but when dealing with relatively low polarisation materials, or around interfaces and boundaries it is important to make a clear distinction.

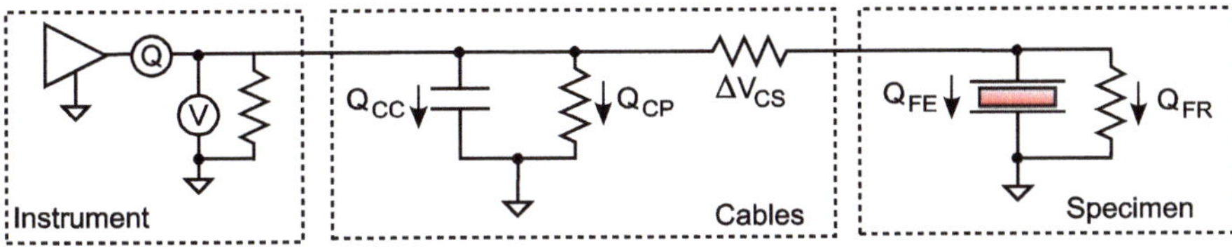

Fig. 5 Sources of P-E loop measurement errors

Fig. 6 Schematic of a Sawyer-Tower circuit for P-E loop measurements

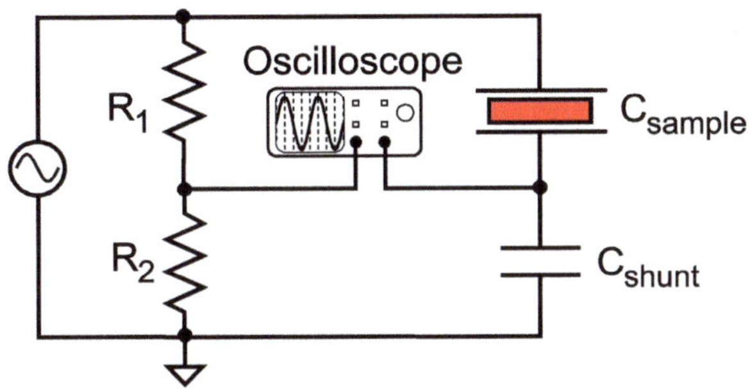

In practical laboratory measurements the idealisation that all charge flow relates to changes in D (and therefore P) is never realised. Figure 5 shows some of the sources of error that can arise in actual measurements.

There can be charge leakage through the impedance of the voltage measuring instrument, cable capacitance, and cable insulation. Because no material is a perfect insulator current will also flow through the ferroelectric sample. This is a particular problem at elevated temperature. These non-ideal current paths mean that Eq. 8 no longer holds and the experimental arrangement should be designed to minimise their effects.

4.1 Ferroelectric Measurement Techniques

The most often quoted method of hysteresis loop measurement is based on a paper by Sawyer and Tower [7] which included some seminal measurements on Rochelle salt. A schematic of the experimental setup using an oscilloscope is shown in Fig. 6. The voltage applied across the sample is attenuated by a resistive divider, R_1, R_2. The sample capacitances, C_{sample}, is connected in series with a shunt capacitor, C_{shunt}, so that the charge flowing onto both is the same. The oscilloscope is connected to measure the voltage across the shunt capacitor. A high quality low leakage shunt capacitor provides a stable and linear relationship between charge and voltage so that the measured voltage can be converted to a charge by multiplying by the capacitance. To a good approximation this will be equal to the charge on the sample. The shunt capacitor is usually chosen to have a capacitance much larger than that of the sample.

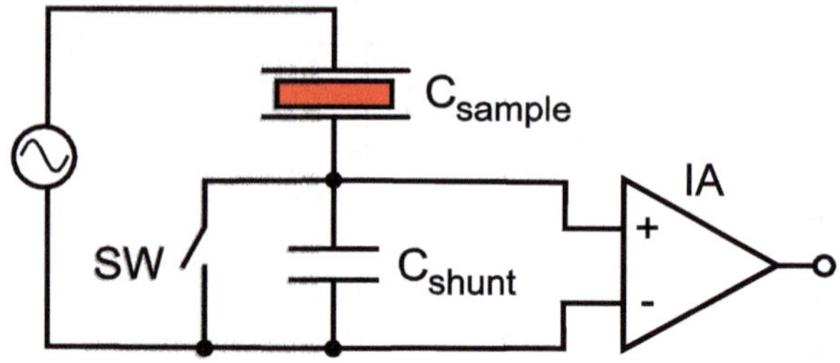

Fig. 7 Sawyer-Tower method implemented with an instrumentation amplifier

This is to ensure that the voltage across the shunt is much smaller than that across the sample. A large shunt voltage would reduce the voltage applied to the sample. In the original work the voltages from the resistive divider and the shunt capacitor were connected to the X and Y plates of a cathode ray tube oscilloscope to generate the P-E loop. With modern digitising oscilloscopes the two signals are recorded separately as functions of time and the P-E loop reconstructed digitally.

A variation of the Sawyer-Tower method, shown in Fig. 7 replaces the oscilloscope with an instrumentation amplifier for measuring the shunt capacitor voltage. This has the advantage of a much higher input impedance than an oscilloscope which reduces drift—particularly important for low frequency measurements. It is also much easier to interface with a PC controlled data acquisition system for automation of experiments. Drift still occurs primarily due to input bias currents of the instrumentation amplifier. If a low input bias current amplifier is used, then this drift will be quite small, but can cause problems at very low frequencies (<1 Hz) or where measurements are made over a period of time. The drift can be measured by disconnecting the sample and compensation made in the analysis. Eventually the drift will move beyond the measuring range of the amplifier, so the circuit shown in Fig. 7 is usually equipped with a reset switch to discharge the capacitor.

The sense capacitor for the Sawyer-Tower Method should be made from a low loss dielectric material such as a polypropylene, polyester or polyphenylene sulphide. This is particularly important for the lower frequency measurements, where leakage currents across the capacitor could degrade the results. The size of the capacitor depends on two factors, the capacitance of the device under test (DUT), and the expected polarisation of the DUT. An important consideration is that the voltage generated on the sense capacitor should be large enough to be easily measurable, usually greater than a few tens of millivolts, but not so large that it introduces a significant error on the voltage applied across the DUT. The polarisation is calculated simply as the product of the sense capacitor capacitance and the electrometer voltage reading, divided by the area of the sample electrodes.

Configuring the circuit as a charge amplifier has the advantage of eliminating the voltage divider effect of the shunt capacitor in the Sawyer-Tower method. A simplified charge amplifier circuit is shown in Fig. 8. The non-inverting input is connected to ground. The sample is connected to the inverting input of the amplifier and the charge is routed to a feedback capacitor. The amplifier adjusts the output to

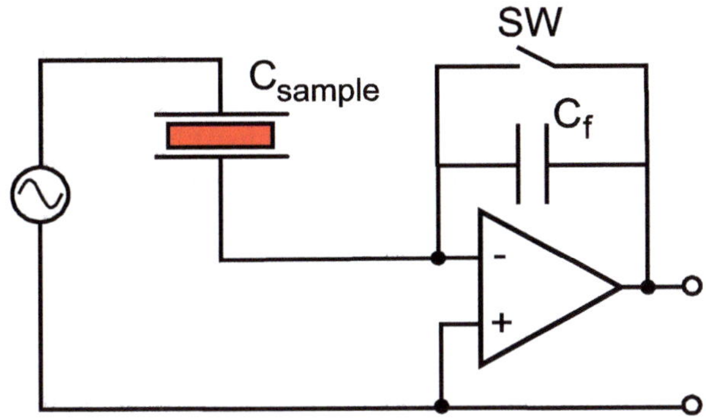

Fig. 8 Charge amplifier for measuring polarisation changes

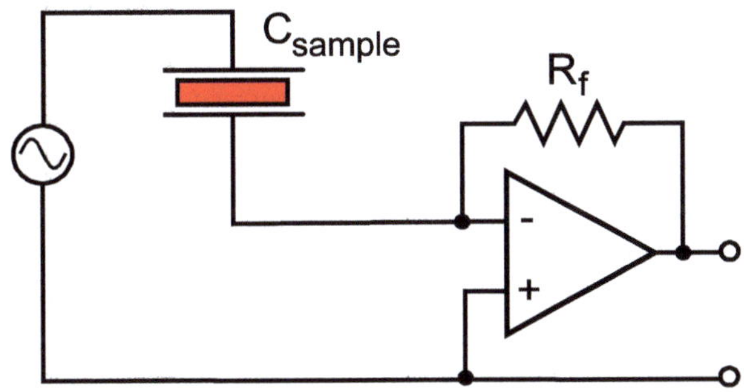

Fig. 9 Current to voltage converter

maintain the two inputs at the same voltage i.e. 0 V. The amplifier therefore presents a "virtual earth" to the sample, and the shunt voltage losses are eliminated. A potential disadvantage of this technique is that, if a commercial charge amplifier is used, the choice of feedback capacitor may be restricted and it will not be possible to place the feedback capacitor close to the sample—sometimes required to reduce noise. Similar considerations regarding drift also apply to this configuration. For low frequency measurements commercial electrometer instruments provide very low drift virtual earth charge measurements . The circuit shown in Fig. 8 has a switch to reset the amplifier to zero. This is to allow dc measurements to be made if it is required to record changes in polarisation over a long period of time. It is common practice in charge amplifier design to use a resistor in parallel with the feedback capacitor to obviate the need to reset. This introduces a low frequency roll-off in the amplifier response and no response at very low frequencies.

Another approach is to measure the current flowing through the sample. An electronic circuit implementation of this shown in Fig. 9. This circuit is a virtual earth current to voltage converter. It works in a similar way to the virtual earth charge amplifier—the op amp adjusts the output voltage to maintain zero volts at the inverting input—only this time the output voltage is proportional to the current through the feedback resistor R_f. Because of the high input impedance of the amplifier this is equal to the current flowing through the sample. Because it is possible to measure very small currents (fA range is readily achievable using commercial instruments) this technique is useful for low frequency measurements. To obtain the P-E loop, it is necessary to integrate the current into a charge, and this can be done digitally after recording the current signal. Care must be taken to control numerical errors in the integration (numerical drift). With the charge measuring techniques described above, all the charge from the sample is integrated onto the measuring capacitor, irrespective of the response of the amplifier. This is not the case with current measurement techniques. For example, a fast current pulse that is shorter than or comparable to the sampling interval, would not be accurately captured using current measurement.

At high gain noise generation across the feedback resistor can be significant. Although the integration to charge removes some of this, it is often found a capacitor is needed in the feedback loop to act as a filter to remove some of the high frequency noise. The value of this capacitor depends on the gain resistor. Usually the time constant of this filter $C_{st}R_{st}$ must be adjusted to at least two orders of magnitude smaller than the period of the applied voltage. Too large a value of C_{st} can lead to a spurious phase shift in the current waveform, leading to an apparently lossy P-E loop.

The IEEE 180 standard (now withdrawn) comes the closest to defining a standard procedure for making P-E loops but since it is only a definition of terms it is not very explicit. Most of the current ferroelectric workers describe their measurement setup for P-E loops simply as a modified Sawyer-Tower circuit and rarely detail the compensation methods if any are used.

4.1.1 Electric Field Application

The choice of high voltage amplifier must be made based on several factors:

- high voltage specification
- maximum drive frequency that can be maintained at the power amplification stated (thus the transfer characteristics are important)
- maximum current that can be delivered into a resistive and capacitive load
- amplification linearity and distortion
- input and output impedance
- circuitry protection
- cost.

Operational electric fields of ceramic piezoelectric materials are typically up to 2 kV mm^{-1}. Also, for measuring ferroelectric properties such as coercive field (an important for consideration for piezoelectric applications [8]) sufficient electric field to achieve a good level of saturation in both directions of electric field is required. Depending on the sample thickness, voltages in excess of several kV may be required. an amplifier gain of 100–1000 times is therefore needed to permit control by a function generator with an output voltage of order 10 V. Piezoelectric elements commonly have a capacitance in the 10–100's nF or even higher for multilayer devices. To drive this capacitive load at high frequencies would require a high power amplifier. For this reason, drive frequencies rarely exceed some 100's Hz. Care must be taken to ensure safe operation of the equipment, particularly when high voltages and high power are present. If a sample began to draw too much current, clipping the amplifier voltage, output would reduce. This could introduce unwanted harmonics, and is sometimes difficult to spot. It is therefore important to monitor the actual voltage applied to the sample. This can also highlight poor connections to the sample, as with high voltages it is possible to create a small air gap which quickly breaks down.

The operating frequency is usually much lower than the bandwidth of the amplifier. The limiting factor is usually the current limit, consequently there is a possibility of introducing high frequency noise from driving the saturated amplifier. In acoustic emission experiments on piezoelectrics this noise can be a problem as the high bandwidth detector can pass these frequencies on to the measurement system. This high frequency component is removed by adding a resistor in series with the sample (capacitor), with a value chosen such that the time constant is much longer than the driving frequency. For strain experiments the mechanical noise generated by the amplifier is less of a problem than the much lower frequency environmental mechanical vibration. Care must be taken to exclude the effect of this additional component in the measurements, i.e. the voltage drop is measured across the sample only.

4.1.2 Charge Measurement for P-E Loop Measurements

In all the virtual earth configurations, the current flowing through the sample must be absorbed by the operational amplifier as these are usually low power devices, and large current due to breakdown of the sample is likely to damage the measuring circuit. This is less of a concern for the Sawyer-Tower method which would therefore be preferred for high power measurements. To prevent damage to the measuring circuit in any case, a combination of back to back diodes and current limiting resistor may be required (as shown in Fig. 10). The critical characteristics of these diodes are they should have a rapid turn on response, be able to withstand the power dissipated under worst case fault condition (short circuited sample) and that the reverse current leakage should be as small as possible so as not to affect the measured current. It is always best to compare P-E curves with and without protection to establish the effect of the additional components.

A typical P-E loop, after the electric field has been cycled at least once, should be closed, that is the polarisation at the end of the cycle should equal the value at the beginning. An open loop can be a result of a real material effect, or due to instrumental factors, such as sense capacitor leakage, DC offsets on the applied voltage or electrometer drift. Resistive leakage from the sample will cause the voltage on the sense capacitor to continuously increase, and so it must be reset periodically by shorting the sense capacitor with the relay. This is also an important factor in

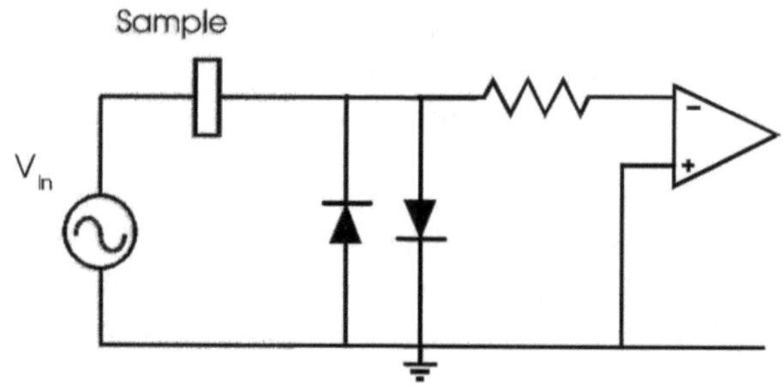

Fig. 10 Protection circuitry for current-to-voltage converter

keeping the sense capacitor voltage well below the Zener threshold of back-to-back diodes, which protect against sample breakdown or other short circuit conditions.

The measured P-E loop is dependent on the history of electrical loading on the sample. If the ferroelectric sample has been previously poled in the positive direction, the application of a negative field will cause domain switching, resulting in a large change in P and therefore a loop that is offset vertically from zero. To eliminate such effects, electrical conditioning of the DUT is required. The application of a cyclical electric field with amplitude decreasing over time should produce a gradual depoling of the materials, with a symmetrical polarisation response on the next measured cycle.

The selection of the best method for P-E loop measurement depends on the nature of the experiment being conducted, and the environment of the experimental setup. The next section illustrates this by means of a case study.

4.2 Case Study: In Situ Measurements of Electric Polarization Within a Synchrotron Diffraction Beamline

X-ray and neutron diffraction can provide valuable insight into structural changes induced by electric fields and the origin of piezoelectric properties at the atomic level. Correlation with macroscopic measurements of strain and polarisation can yield information on the role of crystallographic processes in the piezoelectric or dielectric response, and contributions from extrinsic processes such as domain wall movement. However, this is usually accomplished by means of separate experiments, sometimes on different samples. This is particularly problematic for materials such as relaxors where the history of poling conditions and electrical loading have a significant effect on phase composition and microstructure [9, 10]. For these reasons, comparisons between the crystallography and ferroelectric polarisation is best achieved by measurement of diffraction patterns performed simultaneously with macroscopic polarisation measurements on the same sample. This section describes the implementation of a PE measurement system in situ in the European Synchrotron Radiation Facility (E.S.R.F., Grenoble, France) facility [11].

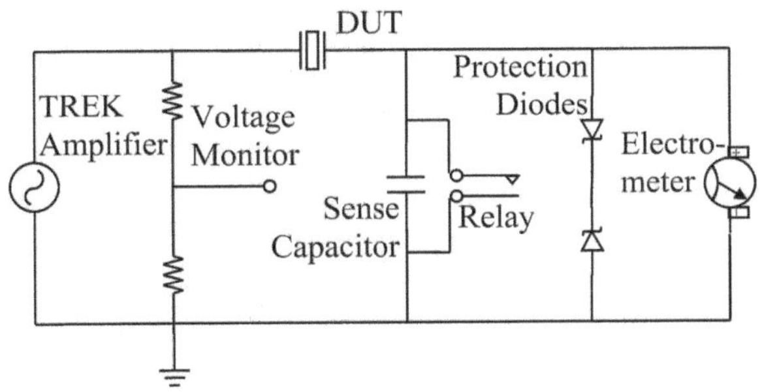

Fig. 11 Circuit diagram of the Sawyer-Tower measurement system implemented on the XMaS beamline. The device under test (DUT) was placed in series with a sense capacitor. The voltage across the capacitor was monitored with an electrometer, and compared to the voltage applied to the DUT via the monitor voltage output from the amplifier. Back to back diodes were placed across the electrometer input for protection in the event of a sample breakdown. The circuit was shorted via a relay at the start of each experiment

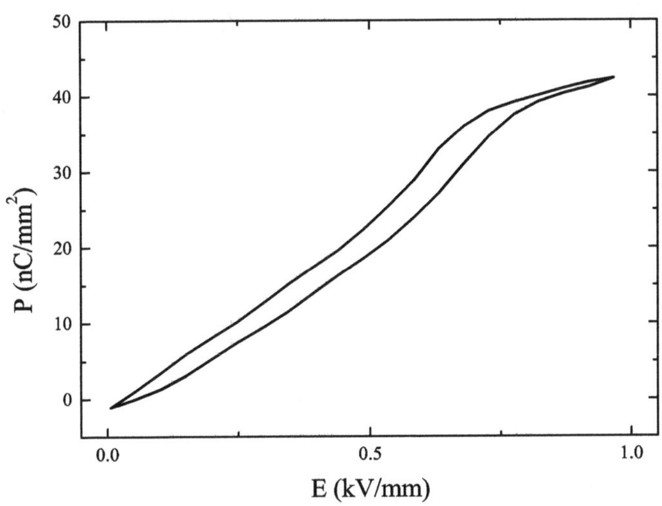

Fig. 12 A unipolar P-E loop measurement of the PMN-PT single crystal collected at 40 points around the electrical cycle at a frequency of 1 Hz and temperature of 25 °C. A change in slope, indicating a change in crystal structure is evident at an applied field of ~0.7 kV/mm

The P-E loop system installed on the beamline used the Sawyer-Tower measurement method based on the schematic shown in Fig. 6. the actual implementation on the beamline is shown in Fig. 11. There were two reasons for the selection of the Sawyer-Tower method in this case. Firstly, the environment surrounding the automated X-ray diffraction setup was electromagnetically noisy because of the many stepper motors used to control the system. This made detection of small currents very difficult, so it was more convenient to integrate the current on a sense capacitor, giving a degree of immunity to the electromagnetic noise. In this case, shielding of the sample from the noise by enclosing in a metallic Faraday cage was not possible because of the requirement to allow access to the sample for the incident X-ray beam.

A sample P-E loop measurement is plotted in Fig. 12. Anomalies such as inflection points and changes in gradient may indicate the presence of structural phase changes that can be determined through the analysis of the diffraction data. The evolution of such features can be tracked with changing physical parameters such as temperature, magnetic field and frequency of the applied electric field.

References

1. Jaffe, B., Cook, W.R., Jaffe, H.L.: Piezoelectric ceramics. ser. Non-metallic Solids. Academic Press, New York (1971)
2. Damjanovic, D.: Ferroelectric, dielectric and piezoelectric properties of ferroelectric thin films and ceramics. Rep. Prog. Phys. **61**, 1267 (1998)
3. Lupascu, D.C.: Fatigue in ferroelectric ceramics and related issues. ser. Physics and Astronomy Online Library. Springer, Berlin (2004)
4. Weaver, P.M., Cain, M.G., Stewart, M.: Temperature dependence of high field electromechanical coupling in ferroelectric ceramics. J. Phys. D: Appl. Phys. **43**(16), 165404 (2010)
5. Scott, J.F.: Ferroelectrics go bananas. J. Phys.: Condens. Matter **20**(2), 021001 (2007)
6. Weaver, P.M., Cain, M.G., Stewart, M.: Temperature dependence of strain-polarization coupling in ferroelectric ceramics. Appl. Phys. Lett. **96**(14), 142905–142905-3, (2010)
7. Sawyer, C.B., Tower, C.H.: Rochelle salt as a dielectric. Phys. Rev. **35**, 269 (1930)
8. Weaver, P.M.: A sensorless drive system for controlling temperature-dependent hysteresis in piezoelectric actuators. IEEE Trans. Ultrason. Ferroelectr. Freq. Control **58**(4), 704–710 (2011)
9. Chen, I.W., Li, P., Wang, Y.: Structural origin of relaxor perovskites. J. Phys. Chem. Solids **57**(10), 1525–1536 (1996)
10. Zhao, X., Dai, J.Y., Wang, J., Chan, H.L.W., Choy, C.L., Wan, X.M., Luo, H.S.: Domain structure and evolution in (PbMg[sub 13]Nb[sub 23]O[sub 3])[sub 0.75](PbTiO[sub 3])[sub 0.25] single crystal studied by temperature-dependent piezoresponse force microscopy. J. Appl. Phys. **97**(9), 094107 (2005)
11. Wooldridge, J., Ryding, S., Brown, S., Burnett, T.L., Cain, M.G., Cernik, R., Hino, R., Stewart, M., Thompson, P.: Simultaneous measurement of X-ray diffraction and ferroelectric polarization data as a function of applied electric field and frequency. J. Synchrotron Radiat. **19**(5), 710–716 (2012)

Piezoelectric Resonance

Markys G. Cain and Mark Stewart

1 Introduction

This chapter is intended to help a user perform resonance spectra measurements on piezoelectric ceramics. The phenomenon of resonance is introduced and linked to the evaluation of the piezoelectric matrix by appropriate choice of sample geometry, and data analysis method. The IEEE method and complex coefficient method is explained and compared with a worked example helping the user understand the stages of measurement and analysis.

1.1 Vibrations

Virtually every system, whether mechanical or electrical, possesses the capability for vibration , and most systems may vibrate freely in a large number of ways. Generally, most systems may freely vibrate at a frequency determined by its size—the smaller the size the faster the periodic motion such as the beating of the wings of a bee, conversely the larger the system the slower the periodic motion such as the ground tremors experienced after an earthquake. If the vibrations are approximated as sine waves (which is physically reasonable and mathematically agreeable) then these oscillations are often described to perform simple harmonic motion. Simple harmonic motion (SHM) may be characterised by an amplitude of motion (displacement for example) and a period (the time between successive maxima). SHM has meaning only when the true steady state of oscillation has been reached—that is long after any transient phenomena have passed. Apart from the system's size, its frequency of free vibrations is determined by other properties such as its stiffness or elastic

M. G. Cain (✉) · M. Stewart
National Physical Laboratory, Hampton Road, Teddington TW11 0LW, UK
e-mail: markys.cain@npl.co.uk

M. G. Cain (ed.), *Characterisation of Ferroelectric Bulk Materials and Thin Films*, Springer Series in Measurement Science and Technology 2,
DOI: 10.1007/978-1-4020-9311-1_2,

modulus. Compared to a compliant material, a very stiff material means that sound can travel much faster from one of its boundaries to another and this means that its free vibrational frequency is correspondingly higher. The resonant frequency is equal to the speed of sound in the material, C, divided by the acoustic length of the device, L. This is the definition of acoustic resonance frequency: $f = \frac{C}{2\pi L}$. At resonance, the standing wave set up in a sample is thus related to the speed of sound within that material. Since the speed of sound is related to the stiffness of the material it is possible to determine the material stiffness using this behaviour as a basis for a measurement technique. For example, by suitably holding a simple bar shaped sample a resonance can be excited by the application of a mechanical impulse (much like striking a xylophone), and the acoustic response can be measured to determine some of the materials elastic properties [1].

1.2 Resonance

When an external force drives a freely oscillating system, then at a certain driving frequency, the system displays a similar—but not identical—maximum in oscillatory amplitude. Resonance in any system can thus be defined as a state of ongoing oscillations that persists for a time much longer than the period for internal motion in that state. Additionally, a resonant system may be excited efficiently by a relatively weak driving force in a narrow band of frequencies close to its resonant frequency.

For piezoelectric materials, since the material can be excited electrically, applying an AC voltage across the device can induce resonance without the need for external mechanical stimulation. Electrically induced resonance occurs in a piezoelectric material because of the elecromechanical coupling that exists between applied field and induced strain, and that is defined by the complete set of piezoelectric equations [IEEE Standard on Piezoelectricity [2] and the more recent IEC standard [3]]. This strain oscillates at the same frequency as the imposed field that act to set up sound waves in the material. In a similar manner to that described above, a resonance is set up in the specimen when dimensions of the specimen match some integral number of sound wavelengths—depending on the mode of vibration. It is clear then that the electrical impedance of a piezoelectric material measured as a function of frequency exhibits peaks that correspond to electromechanical resonance in the specimen. The form of the response is described by two quantities—frequency, f, and its frequency width at half amplitude, δ, (=f/Q) with Q called the Quality factor—that characterise the distinctive properties of a driven system.

The sample response can be measured externally using acoustic or displacement measurements. However, since the current flowing through the device is proportional to the functional response, it is simpler to measure the electrical current. Therefore, the electromechanical resonance of a piezoelectric sample can be readily measured by frequency sweeping a fixed voltage across the sample and measuring the current, i.e. measuring the impedance. One disadvantage of this method is that it can take time to perform a frequency sweep, whereas impulse techniques naturally favour the fundamental and take little time to complete. However, there are a number of

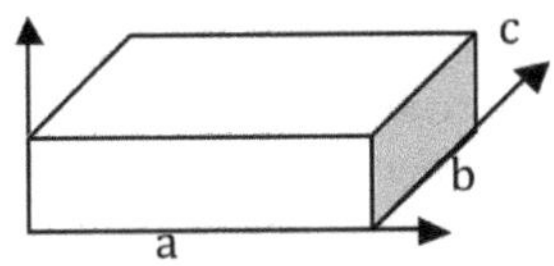

Fig. 1 Cuboid of length a, b, c

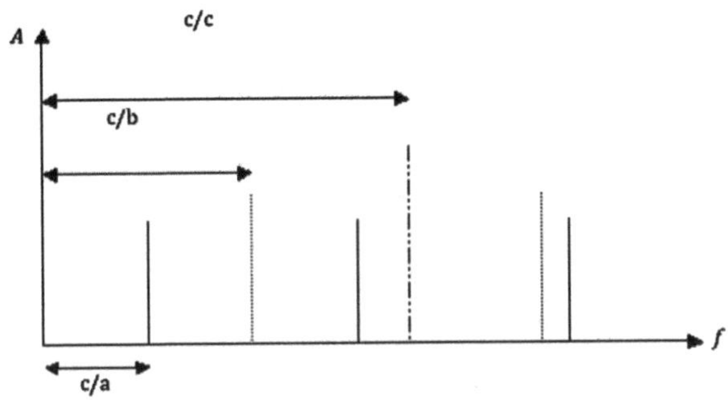

Fig. 2 Progression of higher harmonics from three modes of vibration in cuboid

high speed impedance analysers on the market that can perform a frequency sweep in a matter of seconds or less, so, although it is possible to use less sophisticated instrumentation, it is more convenient to use impedance analysers.

1.3 Modes of Vibration

Any real body will have a very large number of modes of vibration . It is important at this stage to understand the meaning of modes of vibration in real piezoelectric solids and the meaning of *harmonics* . A vibrational mode in piezoelectric materials depends on the sample geometry and the equations of motion governing the piezo-mechanical properties.

Some of these modes are harmonics of other modes, however, the lowest frequency mode will, by definition, be a fundamental mode of some sort. The next maxima may or may not be the first harmonic of this fundamental mode. It may actually be a fundamental mode of a different harmonic series. This is graphically depicted in Fig. 1. Consider a cuboid of length a, b, c.

The harmonic series defined by the lengths a, b, c will be, Fig. 2.

Here, the length b is such that the second mode is not the 1st harmonic of the fundamental. The modes of vibration are clearly related to the physical boundary conditions of the cuboid and the harmonic series associated with each mode simply fits the vibrational forms into the lengths a, b, c.

For example, for a long thin rod the fundamental mode is longitudinal, which is characterised by a fundamental frequency, f_0. The rod will also vibrate at harmonics of this fundamental frequency—$2f_0$, $4f_0$ etc, albeit at reduced intensity. The frequency f_0 is dependent on material parameters and the length of the rod. Of course the rod is free to vibrate in many other modes, although because its length dominates its dimensions their frequencies will be above the fundamental longitudinal mode. If the length of the rod is reduced then other modes of vibration with their own harmonic series—f_1, $2f_1$, $4f_1$, ...—will be superimposed upon the first series, $f_0, 2f_0, 4f_0 \ldots$ where f_1 is dependent upon other geometric and material parameters.

Of course, the frequency of vibration depends on the velocity of sound waves through the material, which in itself depends on material properties such as the stiffness tensor. The factors that affect the shape of the resonance spectra for piezoelectric materials include sample: dimensions, geometry, density, and its elastic, dielectric and piezoelectric constants. The power of this non-destructive technique now becomes clear. It is possible to calculate the various piezoelectric, dielectric and mechanical "constants" of piezoelectric materials by performing resonance measurements on a range of sample geometries. The reason why a range of geometries is necessary is because of the complex coupling that exists between the electrical and mechanical response of piezoelectric materials. When an arbitrary piezoelectric sample is made to resonate a complicated superposition of different resonant vibrations exists within the material. These resonances are made up from the various piezo-electro-mechanical couplings that occur between applied electrical field and boundary conditions arising from the samples fixed surfaces. If samples can be manufactured such that only one resonant vibration exists that can be directly linked to just one of the piezoelectric constants (remember this 'constant' is a 3rd rank tensor) then all the components of the piezoelectric tensor may be extracted from measurement data taken from resonance experiments, on a set of sample geometries. This forms the basis of the (now withdrawn) IEEE standard method [2], which is replaced with the IEC standard [3]. Here, the wave equations for vibration within resonating piezoelectric materials have been formulated based on well-defined boundary conditions of sample geometry. This allows the piezoelectric equations to be uncoupled to give an equation for the impedance of the specimen as a function of frequency—"resonance analysis". These equations have then been solved to yield some of the components of the piezoelectric matrix. Some of the geometries needed to determine the piezoelectric matrix are given in Fig. 3.

So far we have introduced the concept of resonance, mainly as a mechanical phenomenon, but the concept is also familiar in electrical circuits. If the impedance of a piezoelectric device is measured and the results plotted on the complex impedance plane they will describe a characteristic circle, Fig. 4. The three lines OA, OB, OC describe three pairs of frequencies that can be used to denote the resonance behaviour. OA is where the behaviour is purely resistive. OB the frequency of maximum resistance, and maximum conductance, and OC is the maximum and minimum impedance (Z_n and Z_m). Although the three lines have been drawn separately it can be seen that if the circle were symmetrical about the X-axis all three pairs would coincide. This is the case for a material with zero loss, and although strictly speaking most piezoelectric materials exhibit some loss it is assumed that the frequencies coincide for materials with a figure of merit, M greater than 20, Eq. (1):

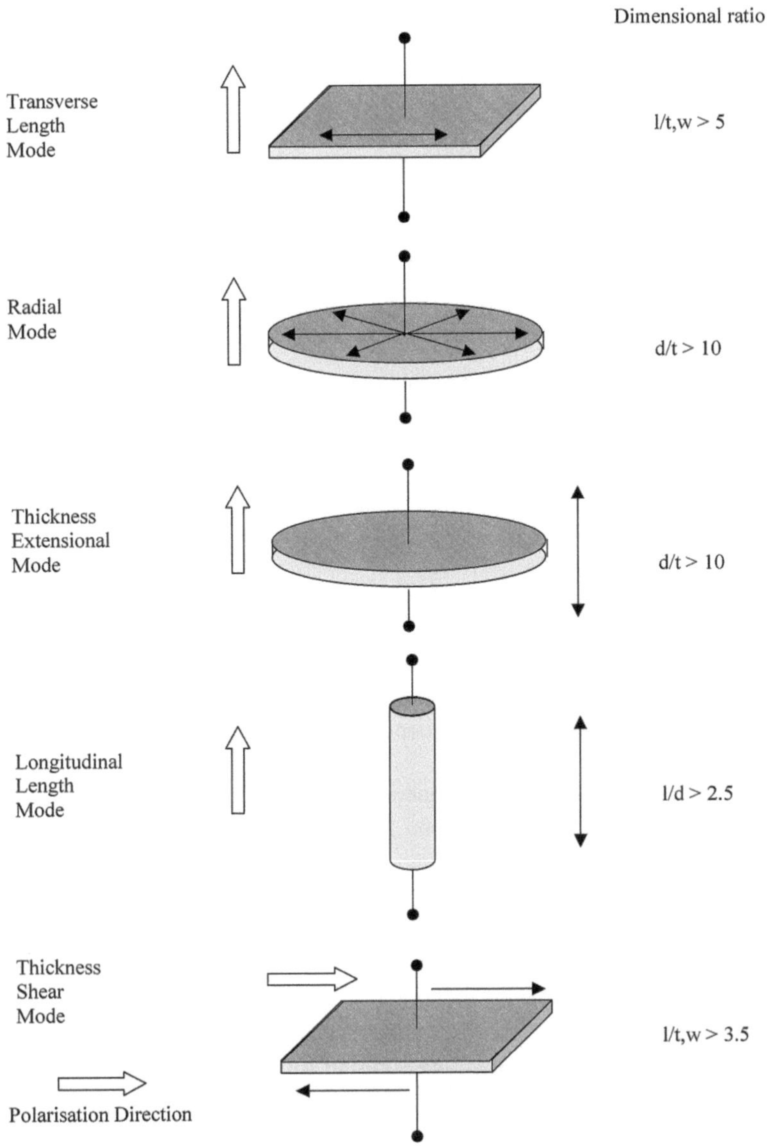

Fig. 3 The BSI [3] standard gives recommended sample geometries for exciting the various modes and these are shown. l, w are length and width, t thickness, and d diameter

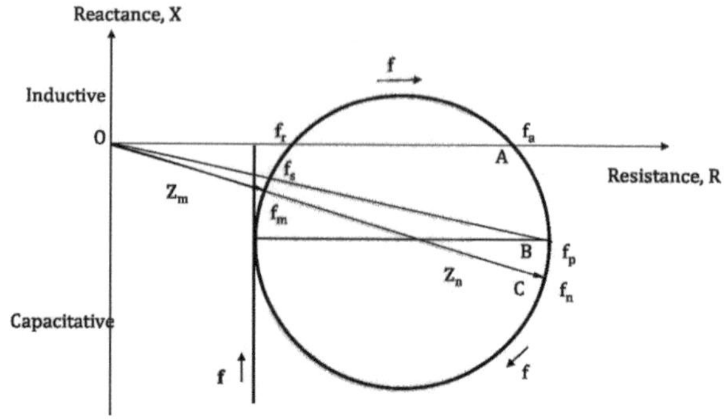

Fig. 4 Representation of the results on the complex impedance plane

$$M = \frac{Q_m k_{eff2}}{(1 - k_{eff2})} > 20 \tag{1}$$

k_{eff} effective electromechanical coupling factor
Q_m mechanical quality factor
f_r resonance frequency, when the susceptance is zero
f_a anti resonance frequency, when the reactance is zero
f_s series resonance frequency, frequency of maximum conductance, (Series arm has zero reactance X=0)
f_p parallel resonant frequency, frequency of maximum resistance
f_m frequency of minimum impedance f_n—frequency of maximum impedance

1.4 Equivalent Circuits

As discussed previously, purely electrical circuits also exhibit resonant behaviour. This concept can help in understanding some of the behaviour in piezoelectric resonators. An equivalent circuit that displays resonant behaviour similar to a piezoelectric is shown in Fig. 5. Here the left hand branch of the circuit represents the behaviour at resonance, and is sometimes termed the motional arm, the resistor determines the resistance at resonance and the capacitor and inductor control the resonant frequency. C_0 is the capacitance of the system without mechanical resonance, and the low frequency capacitance of the device is the sum of C_0 and C_1. In mechanical terms, L represents the vibrating mass, C_1 the elastic compliance, and R the mechanical damping, whilst the right hand arm C_0 is the pure electrical capacitance and has no mechanical significance. The use of equivalent circuits in describing the electrical properties of piezoelectrics or ferroelectrics is commonplace and is a useful tool for property prediction, assuming the materials response is linear. Of course, this

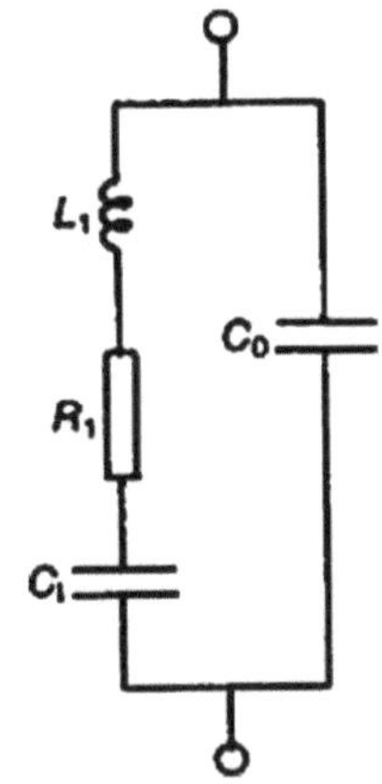

Fig. 5 Equivalent circuit of resonator

is not the case when fields and stresses are increased beyond quite modest values. Methods have been developed to tackle this problem and the work of Holland [4, 5], Smitts [6] and Sherrit [7] have utilised complex quantities for the circuit elements and compared these with the complex descriptions of the piezoelectric, dielectric and mechanical (stiffness) tensors. The complex notation for these tensor quantities has been implemented in order to include the magnitude and phase or loss associated with the driving field/force and response.

1.5 Resonance and Anti-resonance

On first inspection it seems a little odd that a piezoelectric resonating system can be described by two frequencies—resonance and anti-resonance. The differences may be easily explained by noting that at resonance the magnitude of the impedance passes through a minimum value and at anti-resonance the magnitude of the impedance passes through a maximum value. However, this does not explain why these changes occur at all. Remember, the electrical impedance is an indication of current flow or charge generation (since current is the flow of charge). In a piezoelectric material, charge is generated when the piezoelectric effect generates a strain in response to an applied stimulus (voltage). At resonance, the matching of sample dimensions to wavelength of strain amplitude occurs and there is a net charge transfer into and out of the sample during one complete cycle. This is due to the large strains occurring at resonance leading to a maximum change in capacitance of sample and hence charge flow. This amounts to an average current flow and an impedance minimum. At anti-resonance, when the impedance is maximum and hence minimal current is able to flow, the response to the driven field is such that during one complete cycle there is no or little net charge developed because of the symmetry of the field induced vibrations. The symmetry of the strains within the material may be seen as a mix of

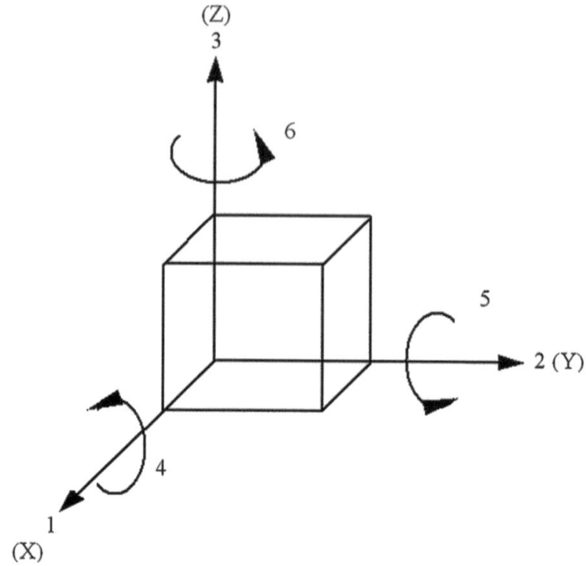

Fig. 6 Axis Notation for piezoelectric crystals

tensile and compressive regions within the sample. On average there is no net change in capacitance and hence minimal current flow [8].

1.6 Tensor Terminology

In order to describe the anisotropy of piezoelectric ceramics the axis notation in Fig. 6 is commonly used. For poled ceramics the material symmetry is ∞ mm and the poling axis is taken to be in the **Z** or 3 direction. [A full description of the tensor notation for piezoelectric materials is beyond the scope of this book, but readers are advised to read the excellent book of Nye [9] where the tensor properties are evaluated in long hand and then appropriately simplified—the starting point for our discussions in this chapter.] All the piezoelectric coefficients have two subscripts, the first identifies the direction of excitation, and the second the direction of the response. So, for example, the coefficient d_{31} in relation to Fig. 6 would refer to an excitation in the 3 (poled) direction, giving rise to a response in the 1 direction. A shear deformation around one of the three orthogonal axes would be represented by a subscript 4, 5 and 6. So the piezoelectric coefficient d_{15} represents a shear deformation around the 2 (Y) axis brought about by an excitation in the 1 direction (perpendicular to the poling direction). Since the 1 and 2 direction are equivalent all occurrences of the 2 axis can be replaced by the 1 axis, and due to the material symmetry conditions many of the coefficients for the ∞ mm case are zero, leading to the following:

$$\begin{pmatrix} \epsilon_{11} & 0 & 0 \\ 0 & \epsilon_{11} & 0 \\ 0 & 0 & \epsilon_{11} \end{pmatrix} \text{ The dielectric matrix}$$

$$\begin{pmatrix} 0 & 0 & d_{31} \\ 0 & 0 & d_{31} \\ 0 & 0 & d_{33} \\ 0 & d_{15} & 0 \\ d_{15} & 0 & 0 \\ 0 & 0 & 0 \end{pmatrix} \text{ The piezoelectric matrix}$$

$$\begin{pmatrix} c_{11} & c_{12} & c_{13} & 0 & 0 & 0 \\ c_{12} & c_{11} & c_{13} & 0 & 0 & 0 \\ c_{13} & c_{11} & c_{33} & 0 & 0 & 0 \\ 0 & 0 & 0 & c_{44} & 0 & 0 \\ 0 & 0 & 0 & 0 & c_{44} & 0 \\ 0 & 0 & 0 & 0 & 0 & \frac{c_{11}-c_{12}}{2} \end{pmatrix} \text{ The stiffness matrix}$$

2 Resonance Measurements

2.1 Impedance Measurement Systems

In making impedance measurements there are several possible terminal configurations but usually the choice comes down to either two or four terminal connections. Although four terminal is usually considered better there are a number of reasons why it is preferable to use two terminal connections when making impedance measurements on piezoelectric resonators. The main reason is the difficulty in making suitable contacts at the nodal point with more than two electrodes. However, having settled on two terminal measurements it is a mistake to carry this reduction in cabling back to the analyser itself, and the two terminal wire pair should be kept to a minimum, and should be shielded as far as possible. It is good practice to use four-wire connection from as close to the sample contacts as possible. Another common error is to assume that the shielding used in the four terminal cabling is actually at ground potential and to ground the shield at several points. Some impedance analysers use a 'driven shield' so that an equal current flows in the opposite direction to the measurement signal, in order to reduce inductive coupling. In this case the shields of the four terminals should only be connected together near the sample, and should not be connected to ground. It is also advisable to have a shield surrounding the sample that is also connected to the driven shield, although the necessity of its use depends greatly on the environment and the measurement frequency. However, it is possible to see measurable changes in spectra when merely moving a finger close to the sample, Fig. 7, when the sample is unshielded [10].

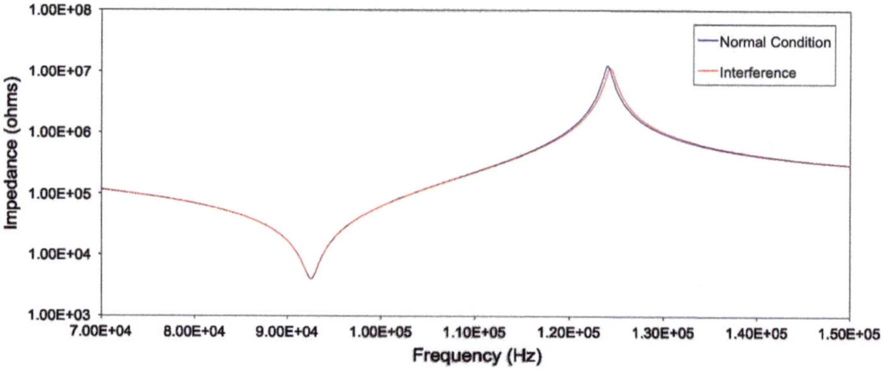

Fig. 7 Changes in spectra when moving a finger in close proximity to the sample, when the sample is unshielded

2.2 Compensation

There is a danger in using sophisticated analyser systems to trust the output implicitly, and perhaps neglect some of the subtleties of these systems. One of these forgotten areas is the compensation for fixtures and cabling. Most analysers come equipped with standard fixtures and fittings that are designed for using with standard electronic components, however these must often be modified for use with piezoelectric shapes or perhaps if used in conjunction with an environmental chamber. These modifications will often have an effect on the impedance measurements since the measured result is a combination of the impedance of the sample and the impedance of the cabling and fixtures. The analysers will often be compensated to account for the impedance of the "as supplied" fixtures and modifications may invalidate this compensation.

The solution is simple; the impedance of the fixture must be measured separately and subtracted from the measurement. Some systems can do this automatically or alternatively there is a relatively simple calculation, (2). To determine the impedance of the fixture and its cabling there are three measurements needed; open, short and load. The short measurement is simply the fixture shorted with a very low resistance artefact, typically gold plated brass. Care must be taken that this is clean and the contact to the fixture is also clean. The open measurement is simply the fixture in the open circuit condition, i.e. no sample in position. For all the compensation measurements it is preferable to have the set up as it is to be used. This means for the open circuit measurements that the contacts are as far apart as they are when the intended sample is in place. The final form of compensation is load compensation, and this involves testing a load of known impedance, preferably similar to the DUT (device under test), but often a simple 100-ohm resistor is used.

$$Z_{DUT} = (Z_{meas} - Z_s)/(1 - (Z_{meas} - Z_s)\, Y_0) \tag{2}$$

Z_s = impedance of short circuit, Y_0 = admittance of open circuit, Z_{meas} = measured impedance and Z_{DUT} = DUT impedance.

The importance of the compensation in the measurement results obviously depends on the impedance of the DUT. For a typical piezoelectric sample the lowest impedance (at resonance) is large enough so that inadequate short corrections do not introduce a large error. However, at anti-resonance the impedance can approach that of the open circuit and thus it is possible to introduce significant errors.

The analyser is an instrument for performing frequency sweeps and so the compensation must also be carried out at or near to the measurement frequency. The more advanced systems measure the compensation artefacts over the entire range and use this to correct the measurements, whereas some of the older systems only measure at a fixed frequency and use this value in the corrections.

2.3 Sample Holding and Contacts

Since we are using the resonance of a sample to determine material parameters, not of a device, an ideal sample holder should have no effect on resonance of the sample. Obviously, electrical connections need to be made, but these should be constructed so that there is a minimal influence on the resonance of interest.

It is difficult to define exact rules for how a sample should be held in terms of clamping load, position etc. since this can depend on the sample and the desired resonance. A simpler approach is to see the effect of changing these parameters on the resonance spectra. For a simple geometry sample with no defects the optimum holding conditions should produce the highest resonance, maximum admittance at resonance and no spurious peaks, and any excessive clamping will produce the converse.

The sample should be held at the nodal points, so for a thin disc resonating in the radial mode it should be held with a point contact at the centre of each face. Any offset from the centre will result in increased impedance at resonance, reduced resonance frequency and spurious peaks around between the resonance and anti-resonance, (Fig. 8).

In order to make simple, repeatable contacts to a sample there is a choice of either soldering wires to the sample, or using some sort of clamping arrangement to force electrodes onto the sample. Some workers have suggested using a mixture of both whereby a low melting point indium/gallium eutectic is used in conjunction with a very low clamping force. Soldering is somewhat invasive in that a heating cycle is needed and the solder presents added mass, although it can provide the lowest contact resistance. Also, for higher E-field experiments where a much higher displacement is produced, it may be the only suitable method since any spring loaded contact will either not be able to hold the sample or will dampen the vibration. Figure 9 shows the effect of increasing the solder mass on a 15 mm long 6.35 mm diameter length resonator. Obviously, the amount of solder in this case has been increased to highlight the effect, but it does illustrate the need to keep the solder mass to a minimum.

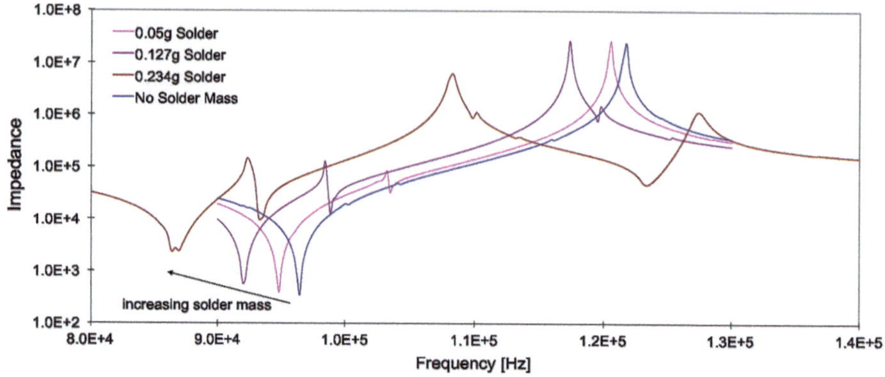

Fig. 8 The effect of increasing the solder mass on a 15 mm long 6.35 mm diameter length resonator

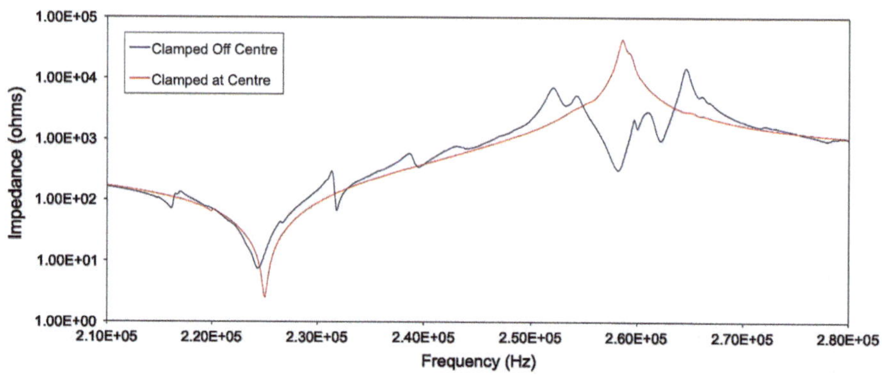

Fig. 9 The effect of holding the sample away from the nodal positions for a thin disc resonator

A method less invasive than soldering is to use spring-loaded contacts. These can either be simple wires bent to provide a small retaining force, or the spring loaded probes that are commonly used in the electronics industry for making test points on printed circuit boards [such as AuPt skt contacts supplied by RS Components, www.rswww.com]. These come in a variety of end shapes to cope with differing conditions, but the pointed types are better than the inverted cone or crown shape for this type of application. Also, as these probes are designed for contact they usually come gold plated to reduce contact resistance. The spring in these probes also have defined spring constants, although the tolerances on these tend to be quite high because of their small size. However, the spring constant can be measured for the individual spring used and a defined clamping load can then be set by the stroke of the spring. These spring probes are designed to apply around 2N, which is sufficient to punch through dirty contact on circuit boards, however this load may be somewhat high for the clamping of resonant piezoelectric devices. Applying a shorter stroke can reduce the load.

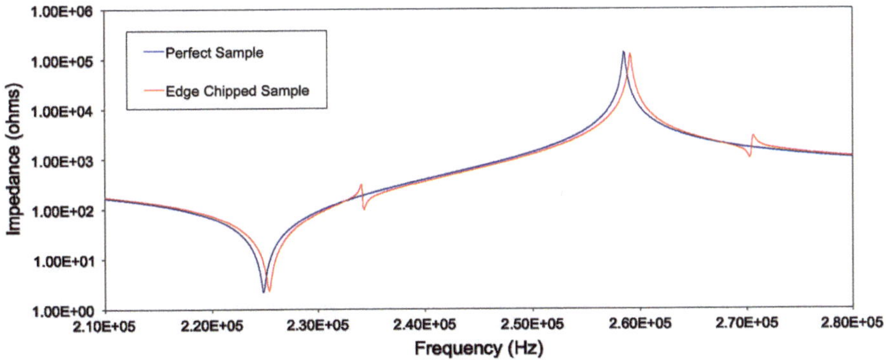

Fig. 10 The effect of edge chips on the radial resonance of a 10 mm diameter disc

2.4 Samples

The problems associated with adding mass to samples by soldering wires to them has been discussed in the previous section. Of course, the sample will not be purely piezoelectric ceramic since in order to make any kind of electrical connection, and to initially pole the material, the ceramic will have some form of electrode. These will introduce deleterious effects on the sample resonance, through the additional mass, the electrode stiffness, and if there is a substantial difference in the electrodes on either side then bending stresses will be introduced. The magnitude of this effect is not as great as for soldering, but could be a problem for very thin discs, where a fired on silver frit electrode could be a substantial fraction of the sample thickness.

In order to produce a clean resonance the sample should be geometrically perfect. Any imperfections in the sample will cause spurious peaks; in effect the sample dimensions have been changed in the area of the imperfection. Figure 10 shows the effect of edge chips on the radial resonance of a 10 mm diameter disc, with the magnitude of the chip shown in Fig. 11. Although the presence of these defects can be detected in the resonance spectra, it does not greatly affect the calculated material parameters since this mainly depends on the resonance and anti-resonance frequencies. However, less noticeable geometric irregularities could affect the results, for instance if the ends of a length resonator are not plane parallel, then this will give rise to a broadening in the resonance peak.

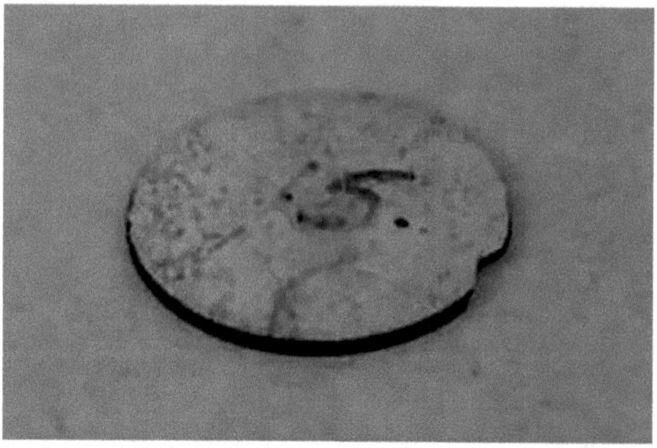

Fig. 11 The chipped disc tested in Fig. 10

2.5 Excitation Voltage

In most of the calculations to determine material parameters from resonance it is assumed that the behaviour is linear with respect to applied field. In most cases, this assumption is true since it is difficult to apply large enough fields at resonance because of the increased current requirements. For most of the commercial impedance analysers the maximum applied field is 1 volt rms and so maximum field levels for typical ceramic samples are restricted to around 1 V(rms)/mm. The US Navy standard (MIL-STD-13767) recommends a field level of less than 0.1 V/mm, and the CEN standard recommends an upper limit of 0.01 V/mm. In practice, it is usually better to use larger signals because of the increased signal to noise, and it is only when fields go above 1 V/mm that it is possible to detect non-linear behaviour in soft PZT [11]. [Although form this work just cited, it is noticeable, even with well aged samples, that there is a detectable variation of properties with electrical field even at the lowest fields used (0.1 $V.mm^{-1}$)]. Figure 12 shows the small change in radial resonance peak in going from 0.02 to 2 V/mm for a 0.5 mm thick soft PZT. The reduction in the magnitude and position of the resonance frequency while the anti-resonance remains unchanged is characteristic of high field non-linearity.

2.6 Modes of Resonance: Sample Geometry

In order to use the resonance behaviour as a technique for measuring material parameters it is important that the sample resonates in a well defined mode with minimal coupling from other unwanted modes. The IEC (BSI) standard [3] gives recommended sample geometries for exciting the various modes and these are shown in Fig. 3.

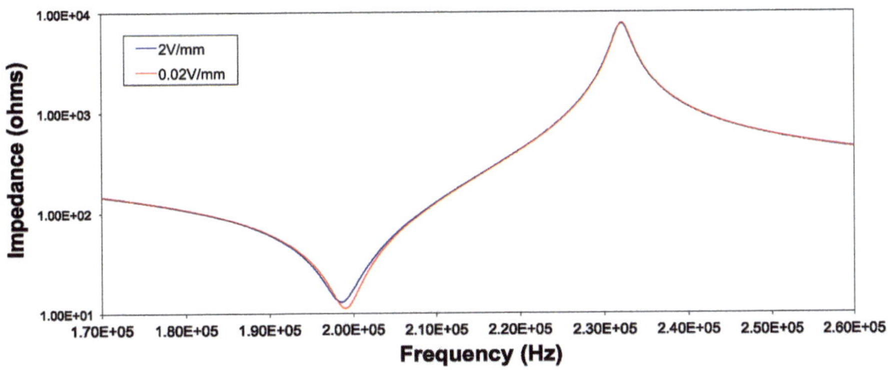

Fig. 12 Small change in radial resonance peak in going from 0.02 to 2 V/mm for a 0.5 mm thick soft PZT

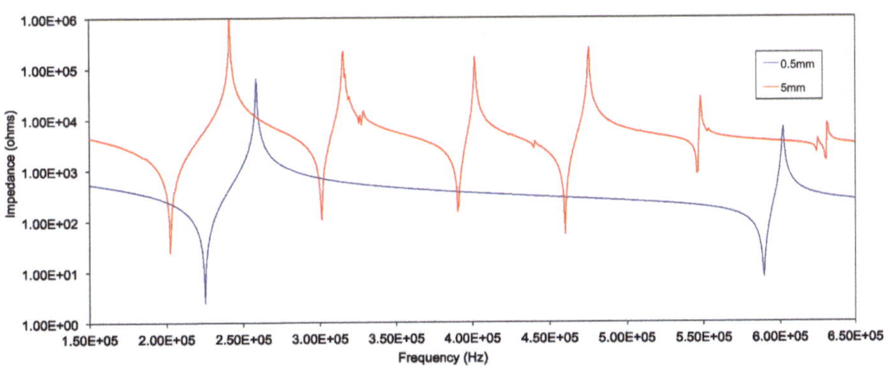

Fig. 13 Comparison of the recommended shape with a non-ideal resonator geometry

Figure 13 shows the consequence of using non-ideal geometry. Here the frequency range covers the first and second series radial resonance of a 10mm diameter disc of a PC 4D material. The 0.5 mm thick sample, which is the IEEE recommended maximum thickness for this diameter, shows only the two expected resonance peaks, whereas the 5 mm thick sample has many more peaks in this range. More importantly, these other modes have depressed the frequency of the first radial resonance. Obviously the 5 mm thick sample is much thicker than the recommended size, and for thinner samples of say 1 mm the multiple resonance peaks are not present. However, it is possible to detect the effect of other modes in the higher order harmonics, and in the second series radial resonance peak for these disc samples. Figure 14 shows the first and second series radial resonance for 0.5 and 1 mm thick 10 mm diameter samples of a hard and a soft composition. Examining the first radial resonance peak shows little difference between the different thickness samples, as any variation is within the variability from sample to sample. However, for the second series radial

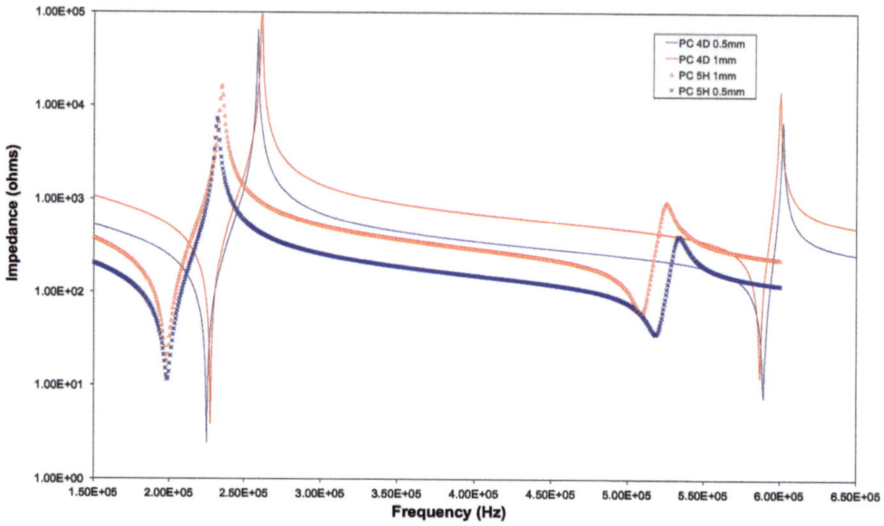

Fig. 14 The first and second series radial resonance for 0.5 and 1 mm thick 10 mm diameter samples of a hard and a soft composition

resonance there is a definite reduction in frequency for the thicker sample, in particular for the soft material. The difference between the first and second resonance peaks is controlled by the stiffness and the Poisson's ratio of the material. The difference in the shift of the second radial resonance with increasing thickness is due to the increased coupling coefficient of the soft PC 5H. This illustrates that the single mode resonance sample geometry restrictions are not only sample dimension dependent but also material dependant, i.e. that some of the geometry restrictions can be relaxed for some material compositions.

2.7 Repeatability

Table 1 gives some idea of the expected repeatability of resonance measurements. The first two rows are the results of 10 measurements on the same sample, where the sample was removed from the measurement fixture, turned through 180 degrees and replaced. The next two rows are for measurements on 10 different samples from the same batch. Clearly, if the measurement fixture is correctly set up and the samples are placed roughly centrally in the fixture the measurements are extremely *repeatable*, although no special precautions were taken to ensure that the sample sat in the same position each time. In comparison, the sample *variability* is over 100 times greater than the measurement *variability*, which is often expected in materials science studies of this nature.

Table 1 Repeatability of radial resonance measurements of a 10 mm diameter 0.5 mm thick PC 4D disc

		First resonance (Hz)	First antiresonance (Hz)	Second resonance (Hz)
10 Measurements	Mean	224854	258514	588975
	Standard deviation	7	7	33
10 Different samples	Mean	227207	258519	594025
	Standard deviation	1510	1321	3692

3 Conclusions

The phenomenon of resonance in piezoelectric crystals has been described. The various modes of vibration depend on sample and electrode geometry, drive frequency and other electro-mechanical constraints.

Materials properties such as elastic, dielectric and piezoelectric coefficients may be calculated from resonance measurements. The method by which the resonance scan is taken can affect the values obtained quite dramatically. If measurements are to be taken in the materials linear regime then excitation fields must not exceed 1 V/mm. Resonance is typically measured using Impedance Analysers. It is important to use quality sample fixings and jigs and cable, and jig compensation is required if accurate property values are required. Effects such as sample off axis location, excess solder mass contacts and sample imperfections have been investigated. All deviations from perfection degrade the signal quality and hence resultant property accuracy. An important issue is that of non-ideal sample geometry and the mixed mode coupling that results. Differences in resonance spectrum peak positions and shapes were more evident for soft materials than hard for small deviations from the IEEE recommendations. Finally, the variation in properties calculated for different samples was consistently higher than the experimental error in single measurements, when adequate precautions were taken in accordance with the recommendations set out in this chapter.

Conditions for materials characterisation using resonance

In order to make meaningful measurements using resonance analysis the following assumptions must be true for your material:

- samples are of geometry defined in IEC (and outdated IEEE) standard
- the material properties may be adequately described by the set of linear piezoelectric equations, (the behaviour is independent of applied field
- all materials properties are constant with frequency
- the materials are poled polycrystalline ceramics with cylindrical symmetry.

Following this chapter we present a laboratory worked example for the student or technician to follow.

Appendix A: Worked Example

The sample is a length mode resonator 15 mm long and 3 mm in dia0meter of a soft PC 5H composition, with silver frit type fired on electrodes. The sample diameter is measured at three points along the length, and the three length readings are made. The average readings are:

```
Diameter = 2.987 mm Length = 15.063 mm
```

The mass of the sample is needed for the density measurement, and although the electrodes are on the sample they are only a small fraction of the overall thickness in this length resonator. The removal of electrodes is a somewhat destructive process, and so for thin discs it is often better to use density values from similar samples. The sample mass is:

```
Sample Mass = 0.79005 g
```

The volume of the sample could be measured by measuring the sample mass in water, however since the geometry of the sample is simple the volume can be calculated from the sample dimensions. This gives a value for the density of:

```
Sample Density = 7485 kg/m^3
```

The sample is now placed in the impedance analyser in the sample jig that has been corrected for open and short circuit conditions (Fig. 15).

An initial sweep from 1 to 500 kHz is made to discover where the fundamental length mode resonance peak is. After this, a refined sweep is performed over the range 70–150 kHz. In this instance, the previous sample measured was a 0.5 mm thick sample that was measured at an applied oscillator voltage level of 10 mV rms. However, when this was used on the 15 mm long sample there was some noise around the anti-resonance curve. This is because under these conditions the high impedance at resonance reduces the current signal. When the oscillator voltage was increased to 100 mV rms the signal to noise ratio improved. The impedance sweep is shown in Fig. 16. The minimum and maximum values for this curve respectively are the values used for the resonance and anti-resonance frequencies. The resonant frequencies are taken as the maximum and minimum values of the measured data, i.e. there is no curve fitting performed, which could of course be carried out. Obviously these values depend on the frequency step of the impedance sweep, in this case 100 Hz, which represents a resolution of around 0.1 %.

```
fr = 92500 Hz fa = 124000 Hz
```

Using the following formula the coupling coefficient k_{33} can be determined:

$$k_{33}^2 = \frac{\pi}{2}\frac{f_r}{f_a}\cot\left[\frac{\pi}{2}\frac{f_r}{f_a}\right] \tag{3}$$

```
k33 = 0.7029
```

Fig. 15 Piezoelectric sample placed between electrodes in the HP sample jig

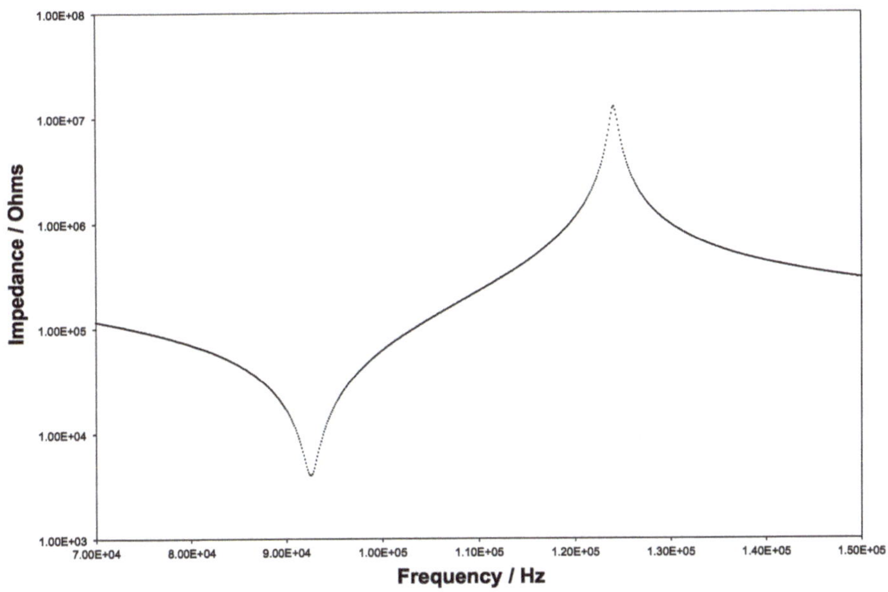

Fig. 16 Worked example for piezoelectric long thin rod resonance and anti-resonance scan

The open circuit elastic compliance s_{33}^D is calculated using the following formula:

$$s_{33}^D = \frac{1}{4\rho f_p^2 l^2} = 9.574\,\exp(-12)(m^2/N) \tag{4}$$

The closed circuit compliance coefficient can now be determined using:

$$s_{33}^E = \frac{s_{33}^D}{1 - k_{33}^2} = 1.89\,\exp(-11)(m^2/N) \tag{5}$$

In order to determine the piezoelectric constant **d**, the dielectric constant under constant (or zero) stress condition is needed. Although the capacitance has already effectively been measured in the impedance sweep these have been carried out near resonance and the condition of zero stress is not satisfied. Normally the 'free' capacitance, C^T, measurement is carried out at 1 kHz leading to the free permittivity through:

$$\epsilon_{33}^{T} = C^{T}\frac{t}{A} = 2.77\exp(-8)(F/m) \tag{6}$$

where t is the thickness and A the area. The piezoelectric coefficients can then be determined using:

$$d_{33} = k_{33}\sqrt{\epsilon_{33}^{T}s_{33}^{E}} = 509\exp(-12)(C/N) \tag{7}$$

and

$$g_{33} = \frac{d_{33}}{\epsilon_{33}^{T}} = 0.01837(Vm/N) \tag{8}$$

Appendix B: Standards

- IEEE Standard on Piezoelectricity, Std 176-1978
- IRE Standards on Piezoelectric Crystals, 1961, Proc. IRE, pp. 1162–1169; July 1961
- IEC standard publication 483, 1976. "Guide to dynamic measurements of piezoelectric ceramics with high electromechanical coupling"
- European Standard, "Piezoelectric properties of ceramic materials and components: part 1 (Definitions and Classifications) and part 2 (Low Power Measurements)", CENELEC, BSI 1999
- IEEE Working Group on: IEEE Standard on Loss in Acoustic Materials, chaired by S Sherritt
- MIL-STD-13768 "Piezoelectric Ceramic Material and Measurement Guidelines for Sonar Transducers".

Acknowledgments The authors wish to express their gratitude to Professor Roger W Whatmore of Cranfield University for many helpful suggestions in the completion of this work. The authors would also like to acknowledge the work of Will Battrick who worked on this project whilst completing his one-year industrial experience for his undergraduate degree, at Manchester University, UK.

References

1. French, A.P.: Vibrations and Waves. CRC Press, M.I.T. Introductory Physics Series (1971)
2. IEEE: IEEE Standard on Piezoelectricity (1978)
3. Cain, M.G.: Piezoelectric Properties of Ceramic Materials and Components. Part 2: Methods of Measurement: Low Power. BSI, Pennsylvania State University (2002)
4. Holland, R., EerNisse, E.P.: Accurate measurement of coefficients in a ferroelectric ceramic. IEEE Trans. Son. Ultrason. **16**(4), 173–181 (1969)
5. Holland, R.: Representation of dielectric, elastic, and piezoelectric losses by complex coefficients. IEEE Trans. Son. Ultrason. **14**(1), 18–20 (1967)
6. Smits, J.: Iterative method for accurate determination of the real and imaginary parts of the materials coefficients of piezoelectric ceramics. IEEE Trans. Son. Ultrason. **23**(6), 393–401 (1976)
7. Sherrit, S., Wiederick, H.D., Mukherjee, B.K., Sayer, M.: An accurate equivalent circuit for the unloaded piezoelectric vibrator in the thickness mode. J. Phys. D: Appl. Phys. **30**, 2354 (1997)
8. Uchino, K., Hirose, S.: Loss mechanisms in piezoelectrics: how to measure different losses separately. IEEE Trans. Ultrason. Ferroelectr. Freq. Control **48**(1), 307–321 (2001)
9. Nye, J.F.: Physical Properties of Crystals: Their Representation by Tensors and Matrices. Oxford Science Publications, Clarendon Press (1985)
10. Honda, M., Hewlett-Packard Company.: The Impedance Measurement Handbook: A Guide to Measurement Technology and Techniques. Yokogawa-Hewlett-Packard (1989)
11. Cain, M.G., Stewart, M., Gee, M., Hill, G., Hall, D.: Electronic Property Measurements for Piezoelectric Ceramics: Technical Notes. NPL Reports (1998)

Direct Piezoelectric Measurement: The Berlincourt Method

Mark Stewart and Markys G. Cain

1 Introduction

Although piezoelectric materials are often used as actuators in order to make small precise movements it can be difficult to measure these displacements in an industrial environment [1]. Consequently simpler methods have been sought to measure the piezoelectric activity, such as resonance methods, and measurement of the piezoelectric coefficient d_{33} using the direct method (often called the Berlincourt method) [2].

This chapter will examine the advantages and disadvantages of the method in detail and with some experimental validation using typical PZT ceramics examine the validity of using the data from this method to predict the displacement of materials in real conditions.

The piezoelectric charge coefficient, d_{ij}, is one of the fundamental parameters defining the piezoelectric activity of a material, basically the higher the d_{ij} the more active the material is. Consequently, manufacturers, designers, and users want to know the d_{ij} coefficient for the material.

Measurement of the d_{ij} coefficient can be realised in several ways varying in accuracy and simplicity. The most reliable method of determining the d_{ij} coefficient is to electrically excite a resonance in a sample, and from the resonance response—given the dimensions of the sample and the density—a d_{ij} coefficient can be calculated, see Chapter "Piezoelectric Resonance", p. xx. One problem with this method is that the geometry of the sample must be such that only a pure fundamental resonance mode is produced, and the calculated d_{ij} parameter relates to this resonance mode. This leaves the problem how to determine the d_{ij} parameter for shapes that don't have an ideal resonance geometry, or where the resonance mode is not the mode that

M. Stewart · M. G. Cain (✉)
National Physical Laboratory, Hampton Road, Teddington, Middlesex TW11 0LW, UK
e-mail: markys.cain@npl.co.uk

M. G. Cain (ed.), *Characterisation of Ferroelectric Bulk Materials and Thin Films*, Springer Series in Measurement Science and Technology 2,
DOI: 10.1007/978-1-4020-9311-1_3,

will be used. For instance, for thin discs poled in the thickness direction it is easy to excite a resonance in the radial direction, and determine the relevant d_{ij} parameter, but to obtain the d_{ij} coefficient for motion in the thickness direction then longer cylinders are needed.

The d_{ij} coefficient is defined as the charge produced for an applied stress, or the strain for an applied voltage, and these are theoretically equivalent [3, 4]. The latter measurement is more difficult to achieve because of the small strains involved, so measurement techniques have concentrated on the former. In this work, initially, the charge was measured in response to an applied static load, but difficulties with thermal drift led to the measurements being performed quasi statically, at a few hundred hertz.

The quasi static method is straightforward; a small oscillating force is applied to the sample and the charge output is measured and divided by the applied force amplitude. The simplicity of the technique has been its downfall, in that anyone can easily build up their own system, and there are a growing number of commercial systems. There are currently no standards for this measurement method, and consequently each system performs the measurement slightly differently. This means that, although the results from these systems are good for measuring within a batch or batch to batch variability, external comparisons usually produce a large variability. This has led to a loss in confidence in these measurement results, and provided the impetus for this study and now this chapter.

Interlaboratory tests have shown systematic differences between results from different laboratories [5]. Figure 1 shows the results from this recent round robin with four participants, illustrating the behaviour for a soft PC 5H (PZT) material. For comparison ±5 % error bars have been drawn on the results for laboratory 2, and it can be seen that many of the results lie within this error band. There is evidence that the errors are systematic, i.e. partner 1 is almost always the lowest and partner 4 the highest. This suggests the measurement method could be improved by closely examining reasons for these deviations.

This chapter aims to answer two questions. Firstly, is the Berlincourt test a valid test method? For example, what are the parameters controlling the accuracy and reliability of the results? Secondly, does the Berlincourt test result predict what displacements can be achieved when the material is excited electrically? Further details can be found in the NPL publication [6].

1.1 History and Physical Basis for Quasi-Static d_{33} Measurement

The phenomenon of piezoelectric (from the Greek word "piezo" or "to press"), was discovered in 1880 during a systematic study of the effect of pressure on the generation of electric charge by crystals [7]. Materials such as quartz and Rochelle salt were some of the crystal structures studied, and due to their non-centrosymmetric crystal structures (i.e. structures lacking a centre of symmetry to their unit cell) an applied stress resulted in the generation of electric dipoles, a process known as

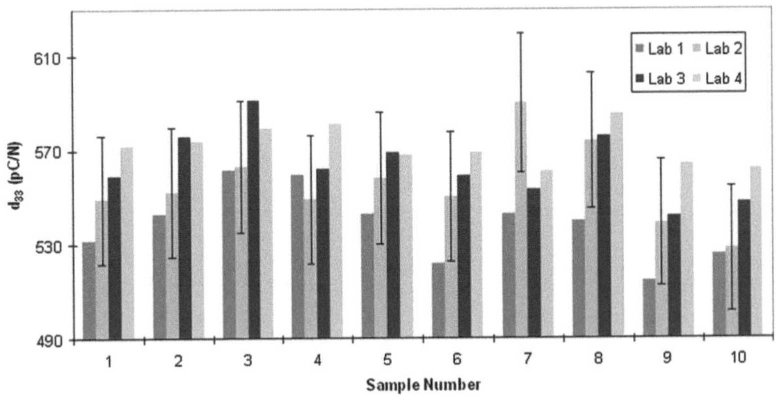

Fig. 1 Round robin results on PC 5H material

polarisation. The applied stress causes the ions in the unit cell to move from their equilibrium positions and distort relative to one another. The necessity for a non-centrosymmetric crystal structure can be rationalised when one considers that a homogenous externally applied stress is centrosymmetric and thus cannot produce a non-centrosymmetric result, such as a vector-quantity like polarisation, unless the material lacks a centre of symmetry.

Such piezoelectric materials also exhibit the converse piezoelectric effect whereby a mechanical deformation is induced on application of an external electric field. The generation of a surface charge on deformation is known as the direct or generator effect, whereas the converse effect of deformation on application of an electric field is designated the motor effect.

The performance of piezoelectric ceramic materials can be quantified either statically (under the influence of a steady strain), quasi-statically or dynamically using the resonance method. The earliest piezoelectric materials investigators determined the piezoelectric constants using static tests, but, due to the difficulty in controlling electrical boundary conditions, static test are now seldom used for piezoelectric materials. Dynamic performance relates to the behaviour of a material when subject to an alternating electric field or mechanical stress applied at frequencies close to the mechanical resonance of the component. The dynamic method, and to a lesser extent the quasi-static method, eliminates the drift due to pyroelectric charges that produce errors in static measurements, and is thus considered to be of higher precision. Additionally, the utilisation of an alternating field allows a more convenient measurement than that carried out at DC. It must be noted that the frequency of the applied signal must be less than the fundamental resonant frequency of the sample and its mounting in order for the equations applied to the static measurements to remain valid and for accuracy to improve in this way.

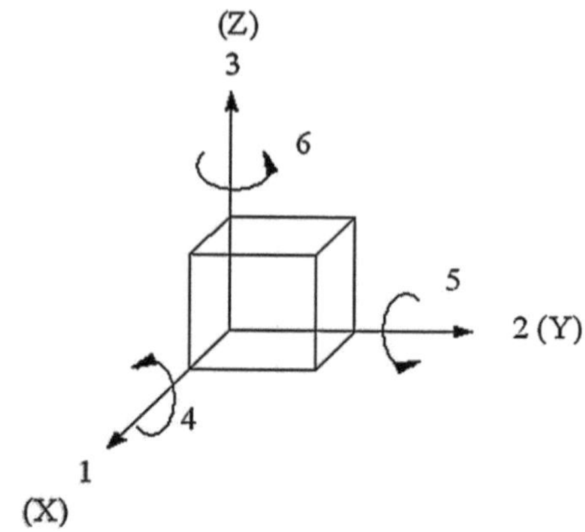

Fig. 2 Conventional axis notation for ferroelectric crystals

A means of quantifying piezoelectric performance is by use of the piezoelectric equations and coefficients. These describe the interaction between electrical and mechanical behaviour and can be summarised as follows:

"d" and "g" are the charge and voltage piezoelectric constants respectively, related by the expression:

$$d = \epsilon^* \epsilon_0 g$$

where, ϵ^* is relative permittivity (dielectric constant), ϵ_0 is permittivity of free space ($8.85\ \mathrm{pF \cdot m^{-1}}$).

Interestingly, the difference between d and g is electrically related to short and open circuit conditions for the piezoelectric materials, respectively.

In summary, the piezoelectric coefficients are defined as follows:

Direct Effect	Converse Effect
$d = \dfrac{\text{charge density developed}}{\text{applied mechanical stress}}$	$d = \dfrac{\text{strain developed}}{\text{applied field}}$
$g = \dfrac{\text{electric field developed}}{\text{applied mechanical stress}}$	$g = \dfrac{\text{strain developed}}{\text{applied charge density}}$

Due to the anisotropic nature of poled ferroelectric ceramics (materials with artificially induced and enhanced piezoelectric properties), and the freedom with which one can select the poling direction, it is necessary to formalise a method for identifying the axes of a component in order to specify its parameters. The direction the material is poled is conventionally taken as the 3-axis (z) with the 1 and 2 axes being perpendicular. Thus, it follows that the subscripts 4, 5 and 6 refer to shear strains associated with the 1, 2 and 3 directions, Fig. 2.

1.2 Mathematical Basis for Measurement Method: Piezoelectric Relations

The Piezoelectric effect is an interaction between the mechanical and electrical behaviour of the material, and to a linear approximation, this interaction can be described by the expressions:

$$S = s^E T + dE \tag{1}$$

$$D = \epsilon^T E + dT \tag{2}$$

where

E = Electric Field Strength
D = Dielectric Displacement
T = Applied Stress
S = Strain
s = Compliance
ϵ = Permittivity

The superscripts E and T are used to indicate values determined at constant field and stress respectively.

The above equations are a generalised form, without the directional notation. If we consider the common case of a ceramic poled in the 3 direction, with electrodes on these faces, and we maintain a constant field, i.e. at zero, then Eq. 1 becomes:

$$d_{33} = [\delta D_3/\delta T_3]_E \tag{3}$$

where, D = electric displacement in the 3-direction (z) and T = applied stress also in the 3-direction (z).

With reference to the measurement of d_{33}, Eq. 3 can be rewritten as:

$$d_{33} = [(Q/A) \cdot (F/A)] = (Q/F) \tag{4}$$

where, F = Applied Force, A = Area force applied over, and Q = Charge developed.

The importance of Eq. 4 is that it gives us a method of determining d_{33} using the direct method, by applying a force on a piezoelectric sample and measuring the charge developed. The areas cancel out (assuming the electroded area is also the stressed area—nearly always the case) so there is no need to measure the area. The only proviso is that the measurement must be performed at a constant field, i.e. under short circuit conditions. This can be achieved by measuring the charge using a virtual earth amplifier or by ensuring that the Device Under Test (DUT) is effectively short circuited by placing a large enough capacitor (much greater than the capacitance of the DUT) across the DUT.

Equation 1 can also be rewritten as

$$d_{33} = [\delta S_3/\delta E_3]_T. \tag{5}$$

So that now the experiment is performed at a constant stress. This gives us a means of measuring the d_{33} in the indirect mode, that is application of a field to produce a strain, all at a constant stress. This gives units in the direct mode of C/N and the indirect m/V, and for PZT materials is usually in pico units. The units are equivalent.

1.3 Measurement Apparatus: Theory of Operation

A piezoelectric coefficient meter consists of two parts, the force head, and the control electronics. The force head incorporates the loading actuator, reference sample, and some means of accommodating different sized samples. The method of applying the oscillating force is usually via a loudspeaker type coil, as this is a relatively cheap and simple means of delivering the force to the sample. The reference sample is in line with the loading train and is used to measure the force applied, so that the reference sample experiences the same loads as the DUT. The reference sample is often a PZT ceramic in order to provide high sensitivity and also a crude form of temperature compensation. The clamping of the samples in the system is usually achieved through a screw thread adjusted vice that forces the sample against the loading actuator. The coil actuator is held in place by some sort of leaf spring arrangement and the stiffness of this spring and the number of turns on the clamping system determines the pre-load put on the sample. This pre-load is required to achieve stable measurements so that the sample does not rattle. Figure 3 shows a schematic of the components in the force head.

The control electronics provide the amplified AC signal to drive the coil at the required frequency, the charge measurement system, the d_{33} calculation system and the readout electronics. The frequency range of these systems are limited at the lower frequencies to around 10 Hz by thermal drifts, and at the higher end, around 1 kHz, by the mechanical resonance of the complete force head assembly. Sometimes the frequency and the applied force can be selected, although the measurements should be insensitive to the applied AC load for the levels used in this experiment (but see later).

The charge measurement electronics must be such that the experiment is carried out under constant field conditions, so either the current is measured using a virtual earth amplifier or the current is shunted through a capacitor of capacitance orders of magnitude greater than the sample (short circuit conditions). The d_{33} measurement is the ratio of charge from the reference material (the force) to the charge developed in the DUT. Either the Root Mean Squared (RMS) values of these signals can be measured and divided to give d_{33} or the two signals could be divided electronically by using a differential amplifier and then the RMS value of this difference signal will be proportional to the d_{33}.

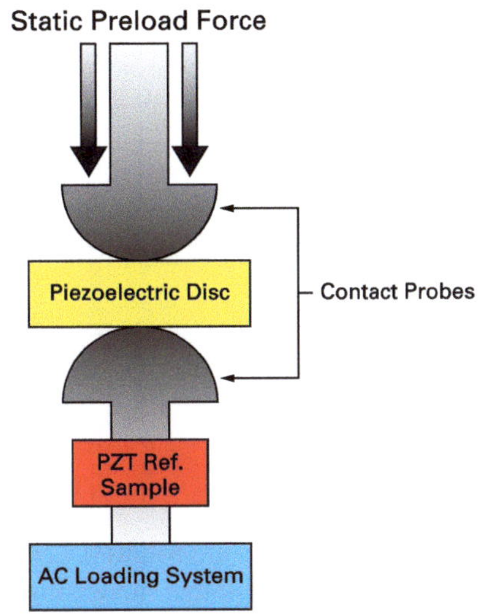

Fig. 3 Schematic of force head

The RMS signals are proportional to the d_{33} and the electronic control system must provide some means of calibration to return the correct value. Usually this is achieved by placing another reference calibration sample as the DUT and then the displayed value is adjusted to give the certified value of the calibration sample.

The Berlincourt Method

The use of the name 'The Berlincourt Method' is somewhat of a misnomer when referring to quasi-static measurement of the piezoelectric d_{33} coefficient using the direct effect, as the skeletal basis for the measurement schema was proposed in 'Piezoelectric Ceramics' by Jaffe et al. [7]. The commonly used name refers to the scientist, Don Berlincourt, who was working for Channel Products during the development of one of the first commercial d_{33} measurement systems. Similar systems are also termed piezometers, d_{33} testers, or piezo d meter e.g. Takecontrol, Piezotest, http://www.piezotest.com/.

2 Experimental Investigation of a Berlincourt System

2.1 AC Measuring Force

The level of the applied AC measuring force can be varied for some systems. The magnitude of the force is not usually important, as long as the stresses introduced into the material mean the piezoelectric is still operating in the linear regime. Typical

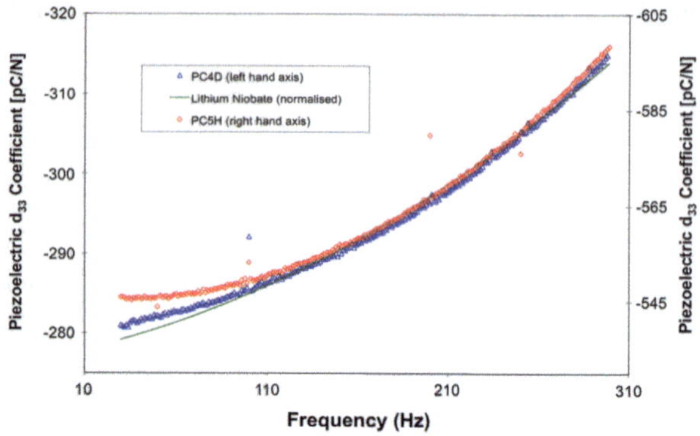

Fig. 4 Frequency sweep for PC4D, PC5H and lithium niobate

operating AC loads are round 0.1 N RMS. If the force can be varied then this can be investigated to check that increasing the AC force has no effect on the measured d_{33}. The obvious advantage of using increased loads is that a larger charge signal will be generated, and this can reduce the signal noise related errors. However, the lower the applied force the greater the likelihood that the material will behave linearly. The level of the AC force can be checked by inserting a load cell in the measuring head, a quartz load cell is useful for providing adequate dynamic response.

2.2 Effect of Frequency

The frequency range of the Berlincourt type instrument is limited to a range between roughly 10 Hz–1 kHz. The lower frequency limit is governed by the charge measurement system, and the stability of the charge over the measurement period. Problems with thermal drift and charge dissipation through the surroundings are one reason that the measurements are not performed at DC (static). The upper frequency limit is governed by the load application method which is based upon a loudspeaker drive coil, and the ability of the system to deliver useful stresses into the sample. Also, it is difficult to design a load application that does not display a resonance peak that interferes with the measurement range. The system is designed so that the peak is far enough away that the measurements are made in a frequency independent region.

Because the method is quick and simple it is not usual to completely shield the instrument electrically from the surroundings, and so there are often measurement anomalies at the mains power frequencies and its corresponding harmonic frequencies. Consequently, it is usual to work just off one of these frequencies, say at 97 Hz, in the United Kingdom and say 110 Hz in the United States of America, and parts of Japan for example.

Figure 4 illustrates some of the issues related to the frequency response of the system, and the effect in different materials. The trace for the PC 4D material shows measurement outliers at 50, 100 and 200 Hz as discussed in the preceding paragraph.

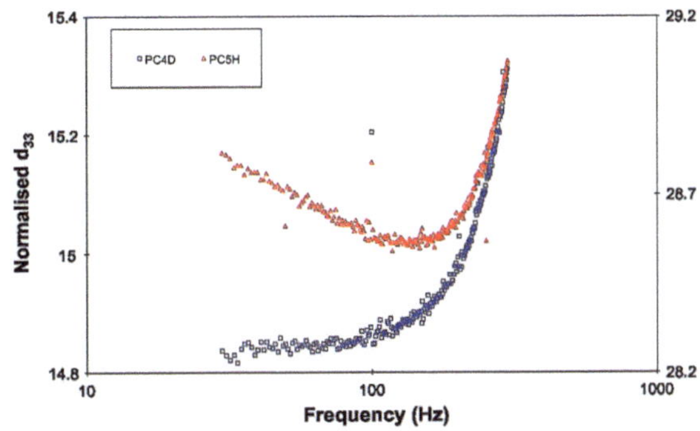

Fig. 5 Frequency sweep for PC4D and PC5H normalised using lithium niobate

To illustrate the material's dependency, the results for the hard and the soft material have been plotted on separate Y axes, and a third line for a fit of measured values on Lithium Niobate has been normalised in a similar manner. Lithium Niobate has a d_{33} of around 20 pC/N, and is thought to behave linearly with respect to applied fields etc. Measurements using the Berlincourt meter on a $LiNbO_3$ single crystal increased from 19 to 21.5 pC/N over the frequency range 30— 300 Hz. We can therefore tentatively assume that this is due to the system response and is increasing towards a resonance peak in the kHz region. In contrast, the soft material displays a pronounced up-turn in the results at low frequencies, whilst mirroring the behaviour of the PC 4D and $LiNbO_3$ at frequencies above 150 Hz. This up-turn must therefore be material dependent, i.e. that the d_{33} increases with decreasing frequencies for the soft material. This effect has been seen in various experiments on soft piezoelectric materials and is due to the inhibition of domain movement at increasing frequencies. This behaviour has been modelled by the Rayleigh law [8], due to the parallels between ferromagnetic and ferroelectric materials. One of the observations for materials systems controlled by domain wall motion is a logarithmic dependence of behaviour with respect to frequency. The Rayleigh model is often applied to soft ferroelectric materials when they are operating in a region where domain wall movement is dominant, i.e. at large stresses and for low frequencies. If the results for the hard and soft material are divided by those for the lithium niobate at each frequency then this should effectively cancel out any system dependant effects.

This has been done in Fig. 5 and plotted against the log of the frequency. Both samples show a similar increase above around 100 Hz, but below this the hard material is largely frequency independent, whereas the soft material shows a linear increase with decreasing frequency. The behaviour below 100 Hz is consistent with known behaviour, however above this it would appear that it is still system dependent, since the behaviour for hard and soft is identical. It may be that because the system gain is set by calibration at 100 Hz then the gain is valid here but at the higher frequencies there may be a frequency dependant gain issue.

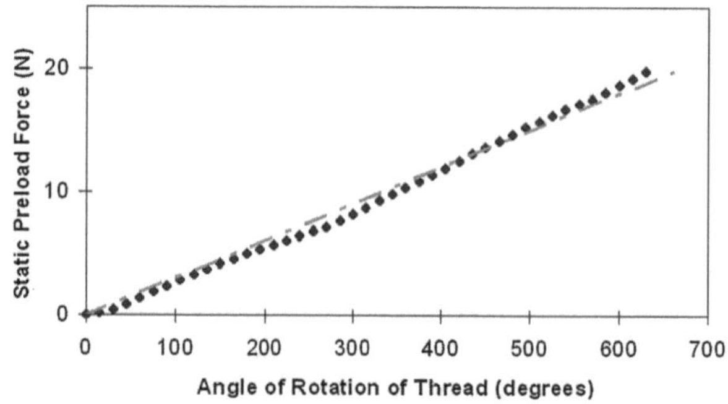

Fig. 6 Calibration of static pre-load. One turn exerts 10.8 N force

2.3 Effect of Static Pre-load

Samples are clamped in position by a small pre-load which is usually applied by turning a screw thread on the measuring head. This forces the sample against a spring, and the stiffness of this spring and the number of turns on the screw thread governs the value of the pre-load. The reason for the pre-load is simply to hold the sample in position while the oscillating measuring force is applied. The pre-load need only be greater than half the amplitude of the measuring load, i.e. so that the probes never leave the sample, but in practice the value is much greater than this.

Figure 6 shows the linearity of the pre-load with number of turns of the pre-load mechanism as measured by a small strain gauged load cell. The slope of this curve is effectively the stiffness of the pre-load spring. From this it can be seen that the value of the applied pre-load might be expected to vary around ±10 % in typical tests.

The value of the pre-load to use in these type of instruments was set historically, based on a value that gave the most consistent results, which is a compromise between enough force so that the sample does not rattle, and not enough to completely clamp the material. That the pre-load can have an effect on the measurements can be seen in Fig. 7. Here, the static pre-load is increased from 0 to over 20 N on a typical hard and soft PZT material. For the soft material, the increasing pre-load has the effect of reducing the d_{33} whereas for the hard material the behaviour is reversed. The slope for the soft material is also much greater showing that the pre-load has a greater effect on the d_{33} than in the hard material. The error bars on the soft material can be seen to reduce as the pre-load is increased, indicating that at very low pre-loads the measurement is less reliable.

In many of the experiments using this kind of instrument it is difficult to separate the change in behaviour due to different experimental conditions and those due to the material itself. The highly piezoelectric PZT ceramics are notoriously dependent on factors such as stress level, temperature, test-time, and it can sometimes be more instructive to examine a less active but more stable material such as quartz. Figure 8 shows the variation of a quartz disc with increasing pre-load, and the zero slope shows that the change that occurs with PZT ceramics is material and not instrumentation dependant.

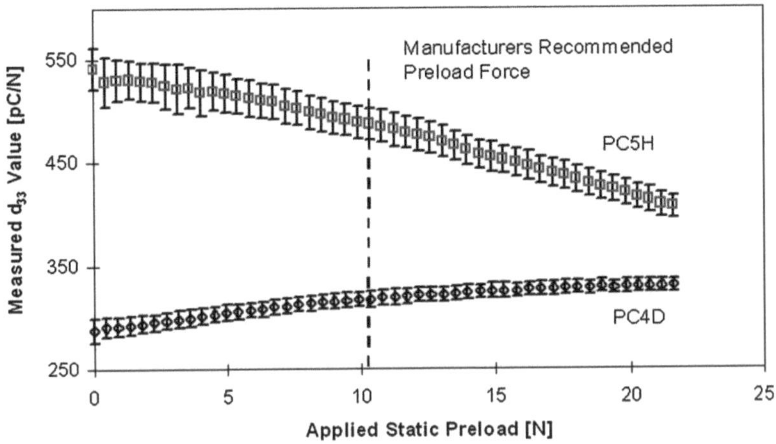

Fig. 7 Effect of static pre-load on PC 4D and PC 5H

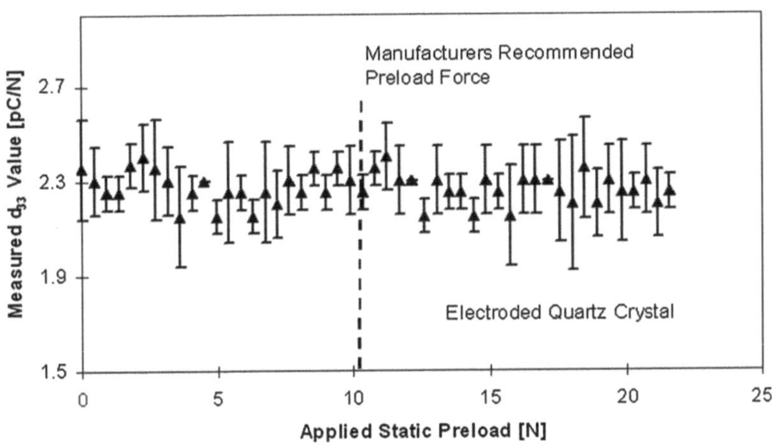

Fig. 8 Effect of static pre-load on Quartz Sample

2.4 Time Dependent Effects

In the investigation of the Berlincourt system it was found that there was a time dependent variation in the measurements. In all cases the measured d_{33} decreased with increasing measuring time under load. Figure 9 shows an example decay curve for a hard PC4D material at several different pre-loads. Initially it was not clear if this was a measurement artefact dependent on the system. However, after a number of experiments it was proven to be related to the static pre-load because of the following observations:

- If a sample is removed from the system and immediately replaced the behaviour continues along the same curve

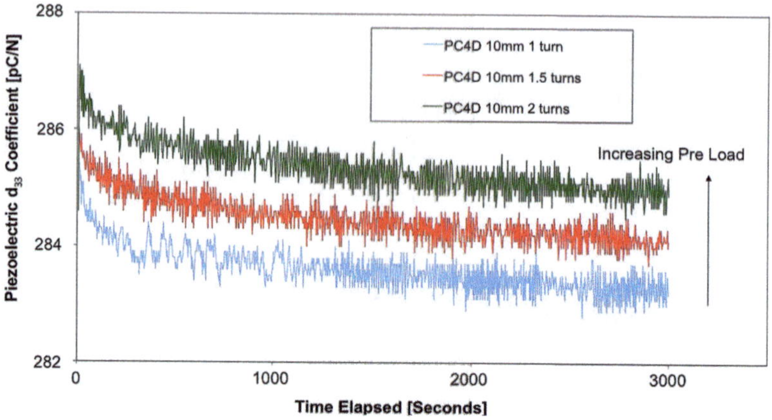

Fig. 9 Effect of static pre-load on relaxation of d_{33} on PC4D

- Inserting a fresh sample at any point initiates a new decay/relaxation curve
- If a sample is removed from the system and left to recover for 24 hours the relaxation curve begins again at time zero
- If a sample is left in the system under pre-load for several hours, but no AC measuring force then the relaxation curve is flat, i.e. most of the relaxation has already taken place
- The behaviour is dependent on material, there is no measurable relaxation for lithium niobate.

The decay of d_{33} with time is difficult to measure, firstly because the changes are close to the resolution of the system and, secondly because the largest change occurs in the first few seconds and since it takes at least 5 s to get a stable reading the change is difficult to quantify. The effect of static pre-load on the decay in d_{33} is dependent on the material, Fig. 10. For a given material the region where the decay flattens out is offset with the initial decay being similar. As seen with the static loading experiments, increasing the pre-load increases the offset for a hard material, and decreases that for a soft composition. The different behaviour of hard and soft material is not just in the offset but also in the rate of relaxation. Figure 10 shows a comparison of a hard and soft sample, where the hard material is stable after 3,500 s, yet the soft material is still relaxing. A general observation was that for equal experimental conditions it took longer for the soft samples to reach a stable region, although some of this could be attributed to the higher d_{33} of the soft material that may be measured more accurately for the same sensitivity.

Further evidence that the decay is due to some form of relaxation in the sample rather than instrumental artefact is seen in Fig. 11, which shows the time decay curves for a thick and thin soft sample. The slope of the 1 mm curve is much greater because of the much higher static stress concentration introduced in the thinner sample. An observation for the 1 mm samples at various pre-loads was that, after an initial rapid

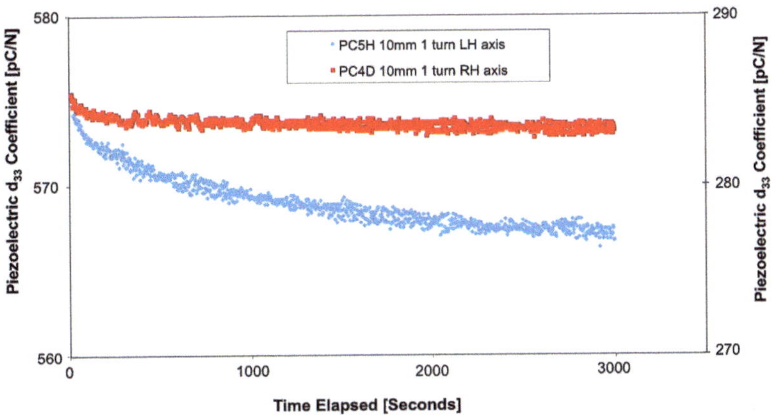

Fig. 10 Relaxation of d_{33} in 10 mm PC 4D and PC5H samples

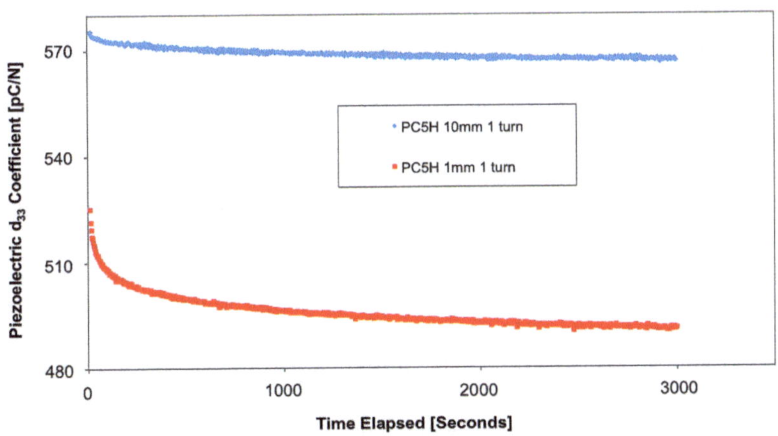

Fig. 11 Comparison of d_{33} relaxation for a 10 and 1 mm PC5H sample

decrease, there was a slight rise at around 250 s before continuing the downward trend. Thus, there maybe two competing mechanism for stress relaxation at the high stresses in the thinner samples.

2.5 *Calibration of System*

In the Berlincourt system there are two calibration points, the zero calibration, and the gain setting. The zero calibration is simply that the reading for a non piezoelectric material should be zero, and the calibration/correction is achieved by placing a non conducting sample in the system and checking that the reading is zero.

The gain calibration issue is more difficult. For measurement instrumentation it is normal for the calibration of the system to be traceable to the national measurement system via the measurement of some electrical quantity or reference artefact. In the Berlincourt system the quantities measured are force and charge. The traceable calibration of the charge is relatively straightforward, but the calibration of force would be more difficult. Consequently, in the Berlincourt system the force is measured by having a piece of reference PZT in the load train. This has the advantage that the charge signal from the internal reference can be compared to the sample charge to give a d_{33} measurement. This simplifies the measurement electronics, but does create a more difficult calibration problem. The solution is to use a reference artefactthat has been calibrated by some other means, and to adjust the gain of the system to agree with the artefact. The normal reference artefact is a sample that has had the d_{33} measured by longitudinal resonance using the IEC standard method [9]. The disadvantage of this is that it is almost impossible to traceably calibrate this, and it also assumes that the value derived from resonance at around 100 kHz is identical to that in the 30–300 Hz range. This poses the question, 'at what frequency should the calibration be carried out?'. We have seen from the discussion on the effect of frequency, that although the calibration may have been carried out at the "standard" operating frequency, once away from this frequency the calibration route becomes more tenuous.

The material used in the reference material should obviously be stable with respect to time and temperature, and show little AC and DC load and frequency dependence. It should also have a high enough d_{33} to be able to calibrate the highest end of the measurement range. Quartz or Lithium Niobate would be ideal candidates but the d_{33} is too low to satisfactorily calibrate the higher ranges. Although the hard PZT compositions have roughly half the value of the most active soft materials, their improved stability with time make them a more suitable choice for a reference material.

Since, for independent measurements of d_{33} the reference sample will invariably be a thickness resonator, the thickness of the sample will be sufficient (probably at least 10 mm long) so that the thickness effects seen previously will not be an issue.

2.6 Sample Geometry

An advantage of the Berlincourt method is that virtually any shape or size sample can be measured in the system, as long as there are electrodes to make electrical contact and apply the load. The method is supposed to be sample geometry independent,and in order to examine this a selection of samples from the same batch of material were examined.

Figure 12a and b shows the effect of sample thickness on the measured d_{33} for a soft and a hard material, respectively. Over the entire range of thickness from 2 to 10 mm there does not appear to be a change in measured d_{33}. However, there is a definite difference between the 2 and 1 mm samples, with the soft PC5H material exhibiting

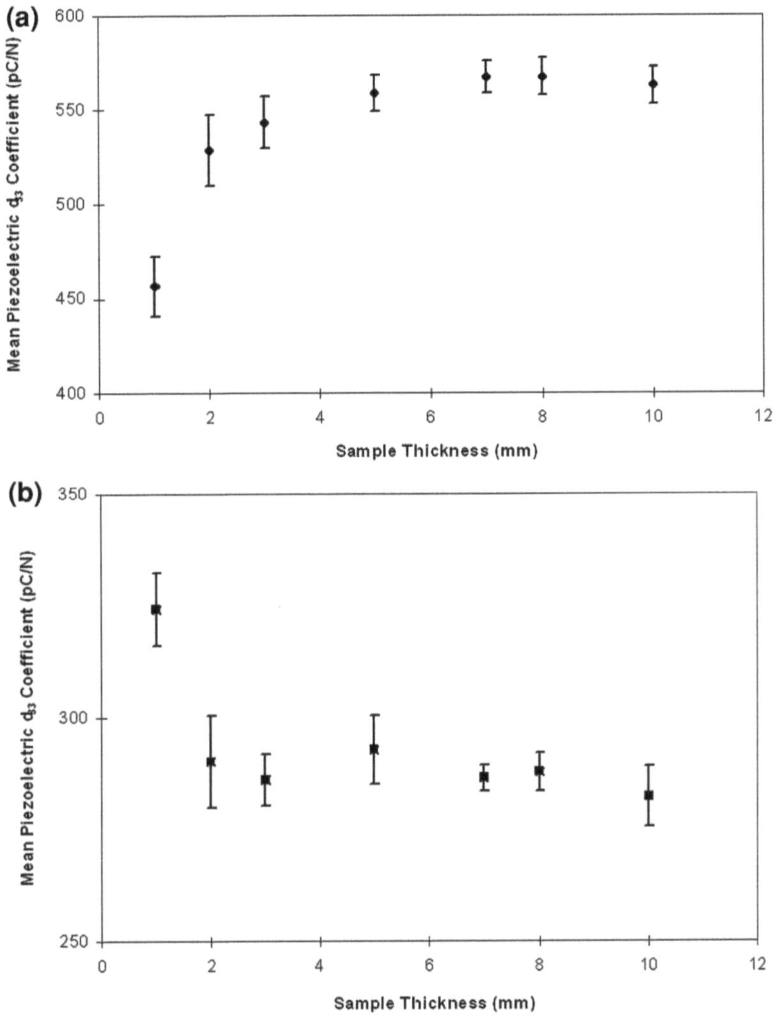

Fig. 12 **a** d_{33} against thickness for a soft PC5H composition, **b** d_{33} against thickness for a hard PC4D composition

a large decrease in the d_{33}, whereas the hard material showing a slight increase. Although the samples all come from the same batch of fired ceramic there could still be slight variations in the poling process for the individual thickness that could lead to differences in piezoelectric activity. In order to verify this, an experiment was devised where thin samples were stacked and compared with their thick counterparts. The results of these measurements are given in Table 1. Assuming linear behaviour the measured d_{33} of a stack of thinner discs should be equal to the average of the individual discs. However, this is not the case for the 1 mm discs, and the d_{33} of two

Table 1 Effect on measured d_{33} of stacking 1 mm discs

Mean d_{33} value [pC/N]		
Stack height/mm	Soft PC 5H	Hard PC 4D
1	456	324
2	555	316
3	577	302
4	559	288
5	577	293
6	565	297
7	571	296
8	572	292
9	562	291
10	568	289

stacked 1 mm samples tends towards the value for the 2 mm samples. This indicates that the different thickness samples have identical piezoelectric activity, and the difference measured with the Berlincourt is instrumental.

As discussed previously, almost any shaped sample that can be inserted in the measuring head can be measured. However, difficulties arise when the samples have a large area. For large samples it can be difficult to position the sample centrally such that the sample is balanced. If the sample is off centre it can introduce bending stresses that will affect the results. For very thin samples the misalignment of the loading pins can lead to considerable shear stresses, again giving misleading results.

For samples with a large surface area compared with the normal 10 mm diameter a separate problem was identified, namely the neutralisation of the charge via the surface. It was noted that the reading on the meter display could be affected instantaneously by breathing over the sample, i.e. the extra moisture in the breath was leading to an increased charge leakage path. The value quickly regains the steady state value, but it is possible that convection currents could have a greater effect on larger surface area samples.

2.7 Loading Geometry

The normal loading contacts for a Berlincourt system are hemispherical contacts of the order 4 mm diameter. This point contact is the preferred mode because firstly the contact points for samples with rough surfaces is more controlled, and secondly it minimises any clamping in the lateral direction by the loading contact. Other types of contact test probes can be used and are more useful for more esoteric shaped samples. For a spherical sample or loading on the perimeter of a disc, a flatter loading probe will provide a more practical loading set-up and still maintain essentially point contact area. For very thin samples and films a flatter electrode will reduce shear due to misalignment, and also reduce the stress levels for a given measuring force.

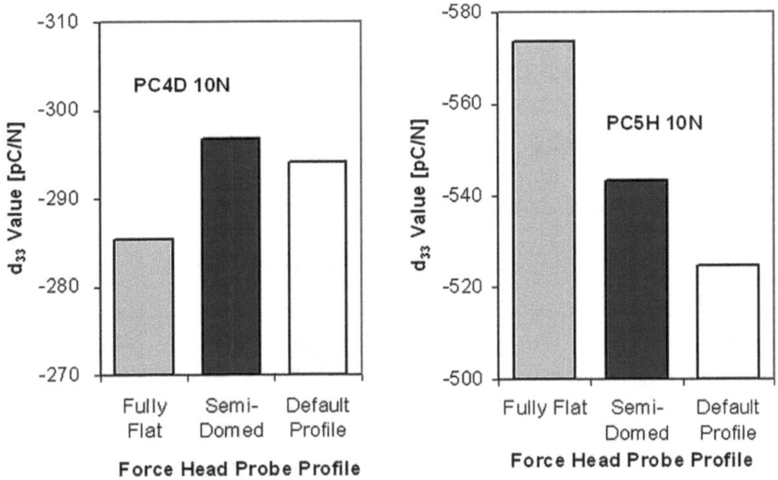

Fig. 13 Effect of different loading probe on the measured d_{33}

The effect of the test probes on the measured d_{33} was investigated by using three type of test probes, the standard point contact, a flatter rounded contact (roughly 16 mm diameter), and a completely flat probe with a diameter of 8 mm. The contact area for the rounded probes was measured by examining the imprint of the area using carbon paper. For the standard contact the imprint was a circle roughly 0.5 mm in diameter, and for the more rounded contact a diameter of around 1 mm. The results for this experiment are shown in Fig. 13 for a soft material with the standard pre-load of 10 N in each case. It appears that the increased contact area probes decrease the static stress, and so for the soft material the d_{33} increases whilst for the hard material the measured d_{33} decreases.

2.8 Second Order Effects

Where it was practically possible the samples were left to recover between each individual measurement so that the change due to one particular experimental parameter could be observed. This was not always practical, for instance the sweep of d_{33} with frequency was performed with a sample continuously in the measurement head. An ideal test would be to measure at each frequency and remove the sample after each measurement and allow the sample time to relax. Another approach would be to leave the sample in the system for several hours before commencing the measurements. However, in the tests performed here there is a mixture of frequency and time dependent information. For instance, if the frequency is swept first up to the high frequency and then back down again the behaviour is hysteretic. Figure 14 shows these sweeps for a soft material of 1 mm thickness. The behaviour is hysteretic

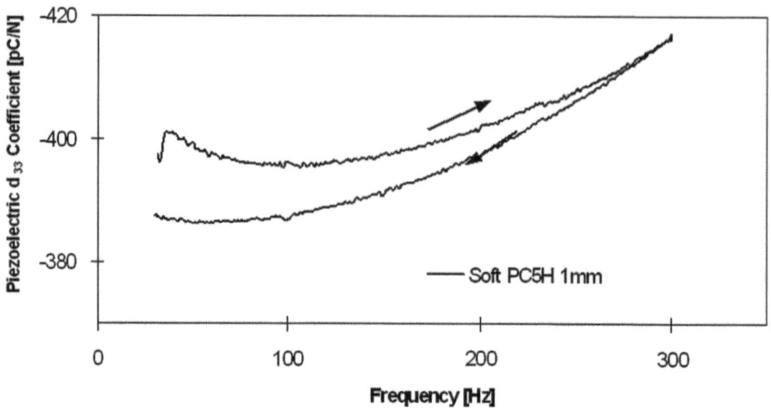

Fig. 14 Up and down frequency sweep on PC5H 1 mm thick sample

because the static load is applied at time zero and by the time the frequency sweep is finished almost an hour will have elapsed, thus showing the time dependent component of this experiment.

2.9 Differences Between Hard and Soft Materials Under Static Load

In most of the experiments discussed here there has been a trend for the hard material to behave in an opposite manner to the soft material. The most probable cause of this behaviour is due to the differing response of the materials to static loads applied parallel to the poling direction [2]. For instance, it is known that the permittivity for hard materials increases under applied stress, whereas the soft materials show a decrease. For the hard material, this can be attributed to a de-aging, where any large mechanical or electrical excursion will initiate a new aging cycle. For the soft materials, the decrease is due to the domain mobility effectively de-poling the material under relatively low stresses.

This explains the difference in the static load behaviour seen in Fig. 7. Also, considering the behaviour under different load contact probes it is apparent that the increased contact area probes decrease the static stress, and so for the soft material the d_{33} increases whilst for the hard material the measured d_{33} decreases. This could also be used to explain the thickness effect, where in very thin samples there are proportionately more regions under a high static stress so the static stress effect dominates. As the thickness of the sample is increased by stacking this high stress region is reduced, thus reverting back to the "true" d_{33}.

3 Finite Element and Theoretical Aspects of Sample Geometry/Loading

It is clear from some of the preceding experimental results, particularly the different response of various sample geometries that the Berlincourt test has some shortcomings. If the response of the piezoelectric is assumed to be linear then this predicts that the results should be independent of sample geometry or method of stress application. If a finite element model of the measurement system is simulated using ANSYS finite element code [10], using linear coupled field elements, and assuming the loading state can be approximated as a static case, then the predicted d_{33} will be equal to the d_{33} used in the input. Although in the Berlincourt measurement a point load on a thin disc can give rise to triaxial stresses and stress concentrations, the linear assumption does not lead to any change in measured d_{33}. In the model each element acts as a linear stress-to-charge converter, and although the stresses may be unevenly distributed throughout the sample, the total charge is conserved.

There are few analytical solutions to the problem that might account for some of the variations seen experimentally. However, in the field of thin films there is an analytical solution to account for the clamping effect of a substrate. There are two solutions [11] to simulate the film d_{33} behaviour in the indirect and direct modes. In the direct mode (the application of a stress to produce a charge) the applied stress is transferred through the substrate to the film, and the film is thought to be held rigidly from moving in the 1 and 2 directions by the substrate. In the indirect mode (the application of a field to produce a displacement) the substrate is passive as the applied field does not induce a displacement, and the film is then clamped by an elastic substrate with a stiffness, Y and Poisson's ratio, σ, where Eq. 6 is for the indirect , and Eq. 7 for the direct mode:

$$d_{33}(ip) = d_{33} - 2d_{31}s_{13}^{E}/(s_{11}^{E} + s_{12}^{E}) \tag{6}$$

$$d_{33}(dp) = d_{33} - 2d_{31}(s_{13}^{E} + \sigma/Y)/(s_{11}^{E} + s_{12}^{E}). \tag{7}$$

For both cases the effect of the d_{31} term is to reduce the effective d_{33} since s_{13} is also negative (and of course d_{31} is negative). The boundary condition for the indirect case represents the extreme case and for the materials used in this work would mean a measured d_{33} of 0.33 times the true d_{33} for the PC5H and 0.45 for the PC4D. The magnitude of this clamping can be confirmed in the finite element model, by preventing the outside of the sample from moving. The reason why the behaviour is now apparently non linear is that a clamping force is needed to prevent the sample moving and this acts against the measurement force leading to a reduced charge output. Of course, the greatly reduced d_{33} predicted by Eqs. 6 and 7 only applies to thin films on substrates, but similar boundary conditions could be envisaged in the Berlincourt experiment.

Consider the case where the force in the Berlincourt is spread out over the whole sample surface, and the friction is such that the sample surfaces are not allowed to

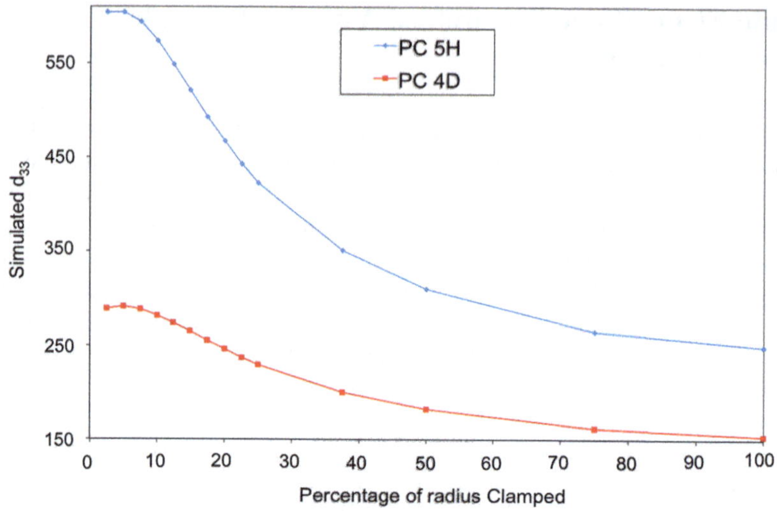

Fig. 15 FEA simulation of direct mode experiment with increasing area of disc loaded and clamped. Disc radius to thickness ratio 10:1

move in the 1-2 directions. This is almost identical to the thin film analytical solution, except that the finite thickness allows some relaxation, giving a slightly higher d_{33} than predicted by Eq. 6. These boundary conditions do not precisely resemble the Berlincourt experiment, and consequently these large reductions in measured d_{33} are not seen.

If the area of load application is reduced then the clamped area reduces, and the clamping forces reduce until, when a point load is used, there is no clamping force and the measured d_{33} is once again the model input d_{33}. Figure 15 shows the change in measured d_{33} with changing the load area for a disc with a 10–1 radius to thickness ratio. The curves go from fully clamped where the d_{33} corresponds to Eq. 6 to the point load which gives the "true" d_{33}. These curves change shape slightly for different radius to thickness ratios, but the end points are fixed. The stress distribution for a partially clamped sample, Fig. 16, is fairly uniform under the loading region and the reduction in apparent d_{33} comes purely from the extra force needed to clamp the displacement at the loading interface.

Another possible mechanism for introducing external boundary conditions could be due to the inertial resistance of the surrounding material under dynamic loading. This would effectively stop the outside of the sample from moving, and show up as a frequency dependence of the measured d_{33} response. However it is unlikely that this effect plays a major role at the frequencies used in the Berlincourt experiment.

In all the preceding modelling of behaviour, the effect of clamping has been to reduce the measured d_{33}, however there are no plausible means by which clamping could increase the charge output, and thus the measured d_{33}. To arrive at a simulation of the Berlincourt experiment where the measured d_{33} was increased some other

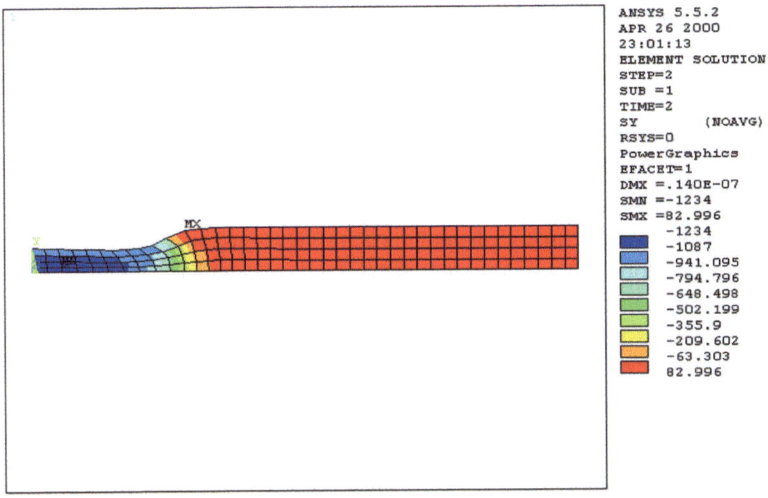

Fig. 16 FEA simulation direct mode loading of 10:1 radius to thickness ratio disc, loaded and clamped over a quarter of the radius

theory is needed. In previous work [12] a methodology was developed to use ANSYS to model systems where the d_{33} was dependent on the stress level. Normally, d_{33} is given as constant, but generally d_{33} is dependent on applied stress or field levels. In the previous work, the d_{33} was assumed to vary linearly with applied AC stress level, based on some independent experimental measurements. This can lead to either higher or lower measured d_{33} values depending on the variation of d_{33} with applied AC stress levels, and could be extended to deal with a d_{33} dependant on the level of DC stress.

3.1 Summary of Finite Element Modelling

In the Berlincourt test, if the material is linear then the measured d_{33} should be independent of sample shape, method of loading etc:

- unless there are experimental conditions which lead to clamping of the sample leading to a reduction in the load—for example:
  - clamping in XY plane by the load applicator
  - dynamic response dependant (frequency dependant)
- Otherwise the changes in response are because:
  - the d_{33} is intrinsically dependant on AC stress levels
  - the d_{33} is intrinsically dependant on DC stress levels.

4 Comparison of Low Field Berlincourt Measurements with High Field/Stress

4.1 Measurement of d_{33} in the Indirect Mode

As previously mentioned, the main advantage of the Berlincourt method lies in its simplicity. However, for many piezoelectric applications, materials are used in the indirect mode—applying a field to achieve a displacement for actuation. This poses the question, is the value of d_{33} determined by the Berlincourt method the same as that determined using the indirect method? For a discussion of how to perform d_{33} measurements of PZT materials using the indirect mode there is an NPL measurement good practice guide available [1].

Results for the indirect d_{33} measurement for a number of PC 5H samples are shown in Fig. 17, where d_{33} is plotted against the applied electric field. In the linear theory of piezoelectricity d_{33} is a constant, i.e. it does not vary with the applied field or stress level. For materials such as lithium niobate this is a valid assumption, but for the soft PZTmaterial measured here the d_{33} has doubled over the applied field range. This graph illustrates one of the fundamental differences between the direct and indirect measurement methods. In the indirect method because of the small displacements involved large fields are applied in order to produce more measurable displacements. The applied field/stress level is therefore a large AC field with zero DC level, and the material experiences tensile and compressive stresses. In contrast with the Berlincourt method the charge from a small AC stress can easily be measured, however a large DC stress bias is needed to hold the sample in place, so the sample is always in compression. This means that for a valid comparison between the measurements the displacement measurements should be extrapolated back to zero applied field levels, and the Berlincourt measurements need extrapolation back to zero pre stress levels.

Figure 18 compares the d_{33} indirect measurement at zero field with the normal Berlincourt measurement (i.e. a 10 N pre-stress). The extrapolation for the displacement measurements assumes that the behaviour is linear all the way to zero field. Considering Fig. 17 this appears valid for the thicker samples, however the thinner samples show a tail off in behaviour as a field of 0.05 kV/mm is approached. One problem with these thinner samples is that the actual displacements measured are smaller for equivalent fields. For the 0.5 kV/mm field a d_{33} of 500 pC/N represents a displacement of only 25 nm, and at these levels system noise and resolution is important. Comparing the two measurement results in Fig. 18 the displacement measurement gives consistently higher results, by about 10 %. However, the trend for the different samples is the same. Two 1 mm samples were chosen for the first of the series of measurements in Fig. 18 that gave different Berlincourt d_{33} values, and this is reflected in the displacement measurements. It is interesting to note that for a given composition the variations in d_{33} measured by the Berlincourt are reflected in the zero extrapolated displacement results, however the gradients of the displacement versus applied field measurements for d_{33} are similar for all the samples.

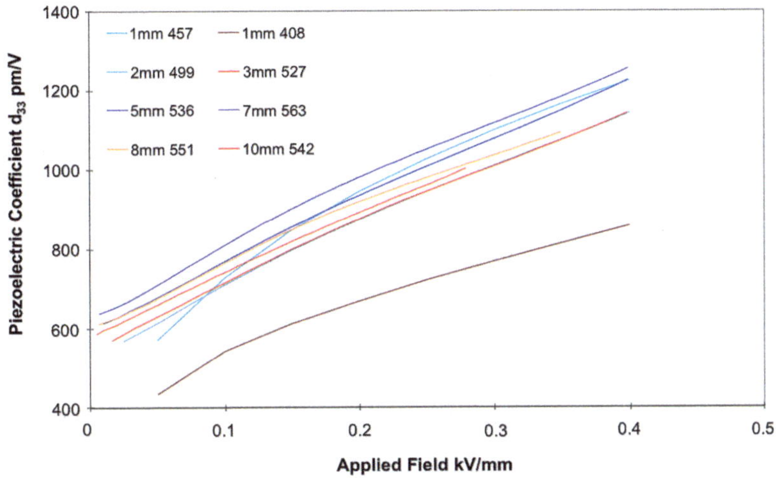

Fig. 17 Indirect d_{33} measurements on PC 5H samples of varying thickness

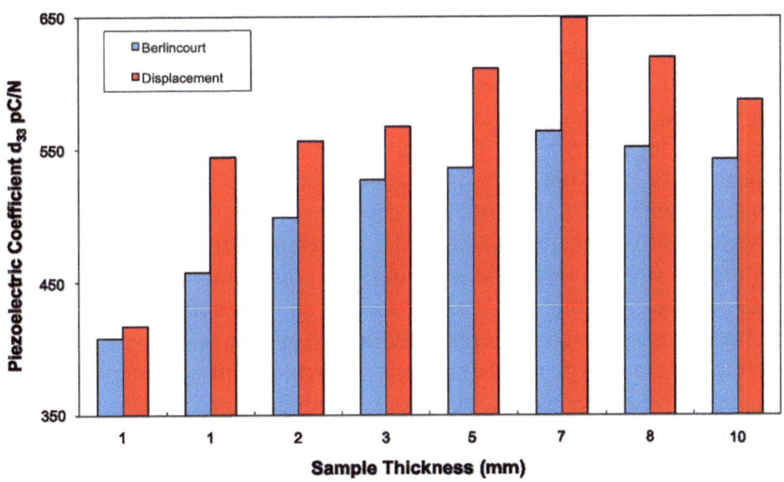

Fig. 18 Comparison of indirect and direct measurements on PC 5H material

The disparity between the two measurement methods could be attributed to the fact that the Berlincourt measurements were taken at 10 N pre-stress. However, the pre-stress should have a greater effect on the thinner samples, not a constant difference as observed. Similar measurements on the PC4D material also gave a higher displacement d_{33} than its corresponding Berlincourt value, but this time only by around 3 %. If the difference was solely due to the pre-stress then the difference should be reversed for the hard material. There are several possible interpretations as to why the measurements are different, including the following:

- Berlincourt measurements are done under a pre-stress
- The calibration of the Berlincourt is at a fixed frequency using a sample measured at resonance
- In the displacement measurement method the sample experiences tension and compression whilst in the Berlincourt method it is always under compression
- The boundary conditions for the two experiments are slightly different.

The discrepancies are not that large considering these issues, and it can be concluded that the Berlincourt results agree with the zero field extrapolated displacement measurements. However, the Berlincourt method gives no indication as to what might happen at higher driving fields.

4.2 Direct Measurements of d_{33} at High Mechanical Stress Levels

In the previous section we have seen that there is equivalence between the Berlincourt measurements and the displacement measurement results when extrapolated back to zero applied field. However, the Berlincourt data cannot be used to predict the displacement of materials at the kinds of fields used in practice. Obviously the Berlincourt result gives us the base level, and some kind of generic rate of change of d_{33} with applied field could be used to give an improved estimate. It may be possible to get an idea of how much the d_{33} changes with increased field levels by using increased stress levels, by using the equivalence of the indirect and direct piezoelectric behaviour. There are two experimental problems with this approach. Firstly, for equivalence with the applied AC field experiments the Berlincourt measurement system would need to generate tensile and compressive stresses. The difficulty with applying tensile stresses could be overcome by comparing with field-strain measurements performed with a DC bias to keep the stresses compressive. The second problem is that, using the conventional Berlincourt system with a voice coil drive, it is difficult to generate large stresses at high frequency. This second problem can be overcome by using a piezoelectric actuator to produce the AC stress.

The results of making direct piezoelectric coefficient measurements at high applied AC stress is shown in Fig. 19. Clearly the behaviour is similar to the displacement behaviour under high applied fields, i.e. there is a near linear increase in d_{33} with applied stress, suggesting this is a possible method for predicting high field displacement behaviour.

5 Summary: Simple Checks for Improved Measurements for Quasi Static Piezoelectric Coefficient (Berlincourt) Measurements

Static Pre-load Think about the load applied to the sample. A 10 N load on a 1 mm area is 10 MPa. Will this affect your sample? Check the effect by changing the pre-load.

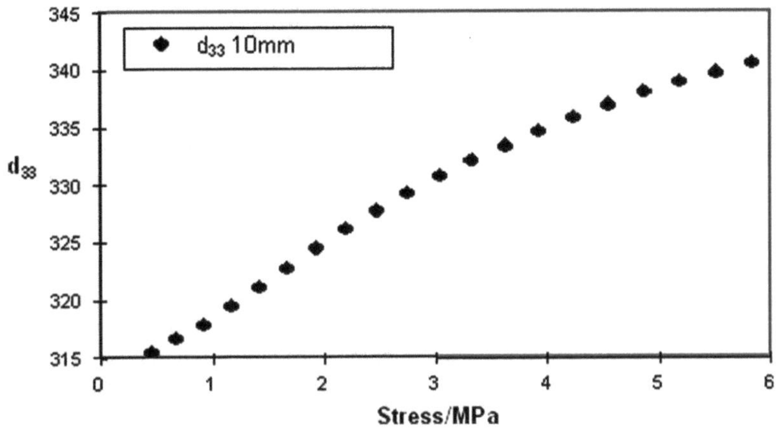

Fig. 19 Piezoelectric coefficient, d_{33} with amplitude (pk/pk) of ac pressure

AC load level Similarly with the AC load level, is this affecting your measurements? If you can change the level, do so, and look at the effect. Generally the lower AC level will give more linear behaviour, but at the expense of decreased signal to noise ratio.

Sample Geometry Thicker samples are always better in terms of spreading the loads, a thinner sample might give increased stress levels. Check the effect on thinner samples by stacking several and looking at the difference. There should be no difference.

Measurement Frequency Don't work exactly on mains frequency, or multiples thereof. Check the resonance behaviour of your system using either Lithium Niobate, or a very hard PZT composition. These should show little change with frequency. Work as far away from the resonance as possible. Work at the calibration frequency, if you want to change frequency, consider recalibration at this frequency.

Calibration How was your reference sample calibrated, and is this calibration valid at the frequency and stress levels used in your system? Use a high sensitivity calibration sample to set the gain of the system, but check the gain independently with a lower activity reference material.

Environment Understand the temperature compensation mechanism of your system, i.e. what is the temperature sensitivity of the internal reference? If you measure a sample with similar properties to the internal reference there will be no variation with temperature. Is this what you want? If the internal reference is a high sensitivity PZT material, check the temperature variation with a more temperature stable material.

Environment What are the effects of moisture and air currents on your system? Look at the measurements as you breathe moist air onto it. Is it stable if you blow dry air over it, for instance from a compressed gas source?

Loading probes Flatter loading contact geometries are better for thinner samples because they reduce the effective stress levels, and also ease misalignment problems. However using completely flat loading contact can lead to bending stresses unless all surfaces are perfectly aligned.

Time Dependant Effects The time dependence of measurements is usually small but measurable. Leave a sample in the system overnight and see the difference. For consistency of results it is good practice to take readings after a fixed time, for example after 5 s.

Acknowledgments Thanks are due to the industrial sponsors of this work. Many of the measurements made in support of this work were carried out using a piezometer supplied by Piezotest Ltd, on PZT samples supplied by Morgan Electroceramics Ltd and Advanced Ceramics Ltd.

Appendix A: Round Robin Tests

As part of the measurements carried for this chapter a small round robin was undertaken to confirm some of the findings. Two laboratories using similar instrumentation measured a set of hard (PC 4D) and soft (PC 5H) samples 10 mm diameter, and thicknesses of 10 and 1 mm. The measurements were all performed at the same frequency, under different levels of pre-load. The results of the round robin confirmed:

1. The variability of the soft material is greater than the hard
2. The variability of thin samples is greater than for thick
3. The hard material is less dependent on the pre-load
4. The d_{33} of the soft material decreases with increasing pre-load
5. The d_{33} of the hard material increases with increasing pre-load
6. The difference in d_{33} for thick and thin materials is consistent with the behaviour under the application of a pre-load
7. Thick samples are less affected by the pre-load.

The spread of results in this round robin has also been reduced as a result of controlling the conditions under which the measurements were performed. Figure 20 shows the results for the current round robin compared with a study on a similar material. The data has been plotted as a percentage deviation from the average for each laboratory, and clearly shows the reduced scatter from close control of the experimental set up. Clearly the control has been achieved by using almost identical piezometers, and further work is needed to maintain this using different makes and models of direct testing machines.

Appendix B: Relevant Standards

1. IEEE Standard on Piezoelectricity, Std 176-1978
2. IRE Standards on Piezoelectric Crystals, 1961, Proc. IRE, pp.1162–1169; July 1961

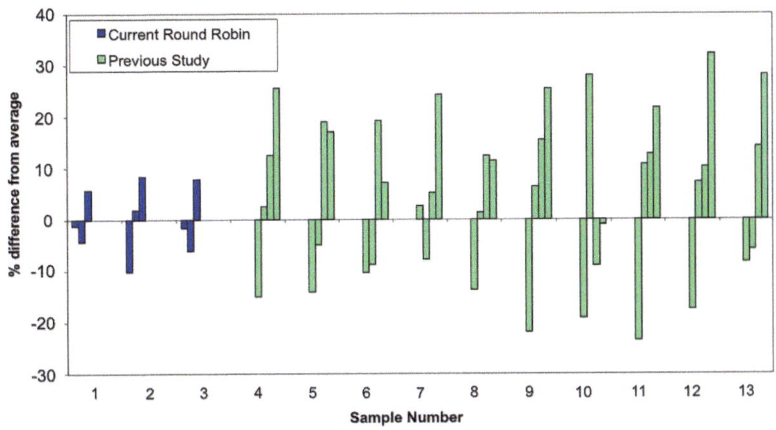

Fig. 20 Round Robin tests using Berlincourt measurements on PC 5H material

3. IEC standard publication 483, 1976. Guide to dynamic measurements of piezoelectric ceramics with high electromechanical coupling
4. BS EN 50324-2:2002, Piezoelectric properties of ceramic materials and components—Part 2: Methods of measurement—Low power

References

1. Stewart, M., Cain, M.G., Gee, M.: Methods of measuring piezoelectric displacement in piezoelectric ceramics. In: Measurement Good Practice Guide, pp. 1–24. NPL, UK (1999)
2. Berlincourt, D., Krueger, H.: Domain processes in lead titanate zirconate and barium titanate ceramics. J. Appl. Phys. **30**(11), 1804–1810 (1959)
3. Damjanovic, D., Demartin, M.: Contribution of the irreversible displacement of domain walls to the piezoelectric effect in barium titanate and lead zirconate titanate ceramics. J. Phys.: Condens. Matter **9**(23), 4943 (1997)
4. Damjanovic, D., Demartin, M.: The Rayleigh law in piezoelectric ceramics. J. Phys. D: Applied Phys. **29**, 2057 (1996)
5. Lodeiro, M.J., Stewart, M., Cain, M.G.: A Round-robin to Measure the Direct Piezoelectric Coefficient Using the Berlincourt Method. VAMAS, NPL, UK (2004)
6. Stewart, M., Battrick, W., Cain, M.G.: Measuring piezoelectric d_{33} coefficients using the direct method. In: Good Practice Guide, pp. 1–34. NPL, UK (2001)
7. Jaffe, B., Cook, W.R., Jaffe, H.L.: Piezoelectric ceramics, ser. In: Non-metallic Solids. Academic Press, New York (1971)
8. Damjanovic, D.: Logarithmic frequency dependence of the piezoelectric effect due to pinning of ferroelectric-ferroelastic domain walls. Phys. Rev. B **55**(2), R649–R652 (1997)
9. Cain, M.G.: Piezoelectric Properties of Ceramic Materials and Components: Part 2: Methods of Measurement: Low Power. BSI, London (2002)
10. Lowrie, F., Cain, M.G., Stewart, M.: Finite element modeling of electroceramics. National Physical Laboratory Report, vol. 150, pp. 1–21. Center for Materials Measurement and Technology (A), NPL, UK (1999)

11. Lefki, K., Dormans, G.: Measurement of piezoelectric coefficients of ferroelectric thin films. J. Appl. Phys. **76**(3), 1764–1767 (1994)
12. Lowrie, F., Cain, M.G., Stewart, M.: Finite Element Modelling of Electroceramics. NPL Report A, vol. 150, pp. 1–29. NPL, UK (1999)

Characterisation of Pyroelectric Materials

Roger Whatmore

1 Introduction

Pyroelectrics form a very broad class of materials. Any material which has a crystal structure possessing a polar point symmetry—i.e. one which both lacks a centre of symmetry and has a unique axis of symmetry—will possess an intrinsic, or spontaneous, polarisation and show the pyroelectric effect. The pyroelectric effect is a change in that spontaneous polarisation caused by a change in temperature. It is manifested as the appearance of free charge at the surfaces of the material, or a flow of current in an external circuit connected to it. The effect is a simple one, but it has been used in a range of sensing devices, most notably uncooled pyroelectric infra-red (PIR) sensors, and has thus come to be of some engineering and economic significance, enabling a wide range of sensing systems, ranging from burglar alarms through FTIR spectroscopic instruments to thermal imagers.

A wide range of material compositions and types are available for potential exploitation, including single crystals, ceramics, polymers, thin films and liquid crystals [1, 2]. An important problem for the materials engineer is how to select the most promising material for exploitation in any given device type. This must be determined by which of the materials properties are relevant to the application being contemplated. Central to the selection decision will be how to accurately measure these properties.

This chapter aims first to describe the main device applications of pyroelectrics and to discuss—on the basis of the physics of how these devices work—the key properties which determine device performance. It will then discuss the main techniques by which these properties can be measured, including some of the potential pitfalls. It

R. Whatmore (✉)
Department of Materials, Imperial College London, South Kensington Campus,
Exhibition Rd, London SW7 2AZ, UK
e-mail: r.whatmore@imperial.ac.uk

M. G. Cain (ed.), *Characterisation of Ferroelectric Bulk Materials and Thin Films*,
Springer Series in Measurement Science and Technology 2,
DOI: 10.1007/978-1-4020-9311-1_4,

will finish with a brief review of the properties of some of the materials which have been applied in practical devices and systems.

2 Basics of Pyroelectric Materials

As noted above, any material whose structure possesses one of the 10 polar symmetry groups (1, 2, m, mm2, 4, 4mm, 3, 3m, 6, 6mm) will be pyroelectric. The reasons for this are fairly simple to appreciate and are illustrated in Fig. 1. This illustrates a very simple polar lattice in which a cation sits within an anion framework, but displaced from the centre of the unit cell (Fig. 1a). This creates an intrinsic spontaneous electrical polarisation (P_S) to the cell. The cation sits in an asymmetric potential well (Fig. 1b), which means that it will find it easier to move in one direction in the cell (back towards the cell centre) than the other. As the temperature of the material is increased, the amplitude of vibration of the cation will also increase, which means that its mean position will be closer to the centre of the cell, and thus the value of P_S will decrease with temperature. Simple examples of real materials which show the effect are ZnO and AlN, which both possess the wurtzite structure, symmetry 6mm [3]. In such materials, the magnitude of P_S decreases with increasing temperature but does not go to zero.

The majority of pyroelectrics which have found practical applications belong to the class of materials known as ferroelectrics [4]. In these materials, the direction of P_S can be switched between equivalent stable states by the application of an electric field of sufficient magnitude. This characteristic, known as ferroelectric hysteresis, is extremely useful, as it means that polycrystalline ferroelectric materials (ceramics and crystalline polymers) can be rendered polar, and hence pyroelectric through a process known as "poling". This usually entails cooling the material from a temperature close to (or even above) T_C while under a sufficiently-high electric field. The field orients the polar axes within the individual crystallites of the material so that they sum together to give a net polarisation. Note that if there are a sufficient number of "options" for the polar axis within any one crystallite, the net polarisation can be close to that achieved in the bulk material (ca 80 % for six or eight options). Examples of ferroelectrics which have been used in pyroelectric applications include:

- $LiTaO_3$ which possesses the ilmenite structure, symmetry 3m at room temperature and for which the direction of P_S can be switched between two opposite directions. This is usually used in single crystal form.
- $PbTiO_3$, which possesses the perovskite structure, symmetry 4mm at room temperature and for which the direction of P_S can be switched between six equivalent directions. (In practical materials, this is usually modified with various other dopants, but the structure remains the same.) This is usually made and used as a polycrystalline ceramic.
- $Pb(Zr_{0.9}Ti_{0.1})O_3$, which also possesses the perovskite structure, but with symmetry 3m at room temperature and for which the direction of P_S can be switched between

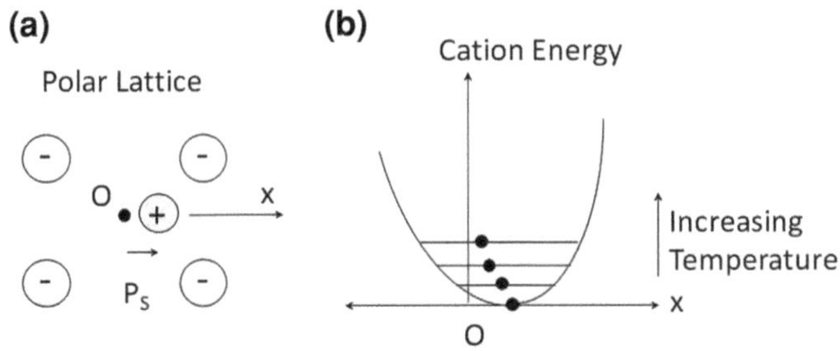

Fig. 1 Schematic diagram illustrating the origins of the pyroelectric effect. **a** Shows a cation sitting off-centre in a cation framework, generating a spontaneous electrical polarisation. **b** Shows the asymmetric potential well within which the cation sits

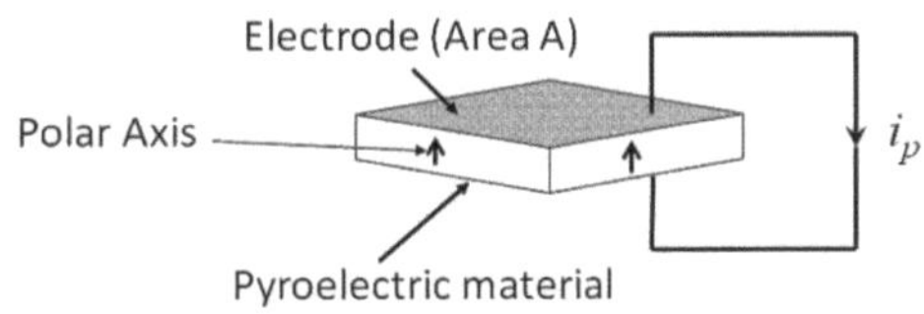

Fig. 2 Schematic diagram illustrating the generation of current by a pyroelectric element

eight equivalent directions. (Again, in practical materials, this is usually modified with various dopants.) This is also usually made and used as a polycrystalline ceramic.

- Poyvinylidene fluoride (PVDF) and its copolymers with trifluoroethylene. These are crystalline polymers with the symmetry mm2.

Several materials have been explored as polycrystalline thin films, usually deposited on silicon, in which the crystallites have a strongly-preferred orientation, which can enhance the pyroelectric effects. These materials will be discussed further below.

In nearly all ferroelectrics, the magnitude of P_S decreases with increasing temperature and goes to zero at the Curie temperature (T_C). In theory, the decrease in P_S is reversible, provided T_C is not exceeded but in practical applications, it is usual to operate the materials well below T_C, as usually P_S will degrade irreversibly unless a DC field is applied to stabilise it. This is discussed further below.

The change in P_S can be detected in the following way. If a piece of pyroelectric material is placed between electrodes, such that the vector P_S is normal to the electrodes and the electrodes are joined with a wire to make an external circuit (see Fig. 2), then a change in temperature will cause a current i_p to flow in the circuit such that:-

$$i_p = A\frac{dP_s}{dT} \cdot \frac{dT}{dt} \tag{1}$$

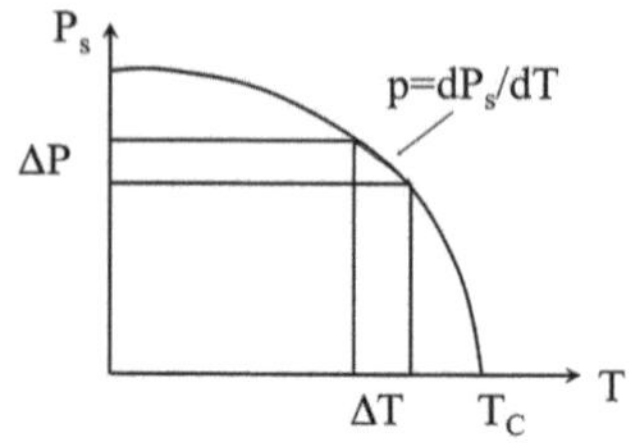

Fig. 3 Illustrating the variation of spontaneous polarisation with temperature in a ferroelectric exhibiting a second-order transition at the Curie temperature T_C

where:
A = area of the electrodes
$\frac{dP_s}{dT}$ = Rate of change of P_S with temperature = p (pyroelectric coefficient)
$\frac{dT}{dt}$ = Rate of change of temperature with time.
This current flow can be used in a variety of ways, as will be illustrated in the next section. In ferroelectrics, the pyroelectric coefficient (p) will increase dramatically as T_C is approached, as shown schematically in Fig. 3. However, there is a strong tendency for ferroelectrics to depole when close to T_C, even if T_C is not exceeded. Nevertheless, the polarisation—and hence the pyroelectric coefficient—can be stabilised close to, and even above, the T_C by the application of an electric field. In this case, the pyroelectric coefficient is given by $p(T, E)$:

$$p(T,\ E) = \frac{dD}{dT} \tag{2}$$

where D is the electrical displacement.

This mode of operation, close to T_C with an applied electric field, is frequently referred to as "dielectric bolometer mode". It has implications for the measurement of the physical properties which determine device operation, which will be discussed in more detail below.

It is important to note that the pyroelectric coefficient is a first rank tensor [5] and thus any crystalline material has a single major component, p.

3 Properties of Pyroelectrics for Radiation Sensing Applications

By far the most common application for pyroelectrics is the sensing of electromagnetic radiation. A typical device structure is shown schematically in Fig. 4. Here, the pyroelectric is shown as thin piece, or element, which is electroded on its major faces. The wafer of the pyroelectric material is usually cut, or poled, so that the polar axis is perpendicular to the electrodes, but there are certain crystals for which an advantage can be gained by having the crystal cut so that p is not normal to the electrodes.

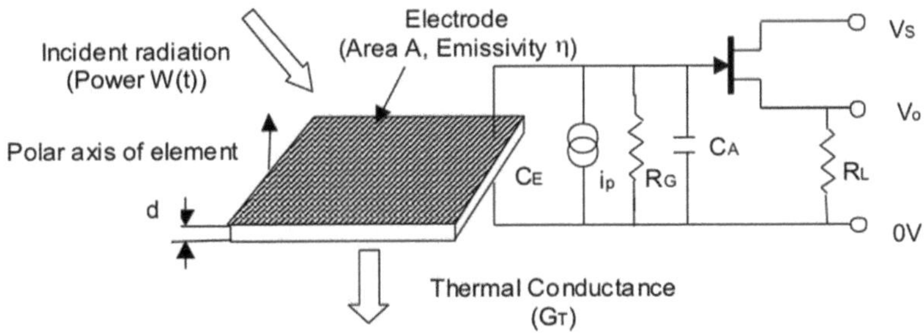

Fig. 4 Showing the structure of a typical pyroelectric detector

The mode of operation is very simple. The radiation to be detected is allowed to fall on, and be absorbed by, the element—causing a change in element temperature. This causes a pyroelectric current i_p to flow through R_G, and a voltage to appear on the gate of the FET. (i_p is shown as a current generator symbol in Fig. 4.) This modulates the transductance of the FET and hence produces an output voltage. Note that this amplifier configuration produces unity gain.

No discussion has yet been made of the wavelength of the radiation to be detected. This is because the detectors simply respond to the energy absorbed from the radiation, and so can be used to detect any wavelength, provided there is sufficient energy present to give a detectable temperature change. Detectors have been demonstrated working at wavelengths from X-rays [6] to microwaves [7]. However, by far the most common uses of pyroelectrics have been for detection of long-wavelength (8–12 micron) or mid-wavelength (3–5 micron) infra-red radiation (usually referred to as LWIR and MWIR respectively). This is because there are few other detectors which can give good performance at such wavelengths without the need for cooling. Resistive bolometers or thermopiles [8] can be used at room temperature, but semiconducting photon detectors generally require some cooling (usually to 77 K for LWIR devices).

The radiation absorption can be achieved by coating the material with a suitable black coating. Examples include gold black [9], platinum black [10] or more-recently layers of carbon nanotubes [11]. Alternatively, the radiation can be absorbed in the material itself, in which case the front electrode should be arranged to be matched to the impedance of free space −377 Ω/square [12]. Other configurations which have been considered include the use of antenna coupling for microwaves or sub-millimetre waves [13].

Generally, pyroelectrics show adequate sensitivity for many LWIR applications, and the fact that they give an output proportional to rate-of-change of element temperature, and hence rate-of-change of radiation intensity, is an advantage in comparison with bolometers and thermopiles. The latter both give outputs proportional to element temperature, and hence radiation intensity. Thus pyroelectrics are very good for detecting the movement of warm (or cold) objects against a static, unchanging background scene, or for comparing radiation intensities from two different sources.

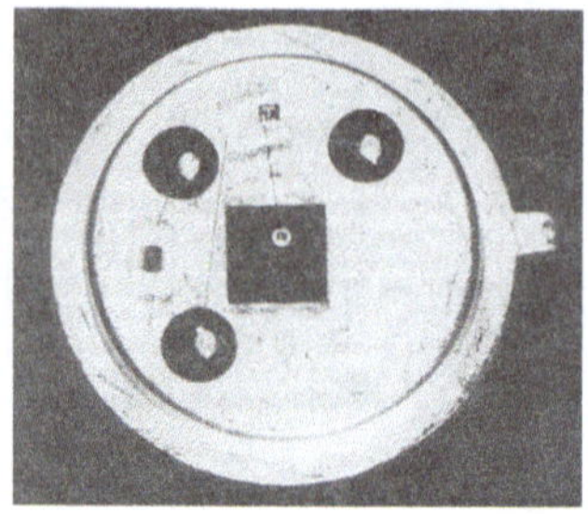

Fig. 5 Photograph of a single-element pyroelectric detector. The element is the *black square* in the *centre* of the package, the FET is the component immediately *above* it and the gate bias resistor is the component on the *left*

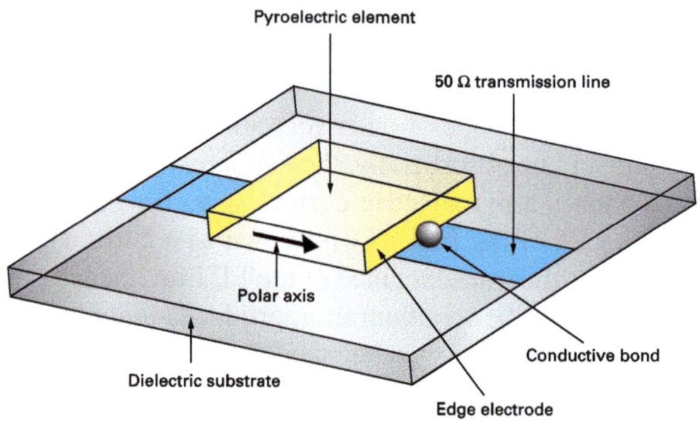

Fig. 6 Diagram of an edge-electroded pyroelectric detector

It is worth considering one further type of pyroelectric detector configuration. The detector configuration shown in Fig. 4 has been widely applied in devices ranging from people sensors (burglar alarms, door openers, automatic light switches) to FTIR spectrometers and gas analysers. An example of a single element detector is shown in Fig. 5. The device shown in Fig. 4 works well at low frequencies (typically 0.1–100 Hz) but, as will be shown below, the performance rolls-off rapidly at higher frequencies. This frequency response is determined by the thermal and electrical time constants (τ_T and τ_E) of the device. Nevertheless, there is nothing intrinsically "slow" about the pyroelectric effect itself, which is mediated by phonons in the material and can be in the picosecond range [14, 15]. Hence, very fast detectors can be made. These are usually edge-electroded, with the polarisation in the plane of the detector (Fig. 6) so that the detector capacitance can be reduced and the device impedance matched to a 50 or 75 Ω transmission line. The sensitivity of such detectors is generally very low, but in applications such as the characterisation of very fast, high power laser pulses, this is not usually a problem. In this case, the lack of a front electrode eliminates the possibility of it being damaged by the pulse.

The physics of operation of the device shown in Fig. 4 has been extensively discussed elsewhere [1, 16–18]. It can be shown that, for radiation of intensity W_0 and modulated at frequency ω, the pyroelectric current i_p, generated per watt of input power (the current responsivity R_i) is given by:

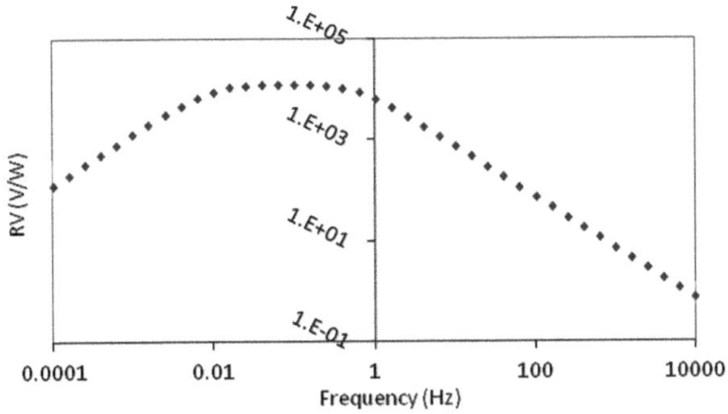

Fig. 7 R_V versus frequency for a pyroelectric ceramic element with area 1 mm^2 and thickness 50 microns, and thermal conductance to the substrate of 50 μW/K

$$\frac{i_p}{W_0} = R_i = \frac{\eta p A \omega}{G_T(1+\omega^2\tau_T^2)^{1/2}}. \tag{3}$$

The form of this response is simple. At low frequencies ($\omega \ll 1/\tau_T$) the response is proportional to ω. For frequencies much greater than $1/\tau_T$ the response is constant, being:

$$R_i = \frac{\eta p A}{H} = \frac{\eta p}{c'd} \tag{4}$$

where c' is the volume specific heat.

Equation 4 gives us the first 'figure-of-merit' which relates the fundamental physical properties of the pyroelectric material with the current responsivity of the device. This is:

$$F_i = \frac{p}{c'}. \tag{5}$$

In this case, we should choose materials with high values of p and low values of c' to maximise the current responsivity.

The voltage responsivity of the detector can be derived from the pyroelectric current i_p and the complex electrical admittance presented to it. The pyroelectric voltage (V_p) generated on the gate of the FET, and therefore the output voltage V_o for a unity amplifier gain and the voltage responsivity (R_V) where:

$$R_V = \frac{i_p}{YW_0} = \frac{R_G \eta p A \omega}{G_T(1+\omega^2\tau_T^2)^{1/2}(1+\omega^2\tau_E^2)^{1/2}}. \tag{6}$$

This function peaks at a frequency $\omega_{peak} = (\tau_E \tau_T)^{-1/2}$, with a value:

$$R_V(max) = \eta p A \frac{R_G}{G_T} \frac{1}{(\tau_E + \tau_T)}. \tag{7}$$

The variation of R_V with frequency is shown in Fig. 7, in this case for a 100 micron square, 25 micron thick ceramic detector element feeding into a typical MOSFET amplifier and with a thermal conductance of 20 mW/K.

Between the two frequencies τ_E^{-1} and τ_T^{-1}, the value of R_V varies by no more than a factor of $\sqrt{2}$ less than R_V(max). Below the lower of these two frequencies, R_V increases as ω^{-1} and above the greater of the two, R_V decreases as ω^{-1}. In the latter region, at high frequencies and for device designs where the element capacitance C_E dominates in Fig. 4, the voltage responsivity is given by:

$$R_V = \frac{\eta p}{c'\epsilon\epsilon_0 A\omega}. \tag{8}$$

This gives us the second figure-of-merit, F_V, where:

$$F_V = \frac{p}{c'\epsilon\epsilon_0}. \tag{9}$$

This is frequently seen in the literature as the most-important figure-of-merit for pyroelectric detectors. However, it should be noted that for most situations, the devices are designed and operated at a frequency to maximise the value of R_V, which is given by Eq. (7). Furthermore, C_E may not dominate over C_A. In this case there is no simple relationship between the material's properties and the voltage responsivity which can be used to give a simple 'figure-of-merit' describing the device's performance, and it is necessary to model the performance of the various available materials in the device structure, and optimise the design accordingly. Maximising the value of F_V is generally a good idea, but it is important to take note of the following points:

- For very small element detectors, where the capacitance of the amplifier may be significantly greater than that of the element, F_i is likely to be the most-useful figure-of-merit. In other words, it may be more useful to maximise p than the ratio p/ϵ. Small detector elements are important for thermal imaging and other array applications, when it is necessary to have a large number of elements in a finite area, defined by read-out IC and optical system implications. Such elements can be as small as 40 microns across [16].
- In general, the best performance will be obtained if the capacitance of the detector element at least matches that of the amplifier. Hence, for small detector elements it can be important to have a high dielectric constant for the pyroelectric material.
- The appearance of G_T in the denominator of Eq. (7) means that the thermal conductivity (k) of the pyroelectric material can be important in determining R_V, although this depends upon the thermal structure of the device. Generally, the thermal diffusivity ($\kappa = k/c'$) is as important as the value of k.

Finally, the importance of noise in a sensor cannot be ignored. There are several sources of noise in a pyroelectric detector [1, 2, 17], including the thermal fluctuation noise (usually negligible), the voltage and current noise in the amplifier, the Johnson noise in the gate bias resistor (which can include a component due to the DC leakage in the pyroelectric material) and the dielectric noise in the pyroelectric. The latter two components usually dominate. In particular, the dielectric noise is proportional to $(\omega\epsilon\epsilon_0 \tan\delta)^{1/2}$, and it is often useful to define a figure of merit which determines the specific detectivity (D*) of the detector:

$$F_D = \frac{p}{c'\sqrt{\epsilon\epsilon_0 \tan\delta}}. \tag{10}$$

Finally, all pyroelectrics are piezoelectric—they will generate charge under mechanical stress. This has two consequences:

Firstly, the pyroelectric coefficient (p^{stress}) which we normally measure under constant (ideally zero) stress consists of two components, a primary component which we would measure at constant strain (p^{strain})—a 'clamped' value—together with a secondary component (p^{sec}) which emerges through a combination of the thermal expansion coefficients of the film (α_{ij}) coupling to the piezoelectric coefficients (e_{ijk}). It is easy to show that:

$$p_3^{stress} = p_3^{strain} + e_{ijk}\alpha_{jk} \tag{11}$$

where the normal repeated suffix summation convention has been used for the tensor coefficients. Generally, the primary coefficient values at constant strain are not important, but the coupling of a pyroelectric thin film to a substrate which has a different thermal expansion coefficient from the film can make a significant difference to the apparent pyroelectric effect (up to $\pm 50\,\mu\mathrm{C\,m^{-2}\,K^{-1}}$—[19]), depending on the relative values of the thermal expansion coefficients of film and substrate and the piezoelectric coefficients of the films. This can make a difference of up to 20 % to the measured values. Generally, the most important of the piezoelectric coefficients to know in determining the magnitude of this effect is e_{31} or $d_{31}/(s_{11} + s_{12})$ [20].

Secondly, from a device performance point-of-view, the piezoelectric effects can be a source of significant electrical noise in a mechanically-noisy environment [21]. This effect is generally termed piezoelectric 'microphony'. Its effects can be minimised by good device and packaging design. However, in order to do this it is again important to know the value of d_{31}, as the largest contribution to the effect comes from the flexure of the substrate to which the element is mechanically connected. The contribution due to self-loading through the thickness of the element (for which the coefficient d_{33} would provide the coupling) is generally small.

Hence, in summary, the material properties which it is important to measure for pyroelectric detector device design are as follows:

- The constant-stress pyroelectric coefficient, p. For ceramic and polymeric materials, it is sufficient to know the magnitude of this quantity, as the principal axis direction will always be parallel to the poling direction, but for single crystals, it is important to know the orientation of this with respect to the crystallographic axes.

- The real and imaginary parts of the dielectric permittivities of the material (e'_{ij} and e''_{ij}). As dielectric permittivity is a second-rank tensor, all the major axis components are needed, together with their orientations with respect to the crystallographic axes. These are essential if a crystal cut is to be used which is not perpendicular to the major axis of the pyroelectric coefficient. The imaginary parts of the dielectric permittivities are also needed to determine the relevant loss tangent, and hence f_D. Dielectric properties (especially loss tangents) are strongly dependent on frequency, and so it is essential to determine these properties at the frequency at which the device will be operated.
- Electrical resistivity (ρ). This will determine the electrical leakage in the device. With care over control of the electrical properties, especially resistivity, the bias resistor R_G shown in Fig. 4 can be "built in" to the material and eliminated as an external component [22–24].
- Volume Specific Heat (c'). This is the product of the mass specific heat (c_p) and the density.
- Thermal Conductivity (k) and Thermal Diffusivity (κ). These will determine the thermal conductance of the device, and hence the thermal time constant. For array-based devices in which there is no physical separation of adjacent elements, the thermal diffusivity will determine the degree of thermal crosstalk between adjacent elements.
- The piezoelectric coefficients (e_{ij}, d_{ij}) are important in determining secondary pyroelectric effects and the level of piezoelectric microphony. e_{31} and d_{31} are the most significant parameters to determine for pyroelectric applications.
- Optical properties: The refractive index and optical loss at the wavelength to be sensed can be important if the absorption of the electromagnetic radiation is to be in the element itself.

4 Measurement of Physical Properties

4.1 Sample Poling

The majority of materials of-interest for pyroelectric applications are ferroelectric, and therefore all the properties should be measured with the material in the poled state. A wide variety of methods have been used to pole ferroelectric materials [4], and for each material it is necessary to determine the optimum poling conditions before definitive property measurements are done. Some of the methods which can be used include:

- Slow cooling of the specimen under a field which is usually a multiple of the coercive field. The specimen is generally held in a bath of an insulating fluid such as a silicone or mineral oil. This is to minimise the risk of surface breakdown and electrical shorting. This method is widely applied to ferroelectric ceramics used for piezoelectric and pyroelectric applications. For example, a modified lead zirconate [25] with a T_C of 230 °C and a value for E_c of ca 10 KV/cm could be reliably poled by cooling from 150 °C to room temperature under a field of 30 KV/cm over a period of ca 30 min. Similar conditions can be used to pole pyroelectric thin films based on a $Pb(Zr_{0.3}Ti_{0.7})O_3$ composition.

- Materials with very high values of T_C (e.g. $LiTaO_3$, T_C ca 620 °C) are often too electrically conductive to pole using the above method because it can lead to electrical breakdown. One technique which can be used in this instance is to slow cool the material through T_C with a pulsed field applied [26]. The field and the duty cycle need to be optimised for the individual material.
- Corona poling [27] is a technique which has been successfully applied to polymeric ferroelectrics such as PVDF. It consists of exposing one, non-electroded, face of the ferroelectric to a corona discharge from a series of metal points held at high voltage (typically >10 KV). The back face of the material is earthed. It has the advantage that, as there is no front electrode, the risks of catastrophic breakdown through a physical defect in the film are much reduced. Recently, it has been demonstrated that this technique can be used successfully for poling ferroelectric thin films [28].

4.2 Pyroelectric Coefficient

Several techniques have been employed for measurement of the pyroelectric coefficient. These are reviewed and discussed below.

Charge Integration

Probably the simplest technique is to heat the material, and integrate the pyroelectric current using a standard charge-integrator circuit, employing a good-quality, FET-input op-amp. An example of such a circuit is shown in Fig. 8. This is the method described by Lang and Steckle [29] and Glass [30]. The FET input is important to give a high input impedance (R_i), which will then give a long integration time-constant (C_i). The charge-integrating capacitor should be sufficiently-large so that the time constant R_iC_i is much greater than the length of time to make the experiment. For $R_i = 10^{12}\,\Omega$ and $C_i = 10\,\mu F$, the time constant would be amply long enough >100 days. The choice of the integrating capacitor is important, as it is essential to minimise leakage. A good quality polyester variety is a good choice, rather than ceramic or electrolytic. If a sample of a typical pyroelectric material is considered (say $LiTaO_3$—see below—for which $p = 200 \times 10^{-6}\,C\,m^{-2}\,K^{-1}$), with an area of $1\,cm^2$, this value of capacitor would give an integration output in the millivolt range for every degree centigrade change in temperature. This can easily be measured using an appropriate digital voltmeter (DVM). The sample temperature can be measured using a thermocouple placed close to the specimen. A Chromel/Alumel thermocouple will produce an output voltage change of approximately 40 μV per °C, although this

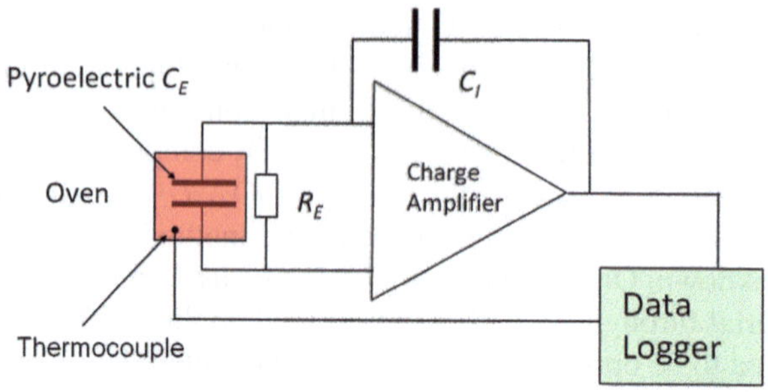

Fig. 8 Charge integration circuit

scale factor is temperature dependent and the appropriate thermocouple calibration curve must be used. The thermocouple voltage can also be measured using a DVM, or with an appropriate digital thermometer. Other devices can be used to measure temperature, such as a Pt resistance thermometer or commercial thermistor. The integrated charge can be logged as a function of temperature, to produce charge integration curves on heating and cooling.

The pyroelectric coefficient at any given temperature can be simply determined from the gradient $\frac{dQ}{dT}$ of this curve and the electrode area (A).

$$p = \frac{dQ}{dT} \cdot \frac{1}{A}. \tag{12}$$

Key issues in making this measurement are:

- As in all temperature measurement, the thermocouple (or other device) must always be at the same temperature as the material being measured. Hence, the design of the heating system is important. Placing the sample on, or within, a large thermal mass within the oven, such as a brass block, can be effective. The thermal mass of the block ensures that the heating and cooling cycles are smooth and steady. The location of the thermocouple in the block, close to the pyroelectric sample being measured, ensures that the measured temperature is as close as possible to the actual temperature of the sample. The sample should be contained within an electrically shielded environment, which reduces the electrical noise. The leads connecting the sample to the charge integrating amplifier should also be fully and individually shielded, with the shields earthed, which reduces stray capacitance (important if the rig is to be used for dielectric measurements see below) and keeps electrical noise to a minimum. If required, channels can be machined into the heating/cooling block to allow the circulation of a coolant fluid. If a measurement below ambient temperature is required, the sample oven must be filled with dry gas possessing a dew point below the lowest temperature to be used, or evacuated, to prevent the condensation of moisture onto the specimen.
- The measurement of pyroelectric currents can easily be confused by the release of trapped charges, causing thermally-stimulated currents (TSC). The trapped

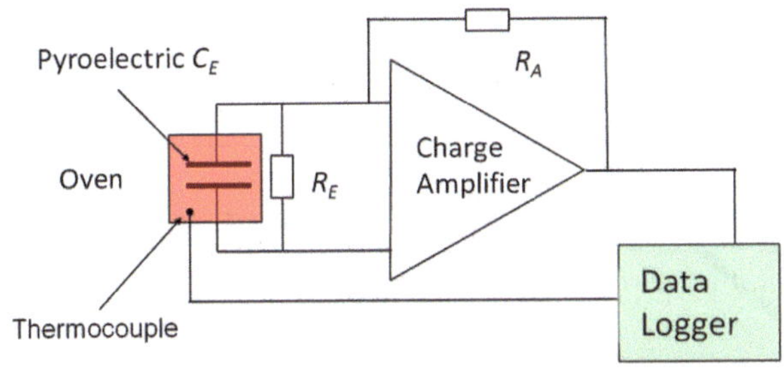

Fig. 9 Circuit for measurement of pyroelectric currents

charges are usually caused by the separation of mobile charge during poling, which become trapped in lattice or structural defects, grain boundaries, electrode interfaces etc. The effect of TSCs is to increase the current released on heating, and reduce it on cooling, so that the gradients of the charge integration curves are greater on heating and reduced on the cooling cycle. The best way to avoid this effect is to wait for a decent length of time (typically 24 h) after poling, before making the pyroelectric measurement. Annealing the sample in an oven for several hours, with the electrodes shorted, can also help to reduce this effect. Clearly the annealing temperature should be well below the Curie temperature. For a typical modified lead zirconate ceramic referred to above, an overnight anneal at 100 °C is usually found to be effective. In any case, it is important to take the pyroelectric coefficients both on heating and on cooling and any residual TSCs taken account of. This can be done as described below.

Pyroelectric Current Measurement

The use of this method has been described by Byer and Roundy [31] and consists of heating or cooling the sample at a known rate while measuring the pyroelectric current—either using a commercial electrometer or using a circuit of the type shown in Fig. 9. The currents can be very small—of the order of nanoamps or less. Hence the circuit needs to be sensitive and well-shielded from sources of electrical noise. There are many excellent examples of commercial devices capable of measuring very low currents.

Figure 10a shows an example of results obtained from this type of measurement, in this case performed on a Cr-doped PZT-PMN ceramic material [32, 33]. The electrode area (A) was 5×10^{-4} m^2. Figure 10a shows the raw data of current and temperature versus time. The heating and cooling rates are plotted in Fig. 10b, together with the corresponding pyroelectric currents. There were significant TSC effects in this specimen. Assuming that the TSC is purely a function of temperature, and not time, then the total measured currents (i_m) on heating and on cooling can be described as:

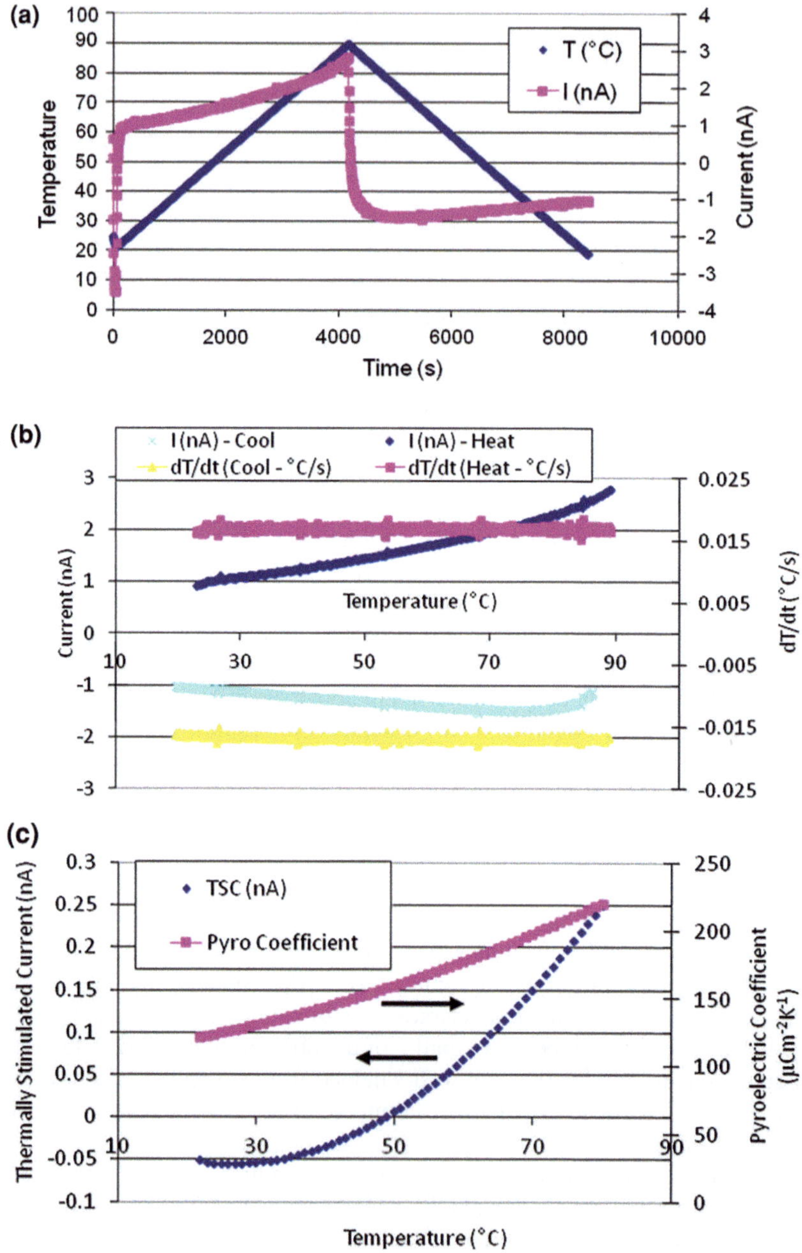

Fig. 10 **a** Pyroelectric currents measured on heating and cooling a Cr-doped PZT-PMN ceramic material **b** Pyroelectric currents and heating/cooling rates as a function of temperature **c** Calculated TSC magnitude and pyroelectric coefficient as a function of temperature (Data from [34])

$$i_m^h = p\dot{T}_h A + i_{tsc} \text{ Heating} \tag{13}$$

$$i_m^c = p\dot{T}_c A + i_{tsc} \text{ Cooling} \tag{14}$$

where $\dot{T}$ is the rate of change of temperature with time and i_{tsc} is the magnitude of the thermally stimulated current at any given temperature. The i_{tsc} can be calculated from:

$$i_{tsc} = \frac{\dot{T}_c i_m^h + \dot{T}_h i_m^c}{\dot{T}_c - \dot{T}_h} \tag{15}$$

Once the TSC is known the reversible pyroelectric coefficients can be calculated from Eqs. 13 and 14. These parameters are plotted in Fig. 10c for this sample. The TSC is small at room temperature (ca 50 pA) but quite significant at elevated temperatures (ca 0.2 nA, or 10 % of the pyroelectric current).

An alternative method for measuring the pyroelectric current consists of making small sinusoidal temperature oscillations about some average temperature, and measuring the current produced [33–36]. An example of the results of this kind of measurement are shown in Fig. 11a. Figure 11b shows three sinusoidal temperature cycles, and the resulting pyroelectric currents.

The pyroelectric coefficient can be calculated using Eq. (16) below [13]:

$$p(T) = \frac{i_{0m}}{\Delta T.A\omega}\sqrt{\left(\frac{R_e}{R_S}+1\right)^2 + \omega^2 R_e^2 C^2} \tag{16}$$

where i_{0m} is the average of the maximum and minimum currents, ω is the angular frequency of temperature oscillation of amplitude ΔT, A is the electrode area, R_e is the resistance of the electrometer (this depends on the instrument range and can be taken from the electrometer specification), and R_S is the resistance of the sample. The effect of TSC on the current plot would be to give a constant DC offset to the measured current (provided the amplitude ΔT is not too great) and hence this method is not badly affected by this perturbation. The value of R_e for a Keithley 6517 electrometer on a 20 nA FSD range would be $<50\,\Omega$ so if $R_S = 10^9\,\Omega$ and $\omega R_e C \ll 1$, the term under the square-root in Eq. (16) is very close to unity. For the data shown in Fig. 11a, $i_{0m} = 2.79$ nA, $\omega = 0.0132\,\text{s}^{-1}$ and $\Delta T = 3.81\,°\text{C}$, giving $p = 110\,\mu\text{C m}^{-2}\,\text{K}^{-1}$. This can be checked by calculating the pyroelectric coefficient from the equation $p = i_m/(\dot{T}A)$, point-by-point through the high gradient parts of the cycle. The values are shown in Fig. 11b for three of the cycles. Close to 26 °C (the mid-point of the thermal cycle) it can be seen that all the cycles give a value close to $p = 110\,\mu\text{C m}^{-2}\,\text{K}^{-1}$. However, the degree of scatter in the measurement increases as the data points move away from this temperature.

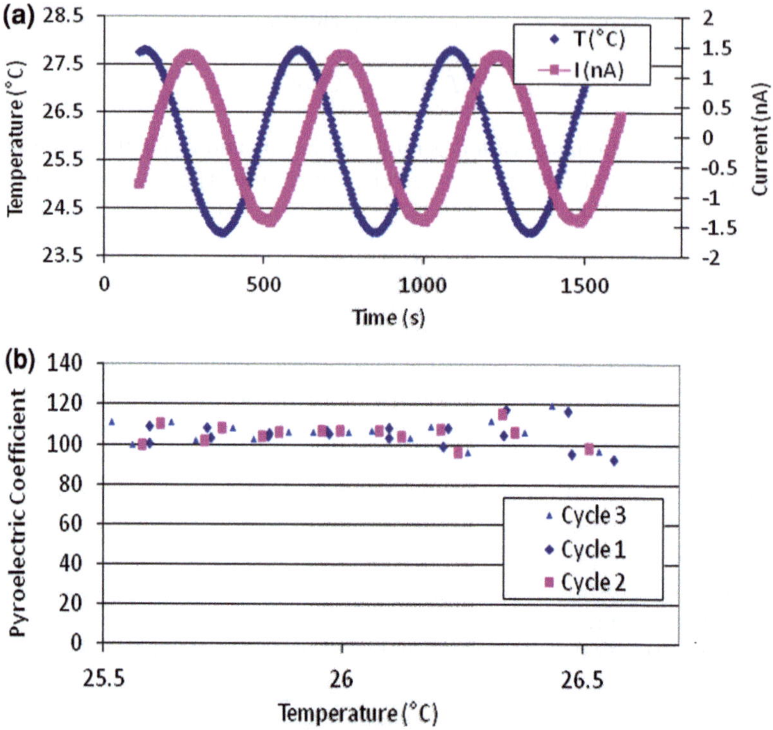

Fig. 11 Alternative method for pyroelectric properties: **a** sinusoidal temperature cycles and measured pyroelectric current and **b** calculated pyroelectric coefficient based on the three cycles in (**a**)

Irradiation Methods

This method, first described by Chynoweth [37], is related to the technique described at the end of the last section. It involves illuminating the specimen with a sinusoidally-varying-flux of infra-red radiation and measuring the resulting pyroelectric current as a consequence of the temperature changes in the material. This is illustrated schematically in Fig. 12.

The method has the great advantage that the temperature of the specimen is continuously cycled by a small amount (usually a fraction of a degree) about an average temperature, which is close to the mode of operation of a pyroelectric detector. The RMS current signal is usually measured using a lock-in amplifier and is unaffected by DC components such as TSC. The disadvantage of the method is that the amount of radiation absorbed is dependent upon the emissivity (η) of the front electrode, so that it can be difficult to be certain of the actual temperature change in the specimen. The radiation can be from a hot body (usually held at ca 500 K), although some authors have used visible or near-IR radiation (e.g. from a gas or solid state laser) for the same purpose. The rms temperature variation (θ_{rms}) for a sample of thermal mass

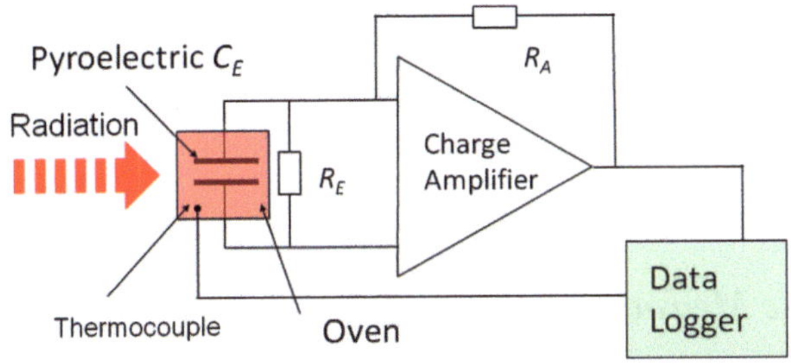

Fig. 12 The "Chynoweth" method for measuring pyroelectric coefficients using AC modulated IR radiation to heat the specimen [37]

H with a thermal conductance G_T to a heat-sink, when illuminated by radiation of intensity W, modulated at an angular frequency ω is:

$$\theta_{rms} = \frac{\eta W}{\sqrt{(G^2 + \omega^2 H^2)}}. \tag{17}$$

As an example, consider a 1 mm square, 50 micron thick piece of a pyroelectric ceramic with $c' = 2.8 \times 10^6\ \mathrm{Jm^{-3}\,K^{-1}}$, bonded to a copper heat sink with a thermal conductance of 1 mW/K (typical for a silver epoxy bond). If illuminated with 5 mW of radiation, modulated at 10 rad/s with $\eta = 0.5$, this would give $\theta_{rms} = 1.4\,^{\circ}\mathrm{C}$ and an rms current response of 11.5 nA if the pyroelectric coefficient is $400\ \mu\mathrm{C\,m^{-2}\,K^{-1}}$. This is readily measurable using a commercial current amplifier, but it indicates the importance of having good shielding and keeping the electrical noise to a minimum. Increasing the thickness by a factor 10 (to 0.5 mm) would reduce the current output to 1.4 nA.

This method relies upon having a good understanding of the operation of the physics of the set-up and the extraction of the basic pyroelectric coefficient from measured pyroelectric current is not straightforward. The 'direct heating' AC method described in the previous section is a much better way to get this parameter. However, the radiation illumination method is a good way to eliminate non-reversible pyroelectric effects and can be a useful method for this reason, if used in conjunction with more direct methods.

Some problems to be aware of with this method:

- If visible radiation is used for illumination, in some cases the results can be complicated by the release of currents due to photoelectric effects.
- The simple theory of pyroelectric devices assumes sinusoidal modulation of the illuminating radiation, so care is required to ensure that the time-waveform of the illuminating radiation is close-to sinusoidal. (This is easy to do with a near-IR solid-state laser, for example). Otherwise, the effects of the higher frequency components of the illuminating waveform must be taken into consideration. In the case of square-wave illumination, the intensity of the higher harmonics can be

significant, considerably distorting the frequency response curve from the ideal. For the example given above, if square-wave modulated radiation, then the higher harmonic terms would boost the apparent response by about 32 % relative to a pure sinusoidally-modulated illumination.

4.3 Dielectric Measurements

The measurement of dielectric constant and loss is well-described elsewhere in this book (Sect. 1 of chapter "Electrical Measurement of Ferroelectric Properties") and so will not be discussed in detail here, other than to mention the key points which pertain to pyroelectric materials. These are as follows:

- It is essential that the dielectric properties are measured under realistic operating conditions. This means that the frequency range of measurement must be chosen to be relevant to the device application. This is particularly important for measurement of tan δ, which can be strongly dependent on frequency, especially at low frequencies. Most pyroelectric devices are operated in the range from 0.1 Hz (for motion sensors) up to about 100 Hz. This poses a problem for many commercial measurement systems (LCR meters and impedance analysers), which are generally designed to work best at high frequencies. Some instruments are specified to low frequencies, however. For example, the Agilent E4980 and Wayne Kerr 6430B are both specified to measure down to 20 Hz, which is within the most-relevant range. There is only one commercial instrument which is specified to go below this frequency, the Solartron Frequency Response Analyser, with 1296a Dielectric Interface will measure down to 10 μHz.
- The dielectric properties of ferroelectric materials are strongly temperature dependent and must be measured over the range of temperatures that the device is to be operated over.
- The dielectric properties of single crystals must be measured for a minimum of three different crystal cuts if the full dielectric constant tensor is to be specified. This is important, as the figure of merit p/ϵ can be optimised for a crystal cut which is not perpendicular to the major axis for the p vector. This was first shown by Shaulov to be the case for deuterated triglycine sulphate crystals [38].

Electrical Resistivity

The bias resistor R_G in Fig. 4 performs an important function in the circuit, in that it both sets the electrical time constant and also biases the gate of the FET. For single element detectors, it is possible to include this as an explicit circuit element, in which case a very high value component is needed (ca $10^{12}\,\Omega$ for a 10 s time constant with C = 10 pF). However, such components are expensive. For large arrays, it is impossible to include a separate resistor for each element. In this case, it is desirable to be able to use the intrinsic DC resistivity (ρ) of the element to provide the required resistance. In this case, we can set the electrical time constant from $\tau_e = \rho\epsilon\epsilon_0$, so that for a ceramic with $\epsilon = 300$, we need $\rho = 4 \times 10^9$ Ωm. There are thus two issues to address. The first is the best method for measuring the DC resistivity in the range 10^8–10^{12} Ωm. The second is the method for controlling it. While many excellent commercial electrometers are available that will permit the accurate measurement of high resistances, the accurate measurement of very high DC resistivities is not trivial. Surface moisture is a major problem and the measurements should ideally be carried out under vacuum, after heating the specimen to expel any adsorbed moisture from the surface. For the most-accurate measurements, guard-ring methods should be used. A further consideration in the measurement of resistivity is the relaxation time associated with the polarisation of immobile charges in the material. This is particularly important for materials that have only moderately-high resistivities. It leads to a peak in current immediately after applying the DC test voltage, followed by an exponential decay in the current to a stable value. The time constant of this decay can be many minutes long and the best way to get the 'DC' value of the resistivity is to measure the current as a function of time and fit it to an equation of the form $i(t) = i_0 \exp 1/t$ where $i(t)$ is the current at time t and i_0 is the current at infinite time. The values of i_0 can be obtained by extrapolation of $\ln(i(t))$ versus $1/t$ and used to compute the DC resistivities at 'infinite' time.

Thermal Properties

As noted above, it is important to have a good knowledge of the volume specific heat and the thermal conductivity/thermal diffusivity. There are many excellent differential scanning calorimeter systems available which can be used to measure the specific heat which, when combined with the sample density can be used to give c'. The thermal diffusivity (which can be important for thermal imaging systems if the target is not reticulated) can be measured directly on a pyroelectric substrate using the laser intensity modulation method (LIMM) described by Lang [39–41] or by using a flash illumination technique [42].

Piezoelectric Coefficients

The techniques used for measurement of piezoelectric properties on bulk materials and thin films are described elsewhere in this book. As noted above, the key coefficients to measure for pyroelectric applications are d_{31} and e_{31}, so that they can be used to model the effects of piezoelectric microphony and the generation of secondary pyroelectric effects due to substrate clamping and thermal expansion effects.

5 Examples of Pyroelectric Materials

There are many different types of pyroelectric, including single crystals, polymers, ceramics and thin films and several reviews [1, 16, 20] have considered the properties of many pyroelectric materials in detail, so the discussion here has been confined to a brief review of pyroelectric ceramics and thin films.

6 Conclusions

This chapter has provided a review of the physics of pyroelectric radiation detectors, the critical physical properties which determine their performance and the various techniques for measuring these. Considerable care is needed, and it is especially important to consider the way in which the material is to be used in the ultimate application the measurement methodologies to be used are selected. This applies especially to the measurement frequencies used for the dielectric properties, which can have a dramatic effect on the results, especially for dielectric loss.

References

1. Whatmore, R.W.: Pyroelectric devices and materials. Rep. Prog. Phys. **49**(12), 1335 (1986)
2. Whatmore, R.W., Watton, R.: Pyroelectric materials and devices. In: Capper, P., Elliott, C.T. (eds.) Infrared Detectors and Emitters: Materials and Devices, pp. 99–148. Kluwer Academic Publishers, The Netherlands (1998)
3. Evans, R.C.: An Introduction to Crystal Chemistry. Cambridge University Press, Cambridge (1964)
4. Whatmore, R.W.: Ferroelectric Materials. In: Kasap, S., Capper, P., (eds.) Handbook of Electronic and Photonic Materials, pp 597–623. Springer, Heidelberg (2006)
5. Nye, J.F.: Physical properties of crystals: their representation by tensors and matrices. In: Oxford Science Publications. Clarendon Press, Oxford (1985)
6. Pontes, W., de Carvalho, A.A., Sakamoto, W.K., de Paula, M.H., Sanches, M.A.A., de Freitas, R.L.B., César, R.B.P., Piubéli, S.L.: PZT for measuring energy fluence rate of x-ray used in superficial cancer therapy. Instrum. Sci. Technol. **38**(3), 210–219 (2010)
7. Lee, T.M., Anderson, A.P., Benson, F.A.: Microwave field-detecting element based on pyroelectric effect in PVDF. Electron. Lett. **22**(4), 200–202 (1986)
8. Kruse, P.W.: Uncooled IR focal plane arrays. In: SPIE's 1995 International Symposium on Optical Science, Engineering, and Instrumentation, vol. 2552, pp. 556–563 (1995)
9. Nelms, N., Dowson, J.: Goldblack coating for thermal infrared detectors. Sens. Actuators, A **120**(2), 403–407 (2005)
10. Lang, W., Kühl, K., Sandmaier, H.: Absorbing layers for thermal infrared detectors. Sens. Actuators, A **34**(3), 243–248 (1992)
11. Parsons, A.D.: Thin-film infrared absorber structures for advanced thermal detectors. J. Vac. Sci. Technol. A: Vac. Surf. Films **6**(3), 1686 (1988)
12. Lehman, J.H., Engtrakul, C., Gennett, T., Dillon, A.C.: Single-wall carbon nanotube coating on a pyroelectric detector. Appl. Opt. **44**(4), 483–488 (2005)

13. Yun, M., Bock, J., Leduc, H., Day, P., Kim, M.J.: Fabrication of antenna-coupled transition edge polarization-sensitive bolometer arrays. Nucl. Instrum. Methods Phys. Res. Sect. A **520**(1–3), 487–489 (2004)
14. Auston, D.H., Glass, A.M.: Optical generation of intense picosecond electrical pulses. Appl. Phys. Lett. **20**(10), 398 (1972)
15. Blackmore, V., Doucas, G., Perry, C., Ottewell, B., Kimmitt, M., Woods, M., Molloy, S., Arnold, R.: First measurements of the longitudinal bunch profile of a 28.5 GeV beam using coherent Smith-Purcell radiation. Phys. Rev. Spec. Top. Accel. Beams **12**(3), 032803 (2009)
16. Porter, S.G., Watton, R., and McEwan, R.K.: Ferroelectric arrays: the route to low-cost uncooled infrared imaging. Proc. SPIE Infrared Technology XXI **2552**, 573 (1995)
17. Putley E.H.: Temperature Noise in Pyroelectric Detectors. Infrared Phys **18**(4), 373 (1978)
18. Putley, E.H.: A method for evaluating the performance of pyroelectric detectors. Infrared Phys. **20**(3), 139–147 (1980)
19. Chang, H.H.S., Whatmore, R.W., Huang, Z.: Pyroelectric effect enhancement in laminate composites under short circuit condition. J. Appl. Phys. **106**(11), 114110 (2009)
20. Muralt, P.: Micromachined infrared detectors based on pyroelectric thin films. Rep. Prog. Phys. **64**(10), 1339 (2001)
21. Shorrocks, N.M., Whatmore, R.W., Robinson, M.K., Porter, S.G.: Low microphony pyroelectric arrays. Proc. SPIE (1985) **588**, 44–51 (1986)
22. Bell, A.J., Whatmore, R.W.: Electrical conductivity in uranium doped, modified lead zirconate pyroelectric ceramics. Ferroelectrics **37**(1), 543–546 (1981)
23. Whatmore, R.W.: High performance, conducting pyroelectric ceramics. Ferroelectrics **49**(1), 201–210 (1983)
24. Stringfellow, S.B., Gupta, S., Shaw, C., Alcock, J.R., Whatmore, R.W.: Electrical conductivity control in uranium-doped PbZrO3-PbTiO3-Pb(Mg1/3Nb2/3)O3 pyroelectric ceramics. J. Eur. Ceram. Soc. **22**(4), 573–578 (2002)
25. Whatmore, R.W., Bell, A.J.: Pyroelectric ceramics in the lead zirconate-lead titanate-lead iron niobate system. Ferroelectrics **35**(1), 155–160 (1981)
26. Herbert, J.M.: Ferroelectric transducers and sensors. In: Electrocomponent Science Monographs. Gordon and Breach Science Publishers, New York (1982)
27. Kumar, A., Periman, M.M.: Simultaneous stretching and corona poling of PVDF and P(VDF-TriFE) films II. J. Phys. D. Appl. Phys. **26**(3), 469 (1993)
28. Marshall, J.M., Zhang, Q., Whatmore, R.W.: Corona poling of highly (001)/(100)-oriented lead zirconate titanate thin films. Thin Solid Films **516**(15), 4679–4684 (2008)
29. Lang, S.B., Steckel, F.: Method for the measurement of the pyroelectric coefficient, dc dielectric constant, and volume resistivity of a polar material. Rev. Sci. Instrum. **36**(7), 929 (1965)
30. Glass, A.M.: Investigation of the electrical properties of sr1xbaxnb2o6 with special reference to pyroelectric detection. J. Appl. Phys. **40**(12), 4699 (1969)
31. Byer, R.L., Roundy, C.B.: Pyroelectric coefficient direct measurement technique and application to a nsec response time detector. Ferroelectrics **3**(1), 333–338 (1972)
32. Whatmore, R.W., Molter, O., Shaw, C.P.: Electrical properties of Sb and Cr-doped PbZrO3–PbTiO3–PbMg1/3Nb2/3O3 ceramics. J. Eur. Ceram. Soc. **23**(5), 721–728 (2003)
33. Molter, O.: Development of new pyroelectric ceramics for thermal imaging applications/Olivier Molter. Dissertation, Ph.D. thesis (M.Sc.), School of Industrial and Manufacturing Science, Advanced Materials, Cranfield University (2001)
34. Whatmore, R.W., Molter, O., Shaw, C.: Electrical properties of Sb and Cr-doped PbZrO3–PbTiO3–PbMg1/3Nb2/3O3 ceramics. J. Eur. Ceram. Soc. **23**(5), 721–728 (2003)
35. Sharp, E.J., Garn, L.E.: Use of low-frequency sinusoidal temperature waves to separate pyroelectric currents from nonpyroelectric currents. part ii: experiment. J. Appl. Phys. **53**(12), 8980 (1982)
36. Garn, L.E., Sharp, E.J.: Use of low-frequency sinusoidal temperature waves to separate pyroelectric currents from nonpyroelectric currents. part i: theory. J. Appl. Phys. **53**(12), 8974 (1982)

37. Chynoweth, A.G.: Dynamic method for measuring the pyroelectric effect with special reference to barium titanate. J. Appl. Phys. **27**(1), 78 (1956)
38. Shaulov, A.: Improved figure of merit in obliquely cut pyroelectric crystals. Appl. Phys. Lett. **39**(2), 180 (1981)
39. Lang, S.B.: Laser intensity modulation method: a technique for determination of spatial distributions of polarization and space charge in ferroelectric materials. Ferroelectrics **78**(1), 129–136 (1988)
40. Lang, S.: Technique for the measurement of thermal diffusivity based on the laser intensity modulation method (LIMM). Ferroelectrics **93**(1), 87–93 (1989)
41. Stewart, M., Cain, M.G.: Spatial characterization of piezoelectric materials using the scanning laser intensity modulation method (LIMM). J. Am. Ceram. Soc. **91**(7), 2176–2181 (2008)
42. Parker, W.J., Jenkins, R.J., Butler, C.P., Abbott, G.L.: Flash method of determining thermal diffusivity, heat capacity, and thermal conductivity. J. Appl. Phys. **32**(9), 1679 (1961)

Interferometry for Piezoelectric Materials and Thin Films

Zhaorong Huang and Glenn Leighton

1 Introduction

Piezoelectricity is the coupling between the mechanical and electric property of materials, which manifests itself by the generation of electric charge upon a pressure or conversely the produce of strain under an electric field. For the converse effect, the displacement related to this kind of strain is generally very small. Even for the best piezoelectric material, for example PMN-PT and PZN-PT single crystals, the longitudinal piezoelectric coefficient d_{33} is no more than 3,000 pC/N. If 1,000 volts are applied on a 1 mm PMN-PT, the resulted displacement is only 3 μm. Other piezoelectric materials, such as PZT the most widely used material, have a d_{33} nearly one order of magnitude smaller. Optical interferometry has long been established as one of the most promising techniques for small displacement measurements, due to its capability of very high resolution and advantages such as no mechanical contact and no need for calibration on the length scale. The development of lasers in the last a few decades has almost eliminated the problems associated with optical path length coherence and beam density, and many techniques based on laser interferometry have been developed for the characterisation of piezoelectric materials. This chapter will review the applications of laser interferometry to the characterisation of piezoelectric bulk and thin film materials and discuss possible issues associated with these techniques.

Z. Huang (✉) · G. Leighton
School of Applied Science, Cranfield University,
Cranfield, Bedfordshire MK43 0AL, UK
e-mail: Z.Huang@cranfield.ac.uk

M. G. Cain (ed.), *Characterisation of Ferroelectric Bulk Materials and Thin Films*, Springer Series in Measurement Science and Technology 2,
DOI: 10.1007/978-1-4020-9311-1_5,

2 Piezoelectric Coefficient Measurement

Piezoelectric d_{ij} coefficients are among the most important parameters for piezoelectric materials. There are generally two approaches to determine them: the direct effect based on charge measurement and the converse piezoelectric effect based on the strain measurement. For bulk ceramic materials, the most commonly used methods include the "Berlincourt" meter based on the direct piezoelectric effect and the sample resonance method based on the converse piezoelectric effect. For thin film samples, things become more complicated. First, it is usually very difficult to produce a homogeneous uniaxial stress on a thin film deposited on thick substrate without also generating a bending effect, which would produce a large amount of charge through the transverse piezoelectric effect. As a result, modifications such as pneumatic pressure rig [1] and sample flexure techniques [2, 3] have been suggested. Piezoresponse force microscopy is a technique based on the detection of local vibrations induced by a testing AC signal applied between the conductive tip of the scanning force microscope and the bottom electrode. Since the radius of the tip apex is in the range of tens of nanometers, the measured piezoelectric response is grain dependent. For some grains, the contribution from d_{15} may be significant and the measured out-of-plane piezoelectric response could be substantially different from the effective value along the poling direction [4]. In this chapter we will discuss the measurement of piezoelectric coefficients using single beam, double beam and heterodyne laser interferometers and the issues related to these techniques.

2.1 Single Beam Michelson Interferometer

Interferometry was used for the detection of piezoelectric and electrostrictive strains nearly 50 years ago [5–8] but was not widely recognised until a feedback circuit was employed to stabilize the optical path length at about the maximum sensitivity point ($\pi/2$ point) [9–11]. The basic principle is that, for a monochromatic light of wavelength λ interfering with a reference beam, the interference light intensity is:

$$\begin{aligned} I &= \left|E_p(t) + E_r(t)\right|^2 \\ &= \left|E_{p0}e^{i\{\omega t - 2k(d+\Delta d)\}} + E_{r0}e^{i(\omega t - 2kd)}\right|^2 \\ &= E_{p0}^2 + E_{ro}^2 + 2E_{p0}E_{r0}\cos(2k\Delta d) \\ &= I_p + I_r + 2\sqrt{I_p I_r}\cos(4\pi\Delta d/\lambda) \end{aligned} \tag{1}$$

where I_p and I_r are the intensities of the probing and reference beams, respectively, Δd is the optical path-length difference between the two beams, and $k = 2\pi/\lambda$ is the wave number. For convenience Eq. (1) can be rewritten as:

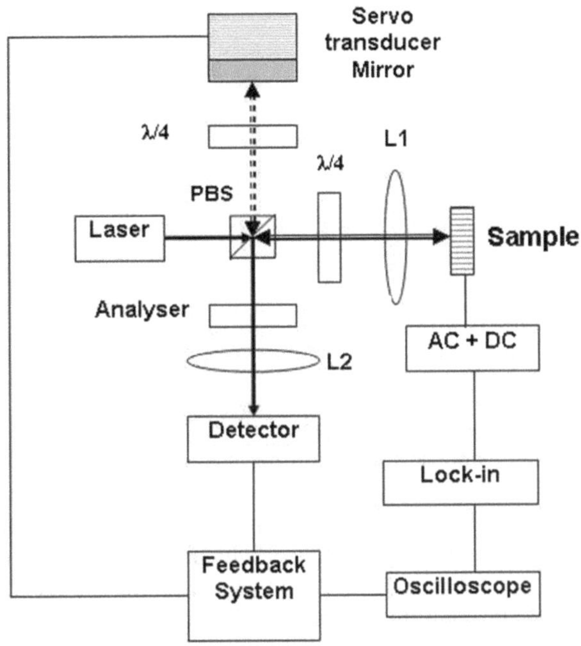

Fig. 1 Schematic of Michelson interferometer for the measurement of piezoelectric strain

$$I = \frac{1}{2}(I_{\max} + I_{\min}) + \frac{1}{2}(I_{\max} - I_{\min})\cos(4\pi \Delta d/\lambda) \quad (2)$$

where I_{max} and I_{min} are the maximum and minimum interference light intensities, which can be measured by a photo-detector.
If $\Delta d = d_{ac} + (2n + 1)\lambda/8$ (so called $\pi/2$ point), the photodiode output can be written as:

$$V_{out} = \frac{1}{2}(V_{\max} + V_{\min}) + \frac{1}{2}(V_{\max} - V_{\min})\sin(4\pi d_{ac}/\lambda) = V_{dc} + \frac{1}{2}V_{pp}\sin(4\pi d_{ac}/\lambda) \quad (3)$$

If $d_{ac}/\lambda << 1$, the sine function can be replaced by its argument. Then the voltage at the photo-detector is $V_{out}(t) = V_{dc} + 2\pi/\lambda\ V_{pp}\ d_{ac}$, where $V_{pp} = (V_{max} - V_{min})$ is the peak-to-peak voltage corresponding to the full fringe displacement and the $V_{dc} = (V_{max} + V_{min})$. From this, d_{33} can be calculated as:

$$d_{33}V = \lambda\delta V/2\pi V_{pp} \quad (4)$$

where V is the driving voltage (rms value) applied to the sample and δV is the corresponding photo-detector output.

Figure 1 shows the schematic drawing of a Michelson interferometer. The operational principle is as follows: The polarized beam from a laser is split into the measuring and reference beams with mutually perpendicular polarization directions

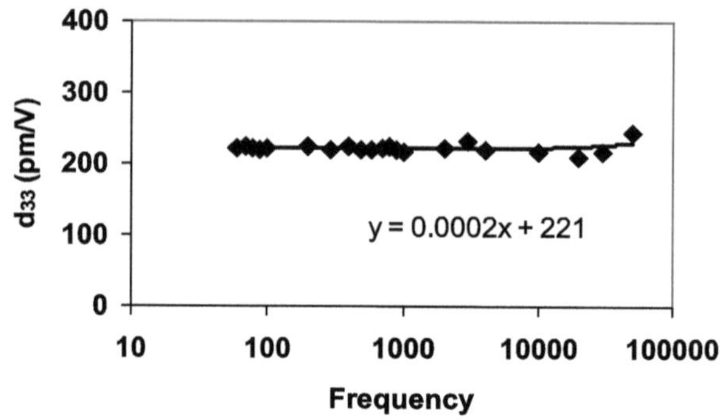

Fig. 2 The piezoelectric coefficient d_{33} for a PZT ceramics as a function of frequency as measured by interferometer techniques. Reprinted with permission from Ref. [41]

by the polarizing beamsplitter PBS. To prevent any reflected beam coming back into the laser and destabilizing the laser a Faraday isolator can be placed between the laser and the PBS. The reference beam travels through a quarter wave plate ($\lambda/4$) then is reflected back by the mirror, and then goes through the PBS. The mirror is attached to a servo transducer which could be a piezoelectric actuator. The probing beam passes through a $\lambda/4$ plate, and is then reflected back by the sample surface, passes through the $\lambda/4$ plate again, to be totally reflected by the PBS. A lens L1 in front of the sample focuses the beam onto the sample surfaces and collects beams reflected from there. Beams reflected from sample surfaces with slightly different tilts will become parallel again due to the lens. After PBS, the probing and reference beams meet at the analyser and the interference pattern is broadened by the lens L2 and the signals are detected by a photodiode and preamplifier detector. The servo transducer feedback system monitors the DC output from the photo-detector and applies a necessary voltage to the transducer to stabilize the optical path length difference between the probing and reference beams at the most sensitive point ($\pi/2$ point). It is also used for calibration. Before an experiment is carried out, the transducer is driven back and forth by an AC voltage to create a full fringe shift so that the V_{max} and V_{min} can be measured. A lock-in amplifier with AC signal generator capability is connected to a PC which can be used to control the experiment. Usually an oscilloscope is employed to monitor the wave forms.

This type of interferometer has been applied to the measurement of piezoelectric and electrostrictive strains by many groups around the world [12–17]. The resolution of the system is limited by noise, which arises due to the fluctuation in the laser intensity and frequency, environmental mechanical and thermal disturbances, shot noise in the photodetector and Johnson noise in the electronics, etc. Resolution as good as 10^{-2} pm has been reported [12] which is close to the theoretical limit by the shot noise in the photodetector [18, 19]. Quartz is often used as a test specimen for the system as its d_{11} (2.3 pC/N) is very stable and well known. Figure 2 shows an example of the measured piezoelectric coefficient d_{33} for a PZT ceramics as a function of frequency. It can be seen that the d_{33} was almost a constant until just before 100 kHz, when resonance started to affect the measured values.

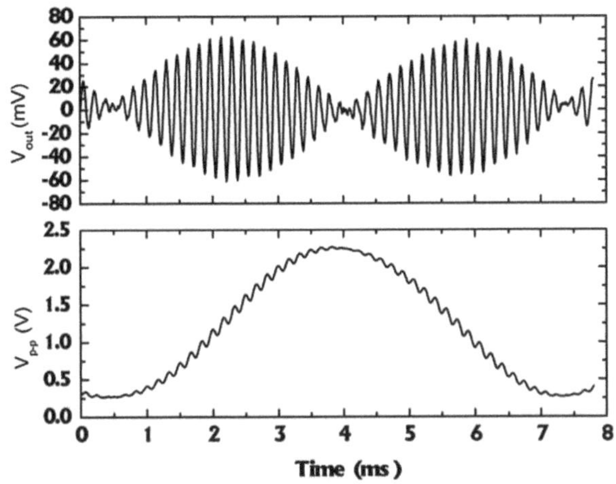

Fig. 3 Simultaneous measurement of Vpp and dV by introducing a periodic movement of the mirror for the reference beam. Reprinted with permission from Ref. [20]

It has been reported recently that a feedback loop is not necessary if a periodic vibration is introduced to the piezoelectric actuator for the reference beam and both the total and the modulated beam intensities are recorded (Fig. 3) [20, 21]. The mirror (or the piezoelectric actuator) travel distance should be larger than $\lambda/8$ so that the total photodiode output varies between the V_{max} and V_{min}. This technique is particularly useful for interferometer scanning measurement because it effectively calibrates the V_{pp} at the same time as the δV measurement, therefore it is less susceptible to the laser intensity instability.

In practice, the use of interferometry for the measurement of piezoelectric coefficients needs to be carried out with great care and patience. Sample shape and mounting, optical alignment, and lock-in amplifier parameters, etc, all can affect the accuracy and even the validity of the measurement. For Eq. (4) to be valid, we have to assume that the back surface of the sample does not move, the applied voltage causes only the front surface of the sample to move, and the sample is able to contract and expand freely along the lateral (perpendicular to the laser beam) direction. This requires that ideally the sample holder and the bonding materials between the sample and the holder to have infinite elastic stiffness constants in the beam direction and the bonding material to have zero elastic constant in the lateral direction. These requirements put some limits to the measurement frequency and specimen shape and dimensions.

The magnitudes of these effects can be analysed using simplified models [9, 10]. Supposing a specimen with length l, width w, thickness h and mass density ρ was bonded to a large sample holder by epoxy glue (Fig. 4a). The epoxy can be regarded as a spring with spring constant $k = clw/h_{ep}$, here c is the elastic stiffness constant of the epoxy and h_{ep} is its thickness. Suppose the front surface moves as $d_0\cos(2\pi ft)$ here f is the frequency, then the mass centre moves as $0.5d_0\cos(2\pi ft)$, and the force to hold the sample is $2\pi^2f^2lwh\rho d_0\cos(2\pi ft)$, and the ratio of back surface motion to the front surface motion is $2\pi^2f^2h\rho h_{ep}/c$. For typical values h = 1 mm,

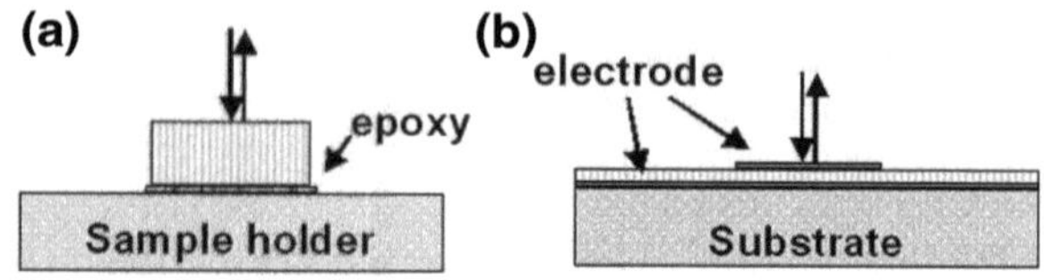

Fig. 4 Schematic of **a** bulk and **b** thin film specimen for laser interferometer measurement

$\rho = 7{,}900\,\mathrm{Kg\,m^{-3}}$, and c = 3 GPa [22]. Usually the epoxy layer thickness h_{ep} is no larger than 0.1 mm, so the ratio $\sim 5f^2 \times 10^{-12}$. Hence, to make the error introduced by the back face motion of the sample less than 1 % requires the measurement frequency to be less than 40 kHz. Since the elastic constant of the epoxy c is about two orders smaller than that of typical piezoelectric material such as PZT, the clamping effect in the lateral direction is negligible. The resonant frequency of the above system is $1/2\pi \times \sqrt{k/m} = 1/2\pi \times \sqrt{c/h_{ep}h\rho} \sim 300\,\mathrm{kHz}$. From this analysis it seems single beam interferometer should be able to measure the piezoelectric coefficient for bulk materials rather accurately as long as the AC signal frequency is not too high, and the back surface is firmly bonded to a large substrate so that sample warp does not happen.

However, if the sample lateral dimension is markedly larger than its thickness, the clamping effect and specimen distortion may not be neglected. Li et al. observed a reduction of nearly a third of the measured d_{11} for a rectangular plate-like quartz specimen, whilst a near-cubical quartz specimen gave a d_{11} very close to its theoretical value [12]. It has been reported that the measured d_{33} varied from 750 to 1,250 pC/N for the same disc specimen PC5H (Morgan Electro Ceramics) simply by changing the way the sample was mounted [23]. For a thin film deposited on a much thicker substrate (Fig. 4b), the bonding is usually regarded as infinitely rigid and the strain at the film/substrate interface is continuous, therefore the clamping effect can not be neglected. Hence the measured d_{33} is a effective piezoelectric coefficient, usually represented as $d_{33,f}$ (If there are no lateral displacements under an applied voltage, the effective converse longitudinal piezoelectric coefficient $d_{33,f} = d_{33} - 2d_{31}\frac{s_{13}^E}{(s_{11}^E + s_{12}^E)}$ where s_{ij} are the mechanical compliances of the piezoelectric film and d_{31} the transverse piezoelectric coefficient [24]. Figure 5 shows the schematic deformation for a specimen under (a) positive and (b) negative applied voltages for a + z polarized piezoelectric film. Assuming the specimen is symmetrical to the central axis at 0 and the bottom electrode is grounded. When a negative voltage is applied to the top electrode (Fig. 5a), the electric field is parallel to the polarization, so the piezoelectric film expands at the thickness direction. At the same time, it contracts in the lateral direction due to the transverse converse piezoelectric effect and this causes the substrate (from 0 to L_1) underneath the top electrode to bend. The substrate outside the top electrode area (from L_1 to L_2) should be the rigid body extension following the slope at the edge of the electrode (L_1). From this we can see that this substrate deformation does not contribute to the front surface displacement at the centre of the top electrode. On the other hand, when a positive voltage is applied to the top electrode (Fig. 5b), the electric field is anti-parallel to the polarization and the piezo-

Fig. 5 Schematic of thin film samples under an external electric field **a** parallel to the polarization and **b** anti-parallel to the polarization direction. Assuming $d_{33} > 0$

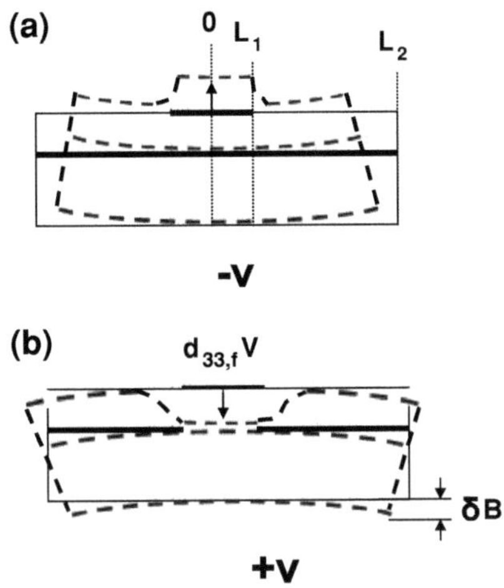

electric material should contract by $d_{33,f}V$ and the substrate bends downward by δB at the edge of the specimen. From Fig. 5b it can be seen that the surface movement is $-d_{33,f}V+\delta B$. So, the measured front surface displacement is $2d_{33,f}V-\delta B$ over a cycle.

We can see from the above analysis it is possible to minimise even eliminate the effect of the substrate bending for the single beam interferometer measurement, if the laser beam is focused at the centre of the top electrode and the applied electric field is **always** parallel to the polarisation, i.e., for the +z poled film, the voltage at the bottom electrode is always higher than voltage at the top. Assuming the bottom electrode is grounded, the top surface voltage can be $-V\{1+\cos(2\pi ft)\}$. For the –z poled film, the voltage at the bottom electrode is required to be always lower than voltage at the top. If the bottom electrode is grounded, the voltage applied to the top surface can be $V\{1+\cos(2\pi ft)\}$. Under these conditions, the induced bending is always upward as shown in Fig. 5a, and the top surface displacement at the centre of the electrode is always equal to $d_{33,f}V\{1+\cos(2\pi ft)\}$ ($d_{33,f} > 0$).

Next we will use a simplified model to estimate δB. Assuming the specimen is a strip and can be treated as a bimorph, which has been studied extensively [25, 26]. The curvature κ of the bending is:

$$\kappa = \frac{6AB(1+B)}{A^2B^4+2A(2B+3B^2+2B^3)+1}\cdot d_{31}\frac{E_3}{t_p} \quad (5)$$

The bending displacement δ at L_1 is:

$$\delta = \frac{1}{2}\kappa L_1^2 = \frac{L_1^2}{2t_p}\cdot\frac{6AB(1+B)}{A^2B^4+2A(2B+3B^2+2B^3)+1}\cdot d_{31}E_3 \quad (6)$$

And the slope α at the L_1 is:

$$\alpha = \kappa L_1 = \frac{L_1}{t_p} \cdot \frac{6AB(1+B)}{A^2B^4 + 2A(2B + 3B^2 + 2B^3) + 1} \cdot d_{31}E_3 \tag{7}$$

The displacement between L_2 and L_1 is α $(L_2 - L_1)$ and:

$$\delta B = \delta + \alpha(L_2 - L_1) \tag{8}$$

where $A = Y_m/Y_p$, $B = t_m/t_p$, Y_m Young's modulus of the substrate layer, Y_p Young's modulus of piezoelectric layer, t_m thickness of the substrate layer and t_p thickness of the piezoelectric layer. For a typical specimen, PZT as the piezoelectric material and Si as the substrate with typical values of: $Y_m = 130$ GPa, $Y_p = 80$ GPa, $t_m = 0.25$ mm, $t_p = 0.001$ mm, $L_1 = 1$ mm, $L_2 = 10$ mm, V = 1 V, and $d_{31} = 100$ pm/V. This gives $\kappa \sim 6 \times 10^{-3}$ m^{-1}, $\delta \sim 3$ nm, $\alpha(L_2 - L_1) \sim 53$ nm, and $\delta B \sim 56$ nm! For a $d_{31} = 100$ pm/V the $d_{33,f} \sim 200$ pm/V therefore the thickness dilatation is 200 pm, meaning the bending displacement is much larger than the longitudinal piezoelectric displacement! This analysis agrees with the conclusions obtained in Ref. [13] where it was assumed that the top electrode covers the whole PZT top surface.

The above analysis is for the quasi-static operation, i.e., the measurement frequency is significantly below the resonant frequency. Treating the specimen as shown in Fig. 5b as a simply supported (free-free boundary condition) beam and neglecting the effect of the piezoelectric film layer on the resonant frequency (since B >>1), the fundamental resonant frequency is estimated as $f_1 \approx \frac{4.73^2}{2\pi L_2^2}\sqrt{\frac{Y_m I}{\rho_{Si} A}} = \frac{4.73^2 t_m}{2\pi L_2^2}\sqrt{\frac{Y_m}{12\rho_{Si}}} \sim 19$ kHz where $\rho_{Si} = 2330$ kg m^{-3}. This indicates that the measurement signal frequency should not be any higher than a few kHz. Strictly speaking, a practical sample is unlikely to be a strip, the Eqs. 5–7 for bimorph beams need to be replaced by plate theory. A plate is stiffer to bend than a strip, so the above calculations over-estimate the bending displacement and under-estimate the resonant frequency.

The above estimation assumed that the specimen was simply supported on the sample holder. However, if the specimen is glued by an epoxy resin for example firmly to a rigid body, the substrate bending can be greatly suppressed [10, 27]. Assuming the substrate of a specimen as shown in Fig. 5b is glued to a rigid sample holder by an epoxy layer, we can estimate how much the epoxy layer would have been stretched upon the application of a voltage to the piezoelectric film. Upon the application of a voltage, an upward bending moment M_{eq} is induced and the specimen moves upward until it is counter-balanced by the stretched epoxy layer. As discussed above, the epoxy layer can be considered as a spring with spring constant $k = cA/h_{ep}$, where A is the contact area of the epoxy with the substrate. For simplicity assuming the maximum displacement at the edge of the top electrode (L_1) is Z_0 and the displacement is a linear function of the distance from the origin (Fig. 6). So the displacement at the position x within the top electrode is xZ_0/L_1.

Considering a small area $w_{ep}dx$ where w_{ep} is the width of the epoxy layer and dx is a small length at the position x. Then the force produced by the epoxy

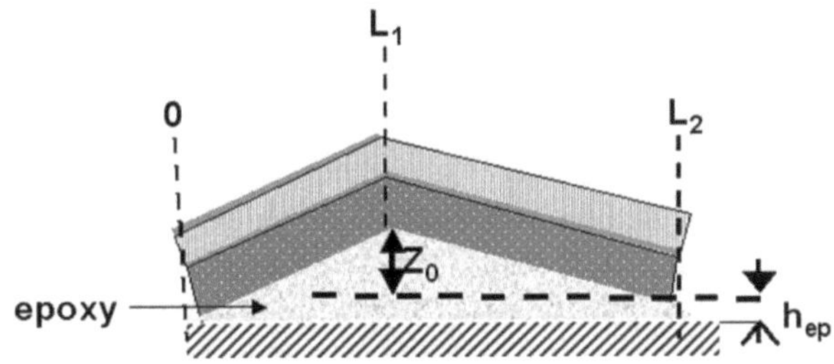

Fig. 6 Schematic of thin film specimen deformation under an applied voltage. The back of the substrate is bonded to a rigid sample holder by an epoxy layer

area $w_{ep}dx$ which acts upon the specimen is $xcw_{ep}Z_0/h_{ep}L_1dx$ and the moment is $x^2cw_{ep}Z_0/h_{ep}L_1$. Between L_1 and L_2, the displacement is $xZ_0/(L_2 - L_1)$, the force is $xcw_{ep}Z_0/h_{ep}(L_2 - L_1)dx$ and the moment is $x^2cw_{ep}Z_0/h_{ep}(L_2 - L_1)dx$. The total moment by the epoxy layer is M and:

$$\begin{aligned} M &= \int_0^{L_1} x^2 \frac{cw_{ep}}{h_{ep}} \frac{Z_0}{L_1} dx + \int_{L_1}^{L_2} x^2 \frac{cw_{ep}}{h_{ep}} \frac{Z_0}{L_2 - L_1} dx \\ &= \frac{1}{3}\frac{cw_{ep}}{h_{ep}} Z_0 L_1^2 + \frac{1}{3}\frac{cw_{ep}}{h_{ep}} Z_0 (L_2 - L_1)^2 + \frac{cw_{ep}}{h_{ep}} Z_0 L_1 L_2 \\ &= \frac{1}{3}\frac{cw_{ep}}{h_{ep}} Z_0 (2L_1^2 + L_2^2 + L_1 L_2) \end{aligned} \tag{9}$$

The equivalent bending moment M_{eq} can be obtained by bimorph theory [25, 28]:

$$\begin{aligned} M_{eq} &= \frac{wt^2 Y_p}{4} \cdot \frac{2AB}{(AB+1)(B+1)} \cdot d_{31} E_3 \\ &= \frac{wY_p t_p^2}{4} \cdot \frac{2AB(1+B)}{(AB+1)} \cdot d_{31} E_3 \end{aligned} \tag{10}$$

where $t = t_m + t_p$. Z_0 can then be obtained by setting $M = M_{eq}$. Then, the ratio of Z_0 and the longitudinal piezoelectric displacement is:

$$\begin{aligned} \frac{Z_0}{d_{33,f} E_3 t_p} &= \frac{wY_p t_p}{4} \cdot \frac{2AB(1+B)}{(AB+1)} \cdot \frac{3h_{ep}}{cw_{ep}(2L_1^2 + L_2^2 + L_1 L_2)} \cdot \frac{d_{31}}{d_{33,f}} \\ &= \frac{Y_p}{4c} \cdot \frac{2AB(1+B)}{(AB+1)} \cdot \frac{3h_{ep} t_p}{(2L_1^2 + L_2^2 + L_1 L_2)} \cdot \frac{w}{w_{ep}} \cdot \frac{d_{31}}{d_{33,f}} \\ &\approx 5 * 10^{-3} \cdot \frac{w}{w_{ep}} \cdot \frac{2d_{31}}{d_{33,f}} \end{aligned} \tag{11}$$

Here we have used the same values as before for the related parameters and a large epoxy thickness value $h_{ep} = 0.1$ mm. The minimum epoxy width is the width of the top electrode w and usually $w_{ep} >> w$, usually $2d_{31} \sim d_{33,f}$, so the maximum value between the maximum back surface displacement Z_0 and the piezoelectric displacement is no more than 0.5 %. If the epoxy width is greater than the width of the top electrode, this ratio is even smaller. This indicates that the sample bending effect can be safely neglected if the specimen is glued to a rigid sample holder. Works reported in the literature support the above estimation. Muensit and Guy were able to avoid the substrate bending excitation and obtained $d_{33} = 2.0 \pm 0.1$ pm/V for GaN thin films by gluing the specimen to a thick brass plate, which was in turn rigidly attached to an optical stage. [29]. Lueng et al. measured $d_{33} = 2.13$ pm/V for GaN thin films with a heterodyne interferometer, [30]. They used silver epoxy to glue the Si substrate to an aluminium block which in turn was rigidly attached to a translation stage.

A distinctive advantage of the laser interferometry technique is its very wide operating bandwidth. The bandwidth for the setup shown in Fig. 1 is mainly limited by the bandwidth of the photo-detector. Tikka et al. constructed a laser interferometer which is capable of detecting vibration amplitude down to less than 10 pm, with a lateral resolution of 300 nm and bandwidth up to 12 GHz [31, 32]. Using a precision controlled sample stage and the laser interferometer system they measured the mode shapes of the vibrations of the thin film bulk-mode resonators which are used as filters of 3G mobile phones at frequencies higher than 1 GHz. At these high frequencies, mechanical movements could not follow the signal variation and the measured vibration is most likely due to the intrinsic piezoelectric contribution.

2.2 Double Beam Mach-Zehnder Interferometer

To overcome the possible substrate bending effect, double beam Mach-Zehnder interferometers have been developed for the measurement of piezoelectric coefficients for bulk ceramics and thin films, [33–41]. In this set up, the probing beam hits both the front and back surfaces of the sample. Figure 7 shows the schematic drawing of a Mach-Zehnder interferometer. The operational principle is as follows: The polarized beam from a stabilised He-Ne laser passes through a Faraday isolator before hitting the polarizing beamsplitter PBS1. PBS1 splits the beam into the probing and reference beams with mutually perpendicular polarization directions. The reference beam is totally reflected by the beamsplitter PBS2, then travels through a quarter wave plate ($\lambda/4$) and then reflected by the mirror M1. The mirror M1 is attached to a servo transducer (usually a piezoelectric actuator). This reference beam is reflected by another beamsplitter PBS4. The probing beam passes through a $\lambda/4$ plate, and is then reflected by the sample surface, passes through the $\lambda/4$ plate again, to be totally reflected by the PBS1. After being reflected by the two 45° placed mirrors M2 and M3, the probing beam is reflected by the PBS2, goes through a $\lambda/4$ plate and is reflected by the back surface of the sample. The two lenses L1 and L2 at either side of the sample focus the beam onto the sample surfaces and collect the reflected beams. Beams reflected from sample surfaces with slightly different axial tilts will become parallel again due to these lenses. After PBS2, the probing beam

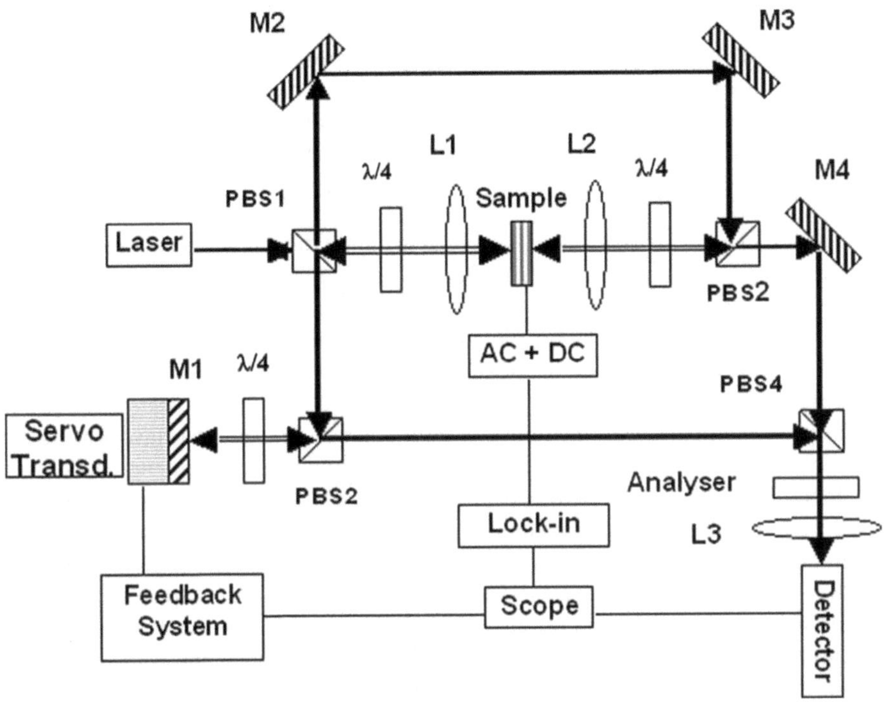

Fig. 7 Schematic of double-beam interferometer

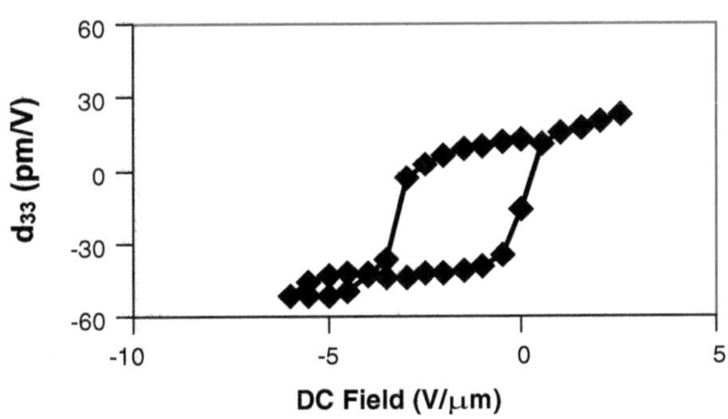

Fig. 8 The piezoelectric coefficient $d_{33,f}$ for a PZT52/48-Mn doped-2 μm film as a function of the DC bias electric field as measured by the interferometer. Reprinted with permission from Ref. [17]

is reflected by a mirror M4 and then goes through the beamsplitter PBS4. After the PBS4, the probing and reference beams meet at the analyser and the interference pattern is broadened by the lens L3 then detected by a photodiode and preamplifier detector. The electric driving and feedback systems are the same as in the single beam interferometer.

Figure 8 shows an example of the $d_{33,f}$ as a function of DC bias for a 2 μm thick Mn-doped PZT52/48 film as measured by this kind of double beam interferometer.

If the front and back beams are perfectly aligned, the substrate bending should not contribute anything to the measured displacement. However, if there is a misalignment of l_s, the measured displacement also includes the bending effect. This can be estimated from Eq. (6). Assuming the front beam is at the centre and the back side beam is misaligned by $l_s = 0.2$ mm, using the parameters the same as in 2.1 and notice B >> 1, then the bending contribution $\delta_s \sim \frac{l_s^2}{2t_p} \cdot \frac{6}{AB^2+4B} \cdot d_{31}E_3 \sim \frac{l_s^2}{t_m} \cdot \frac{3}{AB+4} \cdot d_{31}E_3 \sim 120$ pm, which is more than half of the thickness dilatation! So the precise alignment of the back and front beams is of critical importance to the accurate measurement of the piezoelectric coefficient for thin films using the double-beam interferometer techniques. As for the single beam interferometer, proper specimen mounting such as gluing part of the substrate firmly to a rigid holder can greatly suppress the bending effect.

Various modifications to the above double-beam interferometer have since been reported. The optical fibre version of the Mach-Zehnder interferometers have been developed by Gerber et al. [40] and Fernandes et al. [42]. Like other optical fibre based systems, these interferometers have easier optical alignment, small size and weight, and higher sensitivity. Huang and Whatmore reported a common path laser interferometer which uses two less beam splitters giving a more compact system [17]. In this design, the probing and reference beams share a common path for most of the optical length, LabVIEW was used to control the driving signal and data acquisition so that experiments can be carried out automatically and remotely controlled via a local area network (Fig. 9).

Very compact single beam [43] and double beam laser interferometers [16] have been developed by Nosek et al. These designs enabled the specimen and some key components of the system to be housed within a cryogenic cooling chamber so that the specimen temperature could be as low as 240 K (Fig. 10). Temperature dependence of the piezoelectric coefficients d_{33}, d_{31} and d_h were measured with these systems to study the intrinsic and extrinsic contributions of the piezoelectric effect [16, 40, 44].

2.3 Heterodyne Interferometer and Laser Scanning Vibrometer

The above strain measurement based laser interferometers have very high sensitivity, simple data processing and are relatively cheap to set up. However, they usually require careful optical alignment, very stable work environments with minimum mechanical and thermal disturbances, and samples with very good surface reflectivity which means double side polished wafers for the Mach-Zehnder interferometer. On the other hand, frequency measurement based interferometers, such as the heterodyne interferometer and vibrometer, are more robust and user friendly.

Figure 11 shows the schematic of a heterodyne interferometer. A single frequency laser beam is split into two by the PBS, one being a reference beam and the other the measurement beam. The measurement beam passes through an acoustic-optic

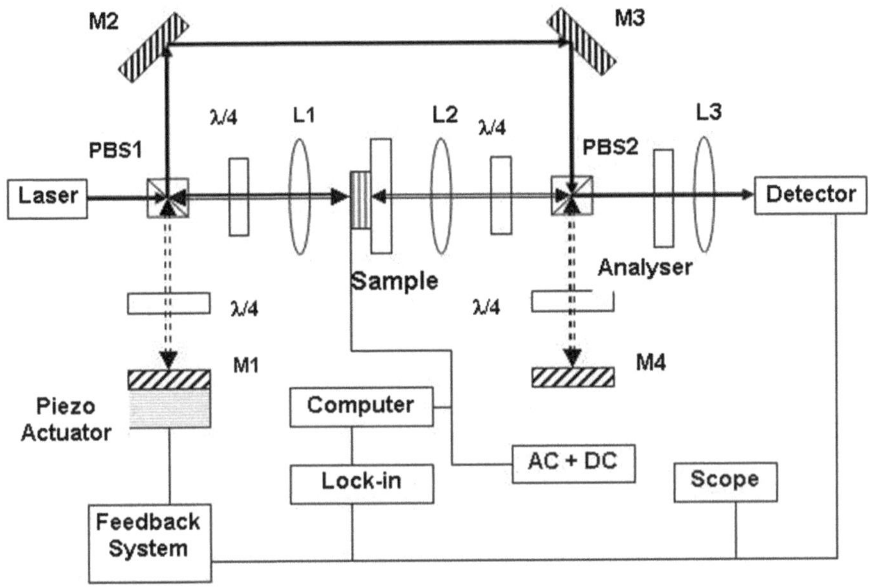

Fig. 9 Schematic of common path double beam laser interferometer. Reprinted with permission from Ref. [17]

modulator (AOM) which produces a frequency shift of $f_R = \omega_R/2\pi$ (say, $f_R = 70\,\text{MHz}$) much higher than the expected motion frequency (or the Doppler frequency shift) of the sample. This measurement beam then passes through another PBS to a quarter wave plate (λ/4) and is focussed by a lens onto the sample surface. After reflection from the surface of the sample the measurement beam retraces its path to the PBS, before it is totally reflected and combined with the reference beam before being incident onto a photodetector. The photodetector output current I(t) is proportional to the light intensity:

$$
\begin{aligned}
I &= \left|E_R(t) + E_p(t)\right|^2 \\
&= \left|E_{R0}e^{i(\omega t+\phi)} + E_{p0}e^{i(\omega-\omega_R)t}\right|^2 \\
&= E_{R0}^2 + E_{po}^2 + 2E_{p0}E_{R0}\cos(\omega_R t + \phi(t)) \\
&= I_p + I_r + 2\sqrt{I_p I_r}\cos(\omega_R t + \phi(t))
\end{aligned}
\tag{12}
$$

where $\varphi(t)$ is the phase shift due to the sample movement, reflection, etc. If the sample surface displacement from its average position is described by z(t), then $\varphi(t) = \varphi_0 + 2kz(t)$, where φ_0 is the phase difference at $z(t) = 0$ and k is the wave number. Assuming a sinusoidal motion for the sample $Z_0\sin(\omega_T t)$, $\varphi(t) = \varphi_0 + 2kZ_0\sin(\omega_T t)$, and $kZ_0 << 1$, then:

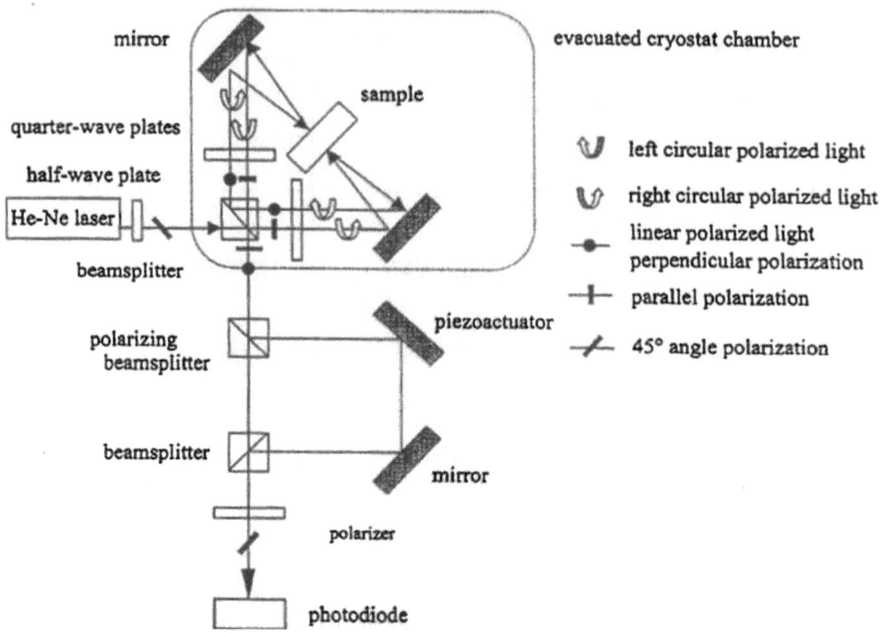

Fig. 10 Schematic of the double beam interferometer for measurement inside a cryostat chamber. Reprinted with permission from Ref. [43]

Fig. 11 Schematic of a heterodyne interferometer

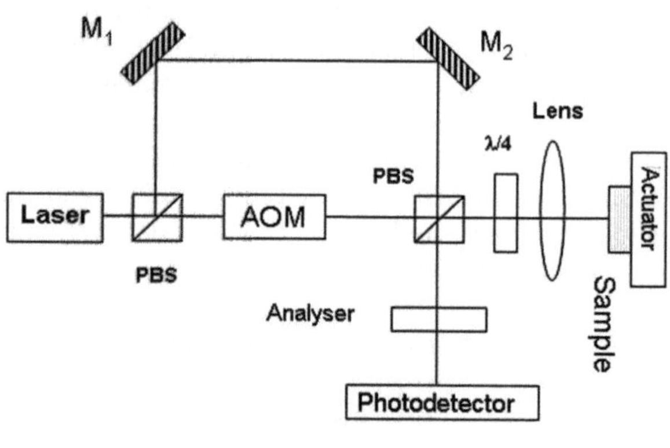

$$
\begin{aligned}
I &= I_p + I_r + 2\sqrt{I_p I_r}\cos(\omega_R t + \phi_0 + 2kZ_0 \sin(\omega_T t)) \\
&= I_p + I_r + 2\sqrt{I_p I_r}\{\cos(\omega_R t + \phi_0)\cos(2kZ_0 \sin(\omega_T t)) \\
&\quad - \sin(\omega_R t + \phi_0)\sin(2kZ_0 \sin(\omega_T t))\} \\
&\approx I_p + I_r + 2\sqrt{I_p I_r}\{\cos(\omega_R t + \phi_0) - \sin(\omega_R t + \phi_0) 2kZ_0 \sin(\omega_T t)\} \\
&= I_p + I_r + 2\sqrt{I_p I_r}\{\cos(\omega_R t + \phi_0) + kZ_0[\cos(\omega_R + \omega_T)t + \phi_0) \\
&\quad - \cos(\omega_R - \omega_T)t + \phi_0)]\}
\end{aligned} \tag{13}
$$

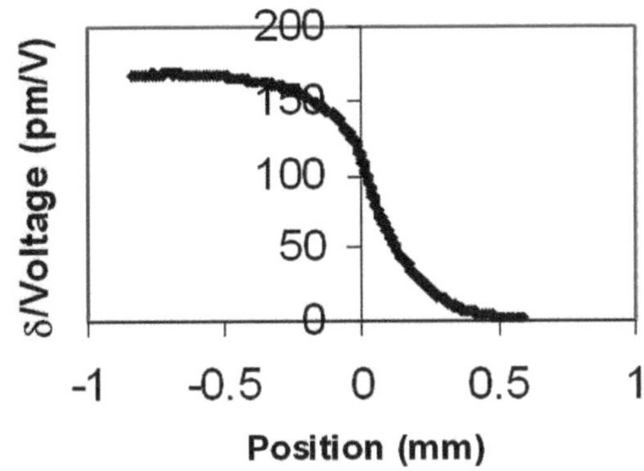

Fig. 12 Displacement profile across the edge of a *square top* electrode for a partially electroded ceramic PZT. Reprinted with permission from Ref. [41]

This shows that, in the frequency domain, the photodetector output will have three peaks: the carrier at ω_R and two sidebands separated from the carrier by $\pm\omega$T. The intensity of the sidebands is proportional to the displacement amplitude Z_0. Usually, only one of the sidebands is sent to the spectrum analyzer.

For laser vibrometers, the Doppler frequency shift f_T from a moving surface at velocity v is given by the Doppler shift equation $f_T = 2v/\lambda$. This Doppler frequency deviation is measured interferometrically and then converted to surface velocity and displacement. Microscope laser scanning vibrometry was particularly useful for the characterization of vibrations in micro-electrical-mechanical systems. One can usually use one-beam or two-beam techniques. For the one beam technique, only one beam i incident onto the sample surface. For the two beams method, apart from the beam focused onto the measuring spot, a reference beam i focused on to the substrate surface, i.e., the sample surface without the top electrode. This beam does not move when the measuring beam i scanning over the sample. In this case, the measured displacement i the displacement difference between the measuring and the reference beams.

Figure 12 shows the displacement δ profile across the edge of a square top electrode for a partially electroded ceramic PZT. Positions <0 were covered by top electrode and positions >0 were without. The reference beam was at the position 0.5 mm, therefore the displacement at 0.5 mm was always 0. It is noticed that the displacement did not reduce to 0 abruptly at the edge, rather it reduced to 0 gradually over a distance nearly 0.5 mm over the un-electroded area. The obtained d_{33} was 176 pm/V at positions inside the electrode and away from the edge (<-0.5 mm for example) at the frequency 10 kHz and this d_{33} value was stable at different frequencies between 2–15 kHz. This compares to the result of 178 pm/V as obtained by the interferometer method, [45].

Figure 13 shows (a) SEM image for a square top electroded PZT(30/70) thin film on SOI wafer; (b) the displacement δ/voltage profile across the square electrode at a particular time with DC bias at 8 V. The top electrode was 1×1 mm, it was electrically connected to a cantilever by a thin line only, the top electrode Au/Cr, PZT and the bottom electrode Pt at areas other than the contact pad and the connecting thin line

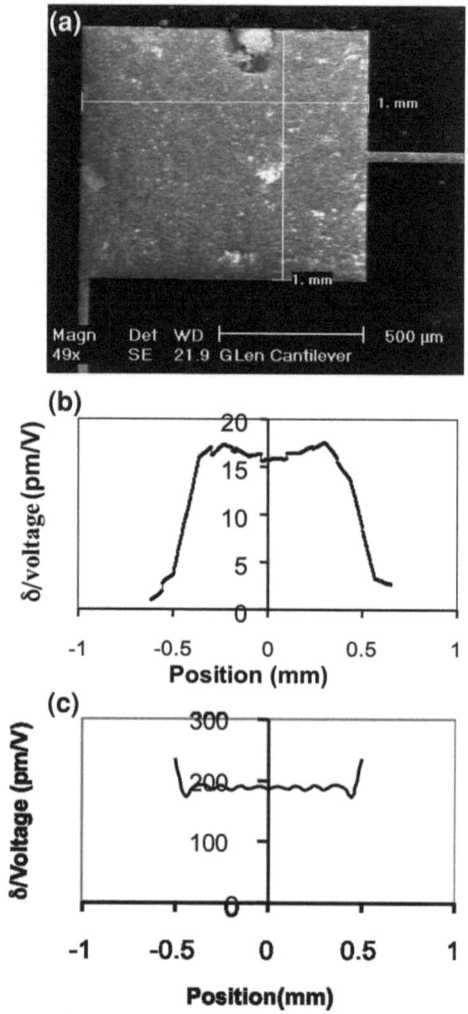

Fig. 13 **a** SEM image of the *top* electrode contact pad consisting of PZT(30/70) on SOI wafer; **b** The displacement d/voltage profile across the contact pad; **c** A finite element analysis simulated displacement line profile across the patterned contact pad. Reprinted with permission from Ref. [45]

have all been etched away. The reference beam was near the position −0.6 mm. It can be seen that the slope had a sharp gradient at the electrode edge and the displacement was a plateau for the centre of the electroded area, also there is a displacement increase toward the edge of the top electrode. FEA was carried out to understand the device behaviour. The PZT property parameters used in the FEA were those for Ferroperm ceramics PZ23 and the Young's modulus for Si was 130 GPa which is the value for the Si<100> [45]. The boundary condition was that the bottom of the Si was clamped. The modelled displacement line profile across the top electrode is shown in Fig. 13c, the increase toward the edge of the top electrode and the similarity between the simulated and the measured results are obvious. Apparently the $d_{33,f}$

should be calculated as the relative displacement between the centre of the electrode and the substrate.

Heterodyne interferometers [4, 30, 46], laser Doppler heterodyne interferometers [47, 48] and laser scanning vibrometers [45, 49, 50] have all been applied successfully for the measurement of piezoelectric coefficients of piezoelectric thin films. As for the strain measurement based interferometers discussed in Sect. 2.1, appropriate sample mounting is of critical importance to suppress the substrate bending and to obtain the accurate piezoelectric strain values.

An advantage of the heterodyne interferometers is its robustness against low frequency environmental mechanical and thermal disturbances, so they can work in the normal laboratory environment. Another advantage is that a spot measurement can be performed in a very short time (<1 s). For these reasons the laser beam can be scanned over a surface area and phase referenced displacement distributions across a sample surface can be obtained. This makes the heterodyne interferometer techniques particularly useful for device characterisation, in which mode shape measurements are necessary. Usually these systems use fibre optics, therefore the required alignment for users is straight forward.

In summary, techniques based on optical interference, such as single beam and double beam interferometers, heterodyne interferometry and laser scanning vibrometry, can all be employed to measure piezoelectric coefficients for bulk ceramics and thin films. However, great care is needed to minimise the sample bending effect. For samples of thin film deposited on a thick substrate, sample bending can easily surpass the piezoelectric thickness dilatation if the samples are simply supported on a surface. Proper mounting of the sample, for example using epoxy to glue the sample back to a rigid holder, can greatly suppress the bending effect to levels that are negligible. There is no reason to believe that the single beam interferometrical techniques, including the Michelson interferometer, heterodyne interferometer and vibrometer methods, should perform inferior to the double-beam interferometer. For the double-beam technique, the bending effect can also easily surpass the real piezoelectric effect if the specimen is not glued to a rigid body and the beam alignment is not precise.

It is straight forward to bond the back surface of a specimen to a rigid sample holder for single beam interferometry measurements. However, for double beam interferometers, the back surface directly opposite the top electrode cannot be bonded to anything in order to provide access for the measurement beam, therefore the precise alignment between the back and front beams must be carried out. From this point of view, single beam interferometry holds a comparative advantage over double beam interferometry due to requiring less optical components therefore minimising energy loss, easier optical alignment, higher resolution [51] and easier sample preparation.

3 Full Field Strain Measurement

In Sect. 2.1 we discussed techniques for spot-measurement, i.e., at any particular time the measurement concerns the strain for that particular spot only, although a

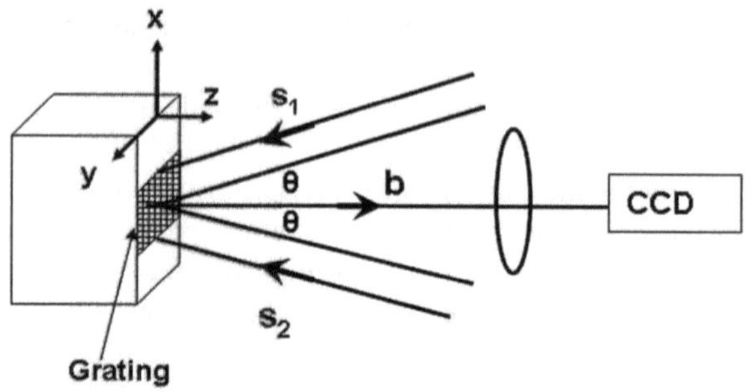

Fig. 14 Principle of moiré interferometer

full field image can be built up by mechanically scanning either the laser beam over the sample surface or move the sample around the fixed laser beam. On the other hand, holographic interferometry offers whole-field displacement and strain measurement capability. In holographic interferometry, two or more wave fields are compared interferometrically, at least one of them must be holographically recorded and reconstructed. For more details please refers to textbooks [52].

The most widely used holographic interferometrical techniques for piezoelectric displacement measurement are moiré and speckle interferometry [53–58]. The operating principle of a moiré interferometer is like the following: highly reflective cross-line diffraction grating (f_g = mm/1200 lines = 0.833 μm) is firmly bonded onto the surface of the specimen, so that deformations of the specimen surface cause displacements of the diffraction grating, which in turn, lead to the phase changes of the reflected beams. These phase changes are detected by the changes of the interference fringes. Figure 14 shows schematically the setup of a moiré interferometer for lateral displacement measurement. A laser beam is collimated and divided into two beams 1 and 2. These two coherent beams are directed to illuminate the set of grating lines perpendicular to the x-z plane at the same but opposite angle such that the grating diffractions the +1 and −1 orders coaxially. The moiré fringe patterns are viewed at normal incidence by a CCD camera.

Let $\mathbf{s}_1 = (-\sin\theta, 0, \cos\theta)$ be the direction for the beam 1 and $\mathbf{s}_2 = (\sin\theta, 0, \cos\theta)$ be the direction for the beam 2, the observation direction is b = (0, 0, 1). Assuming there is a displacement dx along the x direction in the grating. The relative phase change before and after the deformation is:

$$\phi = \frac{4\pi \sin\theta}{\lambda} \cdot d_x \tag{14}$$

For diffraction grating, we also have $f_g \sin\theta = \lambda$.

So,

$$d_x = \frac{\phi}{2\pi} \cdot \frac{f_g}{2} = N_x \cdot \frac{f_g}{2} \tag{15}$$

where N_x is the fringe order in the interference moiré pattern. Similarly, when measuring the in-plane y displacement, two coherent beams in the vertical plane y–z are allowed to illuminate the diffraction grating at the same but opposite angle as for the x measurement. This way, the whole displacement fields u(x, y) and v(x, y) along the x and y directions respectively, can be measured. Generally:

$$u = N_x f_g/2, v = N_y f_g/2 \tag{16}$$

and the in-plane strain components are,

$$\begin{aligned} \varepsilon_x &= \frac{\partial u}{\partial x} = \frac{f_g}{2} \cdot \frac{\partial N_x}{\partial x}, \\ \varepsilon_y &= \frac{\partial v}{\partial x} = \frac{f_g}{2} \cdot \frac{\partial N_y}{\partial y} \end{aligned} \tag{17}$$

In practice, adjustable mirrors were used to direct the beams to form the symmetrical illumination as shown in Fig. 14 and the moiré interferometer system is adjusted to a null field condition prior to measuring the displacement field. Using this technique, Liu et al. studied the fracture behaviours of ferroelectric ceramics during the in-plane and out-of-plane 90° domain switching when the specimen were subjected to a mechanical, electrical and combined electrical and mechanical loads, respectively [54]. Figure 15 shows the moiré patterns (U and V fields) at different electric field loading for a simply supported ceramic beam with a notch at the centre. When the electric field was less than 0.4 MV m^{-1}, the fringes in the u and v fields were nearly uniform and the specimen had shear deformation only (Fig. 15a and b).When the electric field is larger than 0.4 MV m^{-1}, initially around the notch the moiré fringes begin to rotate counter clockwise (Fig. 15b and d), indicating the occurrence of the in-plane 90° domain switching due to the reorientation of the tetragonal crystal. When the field reached 0.9 MV m^{-1} which was above the coercive field strength significant variations in the moiré fringes were observed, indicating intensive 90° domain switching had happened and these domain switching were concentrated in a strip of 45° and (Fig. 15c and f).

As a full-field image technique moiré interferometry has been proved very useful for bulk materials and device strain characterisation. The sensitivity of the technique for the in-plane displacement is half the frequency of the reflective grating, $f_g/2$. The theoretical upper limit of the in-plane displacement sensitivity corresponds to half the wavelength, $\lambda/2$. For more details please refers to textbooks on holographic interferometry, [59].

4 Elastic Coefficients Measurement for Thin and Thick Films

Besides the piezoelectric coefficients, other important basic materials parameters for piezoelectric materials are their elastic stiffness coefficients c_{ij} or compli-

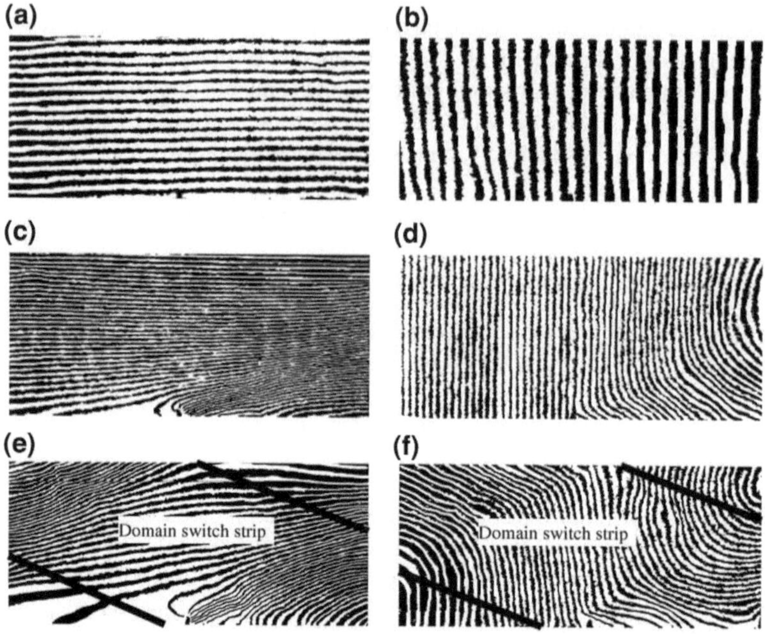

Fig. 15 Moiré patterns of vertical poling specimens subjected to an electric field: **a** 0.3 MV/m (*u field*); **b** 0.3 MV/m (*v field*); **c** 0.6 MV/m (*u field*); **d** 0.6 MV/m (*v field*); **e** 0.9 MV/m (*u field*) and **f** 0.9 MV/m (*v field*). Reprinted with permission from Ref. [54]

ance coefficients s_{ij}. Piezoelectric thin and thick films especially lead zirconate titanate ($PbZr_xTi_{1-x}O_3$, PZT) have received considerable interest for their use as the functional material of a microelectromechanical system (MEMS). Accurate knowledge of the electrical and mechanical properties of these materials is essential for the effective modelling and design of novel MEMS devices and process qualification. Currently, the elastic coefficients obtained from bulk ceramics, with the similar nominal compositions, are used as the values for films in the modelling and design of piezo-MEMS devices [60]. However, given the marked differences in microstructure between the films and bulk materials, and the dependence of these properties on the microstructure, such practices will at best only provide an indication of how a device may respond in real life.

The measurement of elastic coefficients for piezoelectric ceramics has been discussed in chapter "Indentation Stiffness Analysis of Ferroelectric Thin Films", but most of these techniques are not applicable to thin films deposited on thick substrates because the measured properties, such as the resonant frequency, are usually dominated by the presence of the thick substrate. Nanoindentation techniques are widely used for the measurement of the Young's modulus Y_{33} of thin films [61]. However, for PZT thick films which are usually porous, it is difficult to use nanoindentation to determine the averaged Young's modulus since it is a localized measurement technique and will be affected by inhomogeneity in the microstructure. With the help of microscopic

LSV, some techniques usually used for bulk samples such as the resonance method [62] can be applied to thick films, and even thin film samples. Purpose prepared self-supporting cantilever beams can be excited by mechanical or electric activations, the resonant frequencies and the mode shapes can be measured by microscopic LSV.

For a PZT film on Si substrate cantilever (a conventional bending cantilever design used in MEMS applications), as shown in Fig. 16a, the equations relating the resonance frequency and tip displacement of the cantilever with the materials properties are [24, 25]:

$$
\begin{aligned}
f_r &= \frac{3.52t}{4\pi L^2}\sqrt{\frac{Y_p}{3\rho_p}}\left[\frac{A^2B^4+2A(2B+3B^2+2B^3)+1}{(1+BC)(AB+1)(1+B)^2}\right]^{1/2} \\
\delta &= \frac{3L^2}{2t}\cdot\frac{2AB(1+B)^2}{A^2B^4+2A(2B+3B^2+2B^3)+1}\cdot d_{31}E_3 \qquad (18) \\
A &= Y_m/Y_p,\ B = t_m/t_p,\ C = \rho_m/\rho_p,\ t_m + t_p = t
\end{aligned}
$$

Here f_r is the fundamental bending resonant frequency, δ the displacement at the tip, Y_m and Y_p Young's modulus, t_m and t_p the thickness, ρ_m and ρ_p the density, of the substrate and the piezoelectric layer respectively. L is the length of the cantilever and E_3 the electric field across the piezoelectric layer. So, in principle, if the above parameters Y_m, B, C, L and t are known, the PZT film property parameters d_{31} and Y_p could be obtained by measuring the resonant frequency f_r and the cantilever tip displacement δ under the electric field E_3. Huang et al. used this technique to determine the piezoelectric and elastic coefficients of PZT thin films [45]. However, since the thickness of the PZT film (~1 μm) was much smaller than the Si substrate (~20 μm), the resonant frequency was largely determined by the Si substrate, therefore the so obtained PZT Young's modulus was prone to large uncertainty. Furthermore, as in most of the other reported works, the effect of the electrode layers was either ignored, or, the Young's modulus of the bottom electrode (Pt/Ti) layer was assumed to be the same as for the bulk Pt metal (which may not be true due to differences in microstructure). A further complicating factor is the effect of poling a PZT film which will result in different stiffness along or perpendicular to the polar directions.

These problems can be solved using self-supporting PZT cantilever beams prepared by using photolithography techniques [63]. Figure 16 show schematics of the structures required: (b) self-supporting pure PZT; (c) PZT with top and bottom electrode layers, usually Pt. For the homogeneous PZT beam undergoing bending vibration (the bending vibration can be activated by a mechanical shaker), Eq. 18 becomes:

$$
\begin{aligned}
f_r &= \frac{\lambda_i^2}{2\pi L^2}\sqrt{\frac{YI}{\rho A}} \qquad (19) \\
Y &= \left(\frac{f_r 2\pi L^2}{\lambda_i^2}\right)^2\cdot\frac{\rho A}{I} = \left(\frac{f_r 2\pi L^2}{\lambda_i^2}\right)^2\cdot\frac{12\rho}{t^2}
\end{aligned}
$$

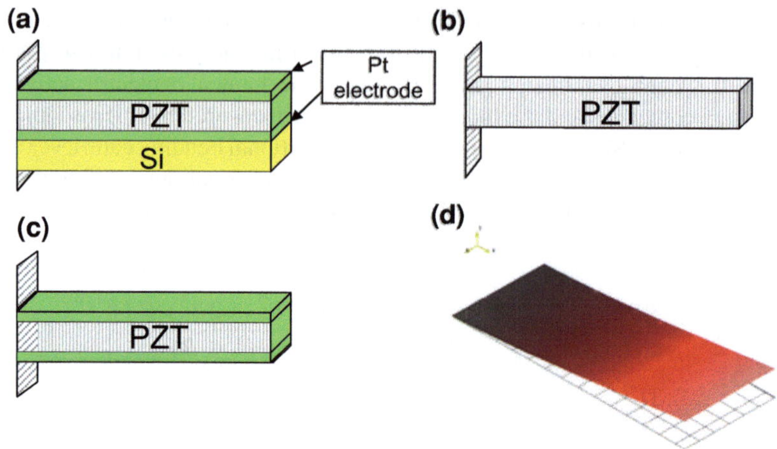

Fig. 16 Schematic of PZT films with different clamping conditions. **a** Electroded PZT on Si substrate; **b** Pure PZT cantilever; **c** Electroded PZT cantilever; **d** LSV image of the PZT cantilever vibration

where λ_i is a dimensionless number relating to mode shape, $\lambda_1 = 1.8751$ and $\lambda_2 = 4.6941$. L is the beam length, Y the elastic modulus and $Y_{11} = 1/s_{11}$ where s_{11} is the compliance coefficient, ρ the material density, A the cross sectional area of the beam and I the moment of inertia. The geometry parameters of the cantilever such as the length L and thickness t can be obtained from SEM observations. The resonant frequency f_r and mode shape can be obtained by using LSV. The LSV focused laser beam size is 2–3 μm and the collecting angle of the objective lens is about 13° allowing very small features to be observed.

To measure the Y_{22}, the cantilever (with the sample mount base) will be rotated by 90° so that the cross section along the width direction is perpendicular to the direction 3 and also the laser beam. By again performing the frequency sweep and displacement image Y_{22} can be determined. Due to the small laser spot size and high sensitivity of the detector vibrations of structures as small as 1–2 μm should be readily detectable provided the reflectivity is high.

For cantilevers with electrode layers as shown in Fig. 16b and c, a sinusoidal voltage applied across the PZT can also induce vibrations via the converse piezoelectric effect. For the device (b), since the top and bottom Pt layers are very thin (0.1 μm) relative to the PZT layer (1 ∼ 10 μm) and have the same thicknesses, the PZT will expand and contract along the direction 3 and contract and expand along the directions 1 and 2 at the same time. By measuring the expansion and contraction vibrations along the directions 1, 2 and 3 at the free end tip (far away from the clamping point) it should enable the piezoelectric coefficients d_{31}, d_{32}, and d_{33}, respectively, to be determined for the unclamped PZT films (ideally, $d_{31} = d_{32}$). From the measurements of the fundamental resonant frequency f_0 along the direction 2 (the resonant

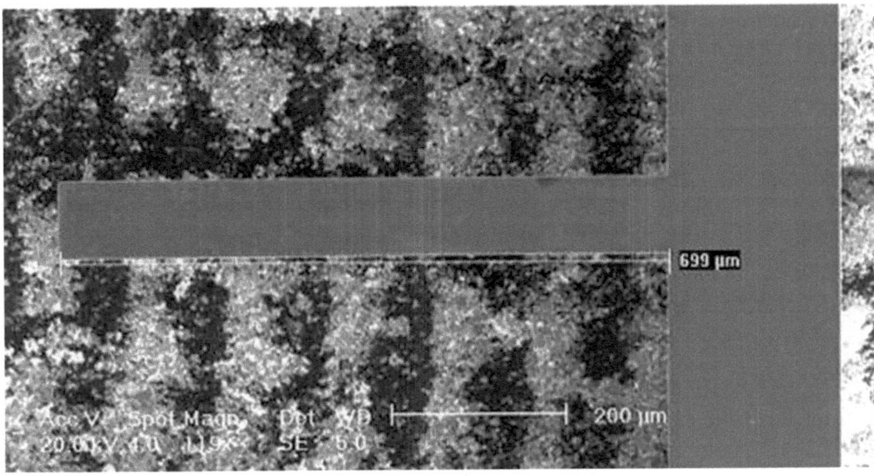

Fig. 17 SEM image of a self-supported PZT cantilever. The cantilever has a *length* 700 μm, *width* 102 μm and *thickness* 2.4 μm

frequencies along the directions 1 and 3 are beyond the vibrometer bandwidth) and the beam length L the elastic compliance constant s_{11} could be obtained for the PZT film. For this vibration,

$$f_0 = \frac{1}{4L}\sqrt{\frac{1}{\rho s_{11}}} \tag{20}$$

where s_{11} is the compliance and ρ the density of the PZT. So, by investigating the resonant frequencies, mode shape and vibration amplitude, parameters such as the PZT compliance coefficients s_{33}, s_{22}, s_{11} with or without substrate clamping, longitudinal piezoelectric d_{33} and transverse coefficient d_{31}, the effective longitudinal piezoelectric coefficient $d_{33,f}$ for the film on thick substrate, and the compliance coefficients s_{Pt} for the electrode layer Pt, can all be obtained.

Figure 17 shows a SEM image of a purpose prepared self-supported PZT cantilever of dimensions 700 × 102 × 2.4 μm. The cantilever was actuated with a shaker in both the directions 1 and 3 and the resonant frequencies of the beam were measured with a LSV. The results of the measurement gave a first mode resonant frequency for the 3 direction of 2258 Hz when measured at 5×10^{-5} mBar. The Y_{33} calculated using the Eq. 20 equals 62.7 GPa. More details are summarized in Table 1, [63].

Similar measurements were carried out for a number of self-supported thick film cantilevers and the results are summarized in Table 1 as well. It can be seen that the elastic constants along the directions 3 and 1, Y_{33} and Y_{11}, equalling 56 and 56.3 GPa respectively, were the same before poling. This indicates that the Young's modulus for thick PZT films are lower than those of the thin films—which is not surprising since there are more voids in the thick films than in the thin films but we assumed all

Table 1 Summary of the measured dimensions, resonant frequencies and the calculated elastic constants for PZT for a few self-supported PZT cantilevers

	Unit	$Y_{33} = 1/s_{33}$				$Y_{11} = 1/s_{11}$	
		Before poling		After poling		Before poling	After poling
Density, ρ	Kg/m^3	7700	7700	7700	7700	7700	7700
Mode constant, λ		4.6941	1.8751	1.8751	4.6941	1.8751	1.8751
Length, L	μm	150	600	700	150	150	150
Thickness, t	μm	10	2.4	2.4	10	10	10
Frequency, f	Hz	12100	3063	2258	12200	48500	46100
Young's Modulus	GPa	56	62.3	62.7	56.6	56.3	50.9

the films have the same mass density. The device was then corona poled at 180 °C with a 20 KV field applied, this field was left on while the hotplate was cooled to 80 °C a process which has showed good poling in previous devices to give a d_{33} of ~70 pC/N.

After the poling the device was re-measured on the vibrometer at 2×10^{-2} mBar vacuum. The elastic constants after the poling, along the directions 3 and 1, were determined to be 56.6 and 50.9 GPa for Y_{33} and Y_{11} respectively. These results show that the PZT film is homogenous prior to poling and that after poling there is a difference in the elastic constants between the directions parallel and perpendicular to the polar direction, [63].

References

1. Xu, F., Chu, F., Trolier-McKinstry, S.: Longitudinal piezoelectric coefficient measurement for bulk ceramics and thin films using pneumatic pressure rig. J. Appl. Phys. **86**, 588 (1999)
2. Dubois, M., Muralt, P.: Measurement of the effective transverse piezoelectric coefficient e31, f of aln and pb (zrx, ti1-x) o3 thin films. Sens. Actuators A: Phys. **77**(2), 106–112 (1999)
3. Southin, J., Wilson, S., Schmitt, D., Whatmore, R.: E31, f determination for pzt films using a conventionald33'meter. J. Phys. D: Appl. Phys. **34**, 1456 (2001)
4. Harnagea, C., Pignolet, A., Alexe, M., Hesse, D.: Piezoresponse scanning force microscopy: what quantitative information can we really get out of piezoresponse measurements on ferroelectric thin films. Integr. Ferroelectr. **44**(1), 113–124 (2002)
5. Fotchenkov, A.A.: Sov. Phys. Cryst. **2**, 643 (1957)
6. Musonov, V.M., Zaitseva, M.P., Chizhikov V.A., Kuznetsov V.P.: Radiotechnics, thin films, computer technics, Institute of Physics, Part I, vol. 1, p. 180. Krasnoyarsk, USSR (1973)
7. Sidnenko, E.V., Zheludev, I.S.: Kristallografiya. Sov. Phys. Crystallogr. **12**, 465–467 (1967) (English Transl. 12, 397–399, 1967)
8. Yamaguchi, T., Hamano, K.: Inteferometric method of measuring complex piezoelectric constants of crystals in a frequency range up to about 50 khz. Jpn. J. Appl. Phys. **18**, 927–932 (1979)
9. Zhang, Q.M., Pan, W.Y., Cross, L.E.: Laser interferometer for the study of piezoelectric and electrostrictive strains. J. Appl. Phys. **63**(8), 2492–2496 (1988)

10. Zhang, Q.M., Jang, S., Cross, L.E.: Highfrequency strain response in ferroelectrics and its measurement using a modified machzehnder interferometer. J. Appl. Phys. **65**(7), 2807–2813 (1989)
11. Pan, W.Y., Cross, L.E.: A sensitive double beam laser interferometer for studying highfrequency piezoelectric and electrostrictive strains. Rev. Sci. Instrum. **60**(8), 2701–2705 (1989)
12. Li, J.-F., Moses, P., Viehland, D.: Simple, high-resolution interferometer for the measurement of frequency-dependent complex piezoelectric responses in ferroelectric ceramics. Rev. Sci. Instrum. **66**(1), 215–221 (1995)
13. Kholkin, A., Wutchrich, C., Taylor, D., Setter, N.: Interferometric measurements of electric fieldinduced displacements in piezoelectric thin films. Rev. Sci. Instrum. **67**(5), 1935–1941 (1996)
14. Tsurumi, T., Ikeda, N., Ohashi, N.: Non-180° domain contribution in electric-field-induce strains of pzt ceramics measured by a mach-zehnder interferometer. J. Ceram. Soc. Jpn. **106**(11), 1062–1066 (1998)
15. Gerber, P., Roelofs, A., Lohse, O., Kügeler, C., Tiedke, S., Böttger, U., Waser, R.: Short-time piezoelectric measurements in ferroelectric thin films using a double-beam laser interferometer. Rev. Sci. Instrum. **74**(4), 2613 (2003)
16. Burianova, L., Sulc, M., Prokopova, M., Nosek, J.: The piezoelectric coefficients in a wide temperature range by laser interferometry. Ferroelectrics **292**(1), 111–117 (2003)
17. Huang, Z., Whatmore, R.W.: A double-beam common path laser interferometer for the measurement of electric field-induced strains of piezoelectric thin films. Rev. Sci. Instrum. **76**(12), 123906 (2005)
18. Sizgoric, S., Gundjian, A.A.: An optical homodyne technique for measurement of amplitude and phase of subangstrom ultrasonic vibrations. Proc. IEEE **57**(7), 1313–1314 (1969)
19. Kwaaitaal, T., Luymes, B.J.: Noise limitations of michelson laser interferometers. J. Phys. D: Appl. Phys. **13**(6), 1005 (1980)
20. Chao, C., Wang, Z., Zhu, W.: Modulated laser interferometer with picometer resolution for piezoelectric characterization. Rev. Sci. Instrum. **75**(11), 4641 (2004)
21. Measurement of longitudinal piezoelectric coefficient of lead zirconate titanate thin/thick films using a novel scanning Mach-Zehnder interferometer. Thin Solid Films **493**(1–2), 313–318 (2005)
22. Lee, K.Y., Case, E.D.: Effects of adhesion on the effective Young's modulus in glass slide/glue laminates. J. Mater. Sci. Eng. **31**(9), 2241–2251 (1996)
23. Moilanen, H., Leppävuori, S.: Laser interferometric measurement of displacement-field characteristics of piezoelectric actuators and actuator materials. Sens. Actuators A **92**(1), 326–334 (2001)
24. Lefki, K., Dormans, G.: Measurement of piezoelectric coefficients of ferroelectric thin films. J. Appl. Phys. **76**(3), 1764–1767 (1994)
25. Wang, Q.-M., Cross, L.E.: Performance analysis of piezoelectric cantilever bending actuators. Ferroelectrics **215**(1), 187–213 (1998)
26. Smits, J., Choi, W.: The constituent equations of piezoelectric heterogeneous bimorphs. IEEE Trans. Ultrason. Ferroelectr. Freq. Control **38**(3), 256–270 (1991)
27. Damjanovic, D., Brooks K.G., Kholkin, A., Kohli, M., Maeder, T., Muralt, P., Setter, N.: Properties of piezoelectric pzt thin films for microactuator applications. In: MRS Proceedings, vol. 360, Jan 1994
28. Smits, J., Choi, W.: The constituent equations of piezoelectric heterogeneous bimorphs. IEEE Trans. Ultrason. Ferroelectr. Freq. Control **38**(3), 256–270 (1991)
29. Muensit, S., Guy, I.L.: The piezoelectric coefficient of gallium nitride thin films. Appl. Phys. Lett. **72**, 1896 (1998)
30. Lueng, C.M., Chan, H., Surya, C., Fong, W.K., Choy, C.L., Chow, P., Rosamond, M.: Piezoelectric coefficient of gan measured by laser interferometry. J. Non-Cryst. Solids **254**(1), 123–127 (1999)
31. Fattinger, G.G., Tikka, P.T.: Modified mach-zender laser interferometer for probing bulk acoustic waves. Appl. Phys. Lett. **79**(3), 290 (2001)

32. Knuuttila, J., Tikka, P., Plessky, V.P., Thorvaldsson, T., Salomaa, M.M.: Recent advances in laser-interferometric investigations of SAW devices, Ultrasonics symposium. In: Proceedings of IEEE, vol. 1, pp. 161–164 (1997)
33. Zhang, Q.M., Pan, W.Y., Jang, S., Cross, L.E.: Domain wall excitations and their contributions to the weaksignal response of doped lead zirconate titanate ceramics. Appl. Phys. Lett. **64**(11), 6445–6451 (1988)
34. Zhang, Q., Wang, H., Kim, N., Cross, L.: Direct evaluation of domainwall and intrinsic contributions to the dielectric and piezoelectric response and their temperature dependence on lead zirconatetitanate ceramics. Appl. Phys. Lett. **75**(1), 454–459 (1994)
35. Kholkin, A.L., Colla, E., Brooks, K., Muralt, P., Kohli, M., Maeder, T., Taylor, D., Setter, N.: Interferometric study of piezoelectric degradation in ferroelectric thin films. Microelectron. Eng. **29**(1), 261–264 (1995)
36. Kloos, G.: Design of a mach-zehnder interferometer for the measurement of electrostrictive strains. Meas. Sci. Technol. **7**, 1027 (1996)
37. Maiwa, H., Christman, J., Kim, S., Kim, D.J., Maria, J.P., Chen, B., Streiffer, S., Kingon, A.: Measurement of piezoelectric displacements of pb (zr, ti) o3 thin films using a double-beam interferometer. Jpn. J. Appl. Phys. **38**, 5402 (1999)
38. Taylor, D.V., Damjanovic, D.: Piezoelectric properties of rhombohedral Pb(Zr, Ti)O[sub 3] thin films with (100), (111), and “random” crystallographic orientation. Appl. Phys. Lett. **76**(12), 1615 (2000)
39. Xu, F., Trolier-McKinstry, S., Ren, W., Xu, B., Xie, Z.L., Hemker, K.J.: Domain wall motion and its contribution to the dielectric and piezoelectric properties of lead zirconate titanate films. J. Appl. Phys. **89**(2), 1336 (2001)
40. Gerber, P., Roelofs, A., Kügeler, C., Böttger, U., Waser, R., Prume, K.: Effects of the top-electrode size on the piezoelectric properties (d[sub 33] and s) of lead zirconate titanate thin films. J. Appl. Phys. **96**(5), 2800 (2004)
41. Huang, Z., Zhang, Q., Corkovic, S., Dorey, R., Whatmore, R.: Comparative measurements of piezoelectric coefficient of pzt films by berlincourt, interferometer, and vibrometer methods. IEEE Trans. Ultrason. Ferroelectr. Freq. Control **53**(12), 2287–2293 (2006)
42. Fernandes, J.R., de Sá, F.A., Santos, J.L., Joanni, E.: Optical fiber interferometer for measuring the d[sub 33] coefficient of piezoelectric thin films with compensation of substrate bending. Rev. Sci. Instrum. **73**(5), 2073 (2002)
43. Nosek, J., Burianova, L., Sulc, M., Soyer, C., Cattan, E., Remiens, D.: About the measurements of the d 33 piezoelectric coefficient of the PZT film-Si/SiO 2 /Ti/Pt substrates using an optical cryostat. Ferroelectrics **292**(1), 103–109 (2003)
44. Sulc, M., Erhart, J., Nosek, J.: Interferometric measurement of the temperature dependence of piezoelectric coefficients for PZN-8Crystals. Ferroelectrics **293**(1), 283–290 (2003)
45. Huang, Z., Leighton, G., Wright, R., Duval, F., Chung, H.C., Kirby, P., Whatmore, R.W.: Determination of piezoelectric coefficients and elastic constant of thin films by laser scanning vibrometry techniques. Sens. Actuators A **135**(2), 660–665 (2007)
46. Royer, D., Kmetik, V.: Measurement of piezoelectric constants using an optical heterodyne interferometer. Electron. Lett. **28**(19), 1828–1830 (1992)
47. Lian, L.: Stress effects in sol-gel derived ferroelectric thin films. J. Appl. Phys. **95**(2), 629 (2004)
48. Lian, L., Sottos, N.: Effects of thickness on the piezoelectric and dielectric properties of lead zirconate titanate thin films. J. Appl. Phys. **87**, 3941 (2000)
49. Yao, K., Tay, F.E.H.: Measurement of longitudinal piezoelectric coefficient of thin films by a laser-scanning vibrometer. IEEE Trans. Ultrason. Ferroelectr. Freq. Control **50**(2), 113–116 (2003)
50. Yao, K., Shannigrahi, S., Tay, F.E.H.: Characterisation of piezoelectric thin films by areal laser scanning. Sens. Actuators A **112**(1), 127–133 (2004)
51. Li, J.-F., Viehland, D.D., Tani, T., Lakeman, C.D.E., Payne, D.A.: Piezoelectric properties of sol-gel-derived ferroelectric and antiferroelectric thin layers. J. Appl. Phys. **75**(1), 442 (1994)

52. Kreis, T.: Holographic Interferometry: Principles and Methods, Ser. Akademie Verlag, Akademie Verlag series in optical metrology (1996)
53. De Nicola, S., Ferraro, P.: Fourier transform method of fringe analysis for moiré interferometry. J. Opt. A: Pure Appl. Opt. **2**(3), 228 (2000)
54. Liu, Z., Fang, D., Xie, H., Lee, J.J.: Study of effect of 90 ° domain switching on ferroelectric ceramics fracture using the moiré interferometry. Acta Mater. **55**(11), 3911–3922 (2007)
55. Watanabe, Y., Tsuda, T., Ishii, S., Goka, S., Sekimoto, H.: Method based on laser speckle interferometry for measuring absolute in-plane vibrational distribution of piezoelectric resonators. Jpn. J. Appl. Phys. **45**(5B), 4585–4587 (2006)
56. Lin, H.-Y., Huang, J.H., Ma, C.C.: Vibration analysis of piezoelectric materials by optical methods. IEEE Trans. Ultrason. Ferroelectr. Freq. Control **49**(8), 1139–1149 (2002)
57. Huang, C.-H., Lin, Y.-C., Ma, C.C.: Theoretical analysis and experimental measurement for resonant vibration of piezoceramic circular plates. IEEE Trans. Ultrason. Ferroelectr. Freq. Control **51**(1), 12–24 (2004)
58. Ma, C.C., Lin, Y.-C., Huang, Y.-H., Lin, H.-Y.: Experimental measurement and numerical analysis on resonant characteristics of cantilever plates for piezoceramic bimorphs. IEEE Trans. Ultrason. Ferroelectr. Freq. Control **54**(2), 227–239 (2007)
59. Jones, R., Wykes, C.: Holographic and Speckle Interferometry, ser. Cambridge University Press, Cambridge Studies in Modern Optics (1989)
60. Zarnik, M.S., Belavič, D., Macek, S.: Evaluation of the constitutive material parameters for the numerical modelling of structures with lead-zirconate-titanate thick films. Sens. Actuators A **136**(2), 618–628 (2007)
61. Chima-Okereke, C., Bushby, A.J., Reece, M.J., Whatmore, R.W., Zhang, Q.: Experimental, analytical, and finite element analyses of nanoindentation of multilayer PZT/Pt/SiO2 thin film systems on silicon wafers. J. Mater. Res. **21**(02), 409–419 (2011)
62. Pardo, L., Algueró, M., Brebøl, K.: A non-Standard shear resonator for the matrix characterization of piezoceramics and its validation study by finite element analysis. J. Phys. D: Appl. Phys. **40**(7), 2162–2169 (2007)
63. Leighton, G.J.T., Huang, Z.: Accurate measurement of the piezoelectric coefficient of thin films by eliminating the substrate bending effect using spatial scanning laser vibrometry. Smart Mater. Struct. **19**, 065011 (2010)

Temperature Dependence of Ferroelectric and Piezoelectric Properties of PZT Ceramics

Paul Weaver and Markys G. Cain

1 Introduction

Temperature induced changes in the properties of piezoelectric and ferroelectric materials have major implications for the design of devices used in wide variety of technological applications. Examples include valves for fuel injection in diesel engines, pneumatics, and gas control in domestic appliances where the flow rate could drift with temperature and compromise valve performance. Similar considerations apply in layered electronic devices employing piezoelectric materials. There is considerable current interest in multiferroic devices for spintronics, sensor and memory applications [1]. Small scale layered devices can provide magnetoelectric functionality through strain coupling between ferroelectric and a magnetostrictive components. Strain coupling will also occur through differential thermal expansion between the layers, so a knowledge of the thermal behaviour of the components is essential. In many of these applications quite large electrical fields are needed to obtain useful energy density, so the temperature dependency of the high field electromechanical coupling is important. It is a significant characteristic of this type of device that the strain or position relative to the value at a reference temperature is an important factor. For instance, in a valve, the position of the sealing member relative to the valve seat is the critical parameter in controlling the flow rate. In a multiferroic device, the strain of the piezoelectric material relative to the magnetostrictive component determines magnetic or electrical state of the device. Thermal expansion may couple mechanically in the same way as the piezoelectric coupling to produce unwanted actuation in response to temperature changes rather than the desired electrical activation. When the piezoelectric material is bonded to a different

P. Weaver (✉) · M. G. Cain
National Physical Laboratory, Hampton Road, Teddington, Middlesex TW11 0LW, UK
e-mail: paul.weaver@npl.co.uk

M. G. Cain (ed.), *Characterisation of Ferroelectric Bulk Materials and Thin Films*, Springer Series in Measurement Science and Technology 2,
DOI: 10.1007/978-1-4020-9311-1_6,

material to form a unimorph or bimorph actuator, the thermal expansion of the partner materials must also be taken into account. In general actuator applications, thermal drift can cause unwanted actuation at temperature extremes, or limit the operating temperature range.

Many of the fundamental characteristics of ferroelectric materials are defined by the temperature dependency of their properties. Changes in permittivity and spontaneous polarisation around the Curie temperature are central to the characterisation and understanding of these materials [2] and have been well studied. The effects of temperature on the electromechanical properties are less well understood. Data are available on the temperature variation of low field properties such as piezoelectric coefficients obtained from resonance measurements [3]. These are not sufficient to provide information on the high field properties which are dominated by non-linear effects, and tell us nothing about thermal drift in the electromechanical performance. Where measurements have been made of large field ferroelectric strain loops [4], these are not related to a datum, so provide no information on the absolute strain. Thermal expansion measurements have been made in PZT ceramics [5, 6], reporting very low values for the thermal expansion coefficient of depoled ceramic ($0–2 \times 10^{-6}$ $°C^{-1}$) and negative values for poled ceramics [5, 6]. Whilst these measurements provide very useful information on phase transitions, they were performed under short circuit conditions, so provide no information on the finite field electromechanical coupling. The need to obtain finite field piezoelectric measurements as a function of temperature relative to a datum has been recognised more recently with regard to the application in diesel injector valves [7]. Direct strain measurements were not reported and the strain relative to the reference was calculated from separate measurements of non-referenced loops, remanent strain, and thermal expansion. The relationship to the polarisation and behaviour above the Curie temperature were not reported.

Thermal expansion in a ferroelectric material is complicated by the coupling between the temperature dependent remanent polarisation and the strain. This coupling is influenced by the polarisation induced by an applied electrical field, leading to a complicated interdependency between temperature, electric field and strain. This coupling can be understood by studying the electrostrictive properties of the material. It is common to neglect the electrostrictive coupling when dealing with small field piezoelectric effects on a poled ceramic [8]. In this case the total remanent polarisation is not changed significantly by relatively small changes in the induced polarisation. However, typical soft PZT materials used at high electric field show a very high degree of induced polarisation. This may be comparable to the remanent polarisation, so the electrostrictive coupling becomes important. It can be shown [9] that the linear piezoelectric effect can be described in the low field limit as an electrostrictive strain with the polarisation equal to the remanent polarisation. The low Curie temperature of soft PZT means that large changes in spontaneous polarisation occur over a temperature range encountered in many practical situations, and this electrostrictive coupling to the temperature dependent polarisation dominates the thermal expansion and high field electromechanical response.

It is often asserted that the electrostrictive coefficients are independent of temperature, but experimental evidence for this is scant. This is particularly true for

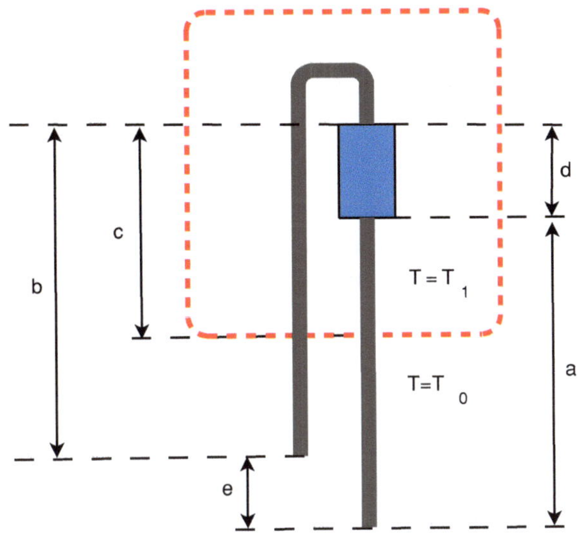

Fig. 1 Schematic diagram of pushrod dilatometer

PZT. This assertion is based on values obtained indirectly by means of a number of assumptions, or measurements over a restricted temperature range for single crystals of materials other than PZT [10].

The objective of this chapter is to obtain a complete mapping of the strain–polarisation–electric field relationship from room temperature to above the Curie temperature for soft PZT ceramics. It was a key objective to relate all the strain measurements to a reference datum so that drift in the actuation position from zero to high electrical fields could be measured. Measurements were performed for both unipolar and bipolar (switching) cycles. We also provide electrostriction coefficients obtained from direct measurements of the strain and polarisation, and show that these are temperature dependent above 100 °C. Analysis of the electrostrictive behaviour is used to provide an explanation for the nearly linear and hysteresis-free strain response when the charge rather than the electrical field is used to control actuation. Recent work on the role of electrostriction in actuation can be found in references [11–14].

2 Experimental Methods

A vertical pushrod dilatometer was used to measure electrically and thermally induced strain. Corrections were made for the pushrod expansion by calibration against a sample of known expansivity. Pushrod dilatometers are capable of high accuracy expansion measurements as long as the expansion of the pushrods themselves is taken into account.

Figure 1 shows a typical dilatometer system. The region inside the box is heated (or cooled) to a temperature T_1 while the area outside the box is maintained at a constant temperature T_0. The sample, of length d, is held between two pushrods with dimensions a and b. Note that only the length of pushrod b from the end of the sample is taken into account as the loop dimension cancels.

In the equations below a suffix 0 indicates the dimensions at ambient temperature i.e. when $T_1 = T_0$.

$$At \quad T_1 = T_0 \quad e = e_0 = a_0 + d_0 - b_0 \tag{1}$$

when the box is heated (or cooled) the measured expansion will be the expansion of the sample plus contributions from the portions of the pushrods inside the box. At $T_1 = T$

$$\begin{aligned} e = e_T &= a_0 + (c_0 - d_0)S_R(T) \\ &\quad + d_0 + d_0 S_S(T) \\ &\quad - b_0 - c_0 S_R(T) \end{aligned} \tag{2}$$

where $S_R(T)$ is the strain of the pushrod material and $S_S(T)$ is the strain of the sample material at temperature T relative to the strain at T_0. The measured expansion, $e - e_0$, is

$$e - e_0 = -d_0 S_R(T) + d_0 S_S(T) \tag{3}$$

we are interested in finding the strain of the sample material

$$S_S(T) = \frac{e - e_0}{d_0} + S_R(T) \tag{4}$$

if we define the measured strain $S_M(T)$ as

$$S_M(T) = \frac{e - e_0}{d_0} \tag{5}$$

then

$$S_S(T) = S_M(T) + S_R(T). \tag{6}$$

The correction term $S_S(T)$ is obtained by measuring a calibration curve using a reference sample of known expansion. In this analysis it was assumed that there are only two temperatures for the portions of the pushrod. In the real situation there will be temperature gradients along the pushrods. This equation will still hold as long as the calibration sample is the same size as the measured sample. The same is true of temperature gradients in the pushrods caused by the rate of change of temperature. Again these will be minimised if the calibration is performed under the same rate of temperature change as the sample.

A factor not considered here are changes in proportions of the pushrods inside and outside of the box. This is a second order effect but may be significant for e.g. very thin samples. Samples used for the experiments were 25 mm × 5 mm × 0.3 mm sheets of a commercial soft PZT, with approximately 4 μm of Ni electrode on both

25 mm × 5 mm faces. Strain was measured in the 25 mm length for an electrical field applied across the 0.3 mm thickness (this was also the direction of poling). This is referred to in this chapter as transverse expansion or contraction. Strain in the poling direction is referred to as longitudinal. A positive strain refers to an expansion. Note that results in this paper are mostly plotted as negative strain (noted on the graph labels) to present the familiar direction of the strain loops. Electrical connections were made through gold contacts held against each electrode. A 17.2 μF capacitor was placed in series with the piezoelectric element to measure the charge on the piezoelectric element (33 nF). The voltage across the capacitor was measured by means of a high input impedance unity gain instrumentation amplifier. The small amount of drift due to input bias currents was measured and compensated for in the results. Electric field was applied using a high voltage amplifier under computer control. Simultaneous measurements were made of the applied voltage, charge, extension and temperature. Electric field, polarisation and strain were obtained using the measured dimensions of the sample. Separate experiments were performed for bipolar and unipolar electrical cycles. For the bipolar set a loop sequence of linear ramps 0 to +400 to −400 to 0 V was applied over a period of 40 s. Two loops were taken for each temperature point (approximately every 10 °C) for a heating/cooling rate of 20 °C per hour. The unipolar loop sequence was 0 to 400 to 0 V over 80 s. The heating/cooling rate for the unipolar experiments was 10 °C per hour.

3 Experimental Results

Electrical loops recorded at different points across the temperature range are shown in Figs. 2 and 3 for bipolar and unipolar cycles respectively. Figure 2a shows polarisation–electric field loops. At 19 °C characteristic ferroelectric loops were observed with a remanent polarisation of approximately 0.3 C m^{-2} at room temperature, typical of this type of material [15, p. 147]. At 186 °C the hysteresis has disappeared and the dielectric response becomes paralectric, albeit with a high and non-linear permittivity ($\epsilon_r \approx 15000$). The transition from ferroelectric to paraelectric behaviour occurs approximately between 140 and 150 °C. This will be referred to as the transition temperature. Figure 2b shows the strain–field loops. The familiar 'butterfly' loop is observed at the lower temperatures. Above the transition temperature the strain hysteresis also disappears. However, there is still a strong eletromechanical response well above the transition temperature. This is shown in Fig. 4 which plots the loop maxima, minima, and the loop end point at 0 V (the remanent strain). The maxima and minima of the strain loops show no discontinuity or other anomaly on passing through the transition temperature. The remanent strain shows a step change in strain as well as a change in slope at the transition temperature. The remanent strain closely follows the remanent polarisation (see Fig. 5) except for a gradual strain drift with increasing temperature corresponding to an underlying thermal expansion. It is not possible to specify a unique expansion coefficient for these loops because it

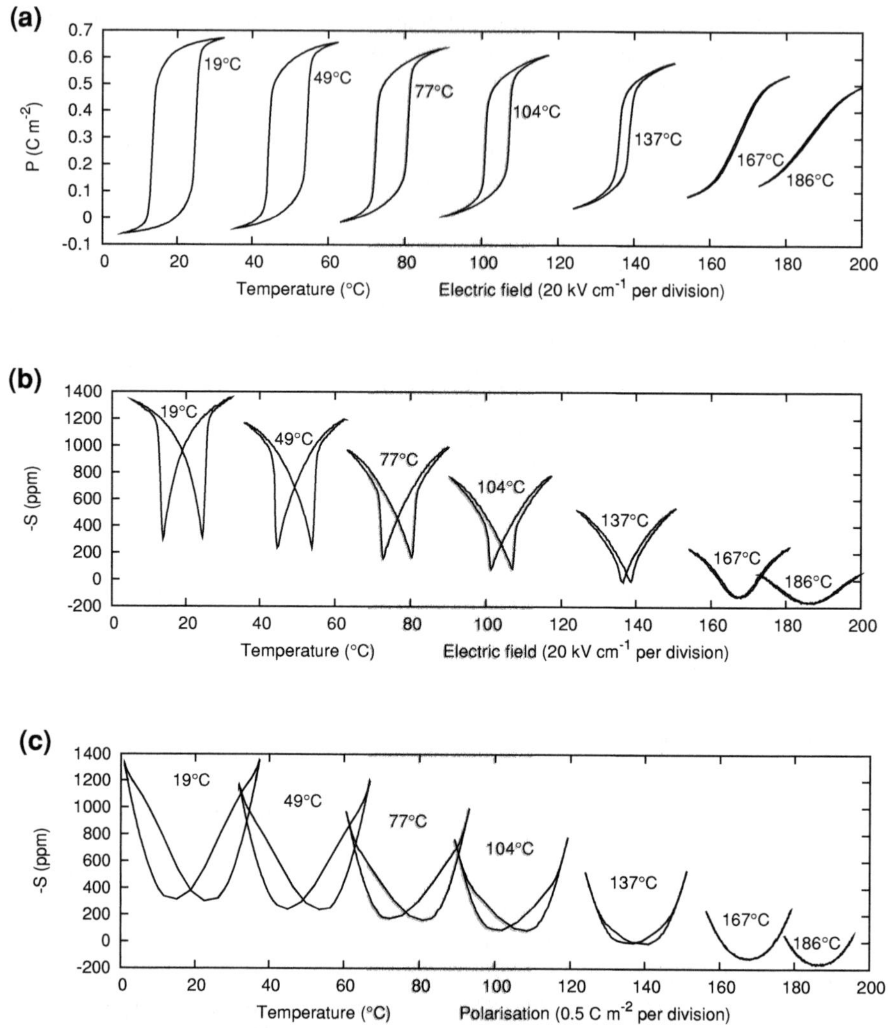

Fig. 2 Composite plots showing bipolar, **a** polarisation-field, **b** strain-field and **c** strain-polarisation loops at different temperatures. The x axis for each loop is scaled (by the same arbitrary factor for all loops in each plot) to allow clear presentation

depends on the point in the loop that is measured i.e. the thermal expansion depends on the electrical conditions. For example, the loop maxima and remanent strain show an expansion of approximately $7.5–8.0 \times 10^{-6}$ °C^{-1} while the loop minima give a value of 2.9×10^{-6} °C^{-1}.

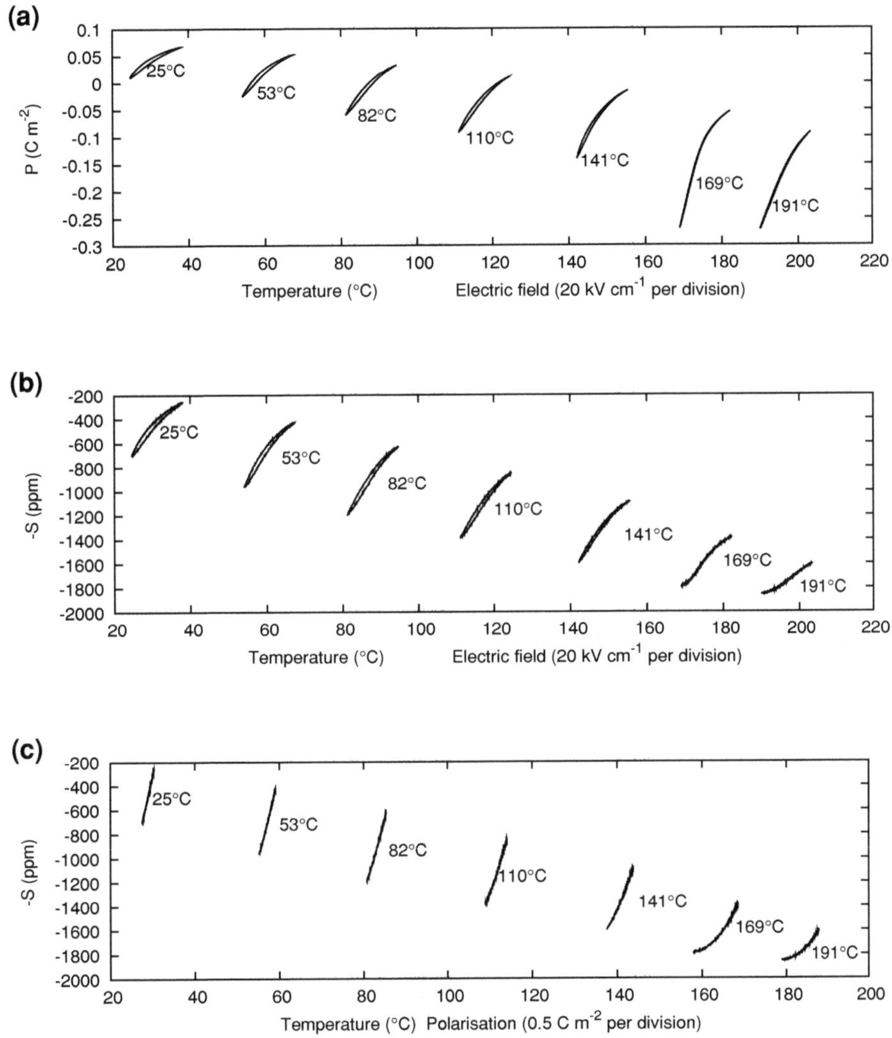

Fig. 3 Composite plots showing unipolar, **a** polarisation-field, **b** strain-field and **c** strain-polarisation loops at different temperatures. The x axis for each loop is scaled (by the same arbitrary factor for all loops in each plot) to allow clear presentation

4 Polarisation Zero

Figure 5 shows the values of the polarisation at different points in the loop, and their variation with temperature. The maximum and minimum polarisation diminished gradually with increasing temperature, and showed no anomaly at the transition temperature, approximately tracking the strain maximum. The remanent polarisa-

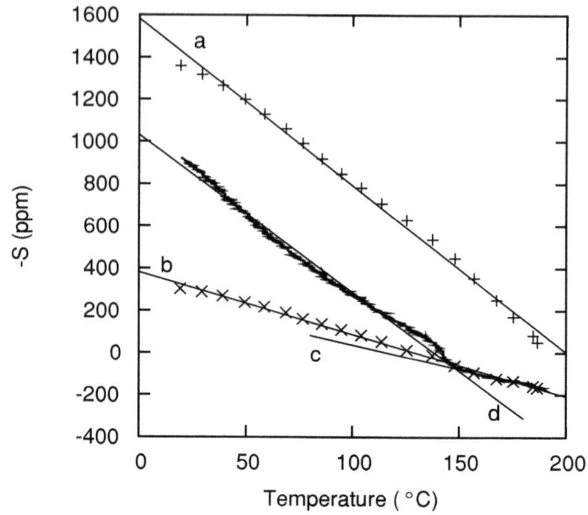

Fig. 4 Thermal expansion during temperature cycling with bipolar loops. **a** is a *straight line* fit to the loop maxima (+) and **b** is a *straight line* fit to the loop minima (×). **c** is a *straight line* fit to the remanent strain (*solid line*) for T > 148 °C while **d** is a fit to the portion of the *curve* T < 138 °C

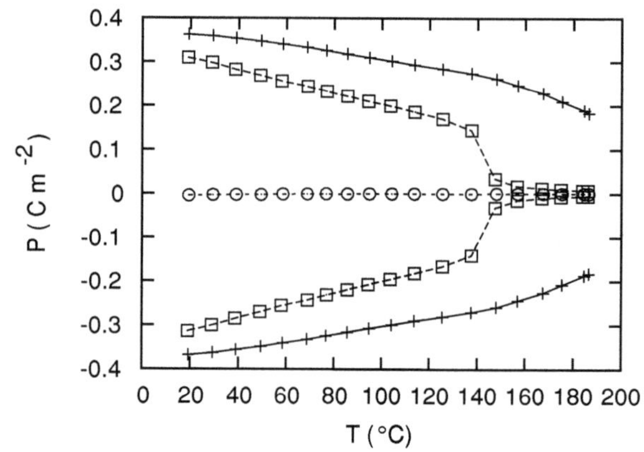

Fig. 5 Variation of PE loop maximum, minimum, mean and remanent polarisation values with temperature

tion decreased somewhat more rapidly, and underwent a step change at the transition temperature. The polarisation averaged over each loop remained approximately constant over the temperature range. The average of the positive and negative remanent polarisations was equal to the same value, and also did not vary with temperature. Above the transition temperature, the remanent polarisation did not equal this value, but approached the same value asymptotically from both directions. The correlation of the different measures of the mean polarisation, and the fact that the remanent polarisation approached the same value, allows the identification of a single unique value for the zero point polarisation P_0. This was set as zero in Fig. 5. The remanent polarisation above the transition temperature appears to be time dependent, indicating a slowly relaxing induced ferroelectric behaviour above the transition temperature.

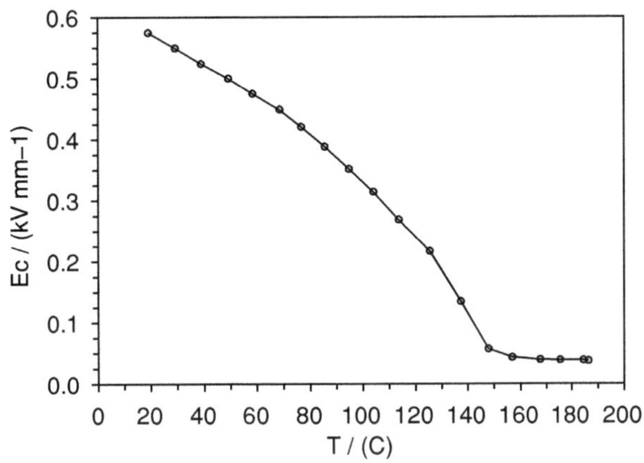

Fig. 6 Coercive field variation with temperature

5 Coercive Field

The coercive field varies with temperature. This field was obtained by measuring the zero crossing of the PE loops and is shown in Fig. 6.

6 Electrostriction and Ferroelectricity

Most research in electrostriction is focussed on materials with zero remanent polarisation but with high permittivity and high electromechanical coupling such as relaxor ferroelectrics [16]. These materials usually exhibit typical ferroelectric behaviour below the Curie temperature, including hysteresis and remanent polarisation. Indeed, materials we consider as typical ferroelectrics, such as Barium Titanate can be considered as electrostrictive in the high field extreme where the applied field is large compared to the coercive field and the remanent polarisation is small [17, 18]. However, for many commonly used piezoelectric ceramics operating at small field levels, the field does not significantly alter the total polarisation and the normal piezoelectric relations apply. However, at high field strengths used in many actuator applications, especially when using 'soft' ceramics with high piezoelectric coefficients, the electrostrictive coefficients are also very high and the polarisation induced by the field is significant compared to the remanent polarisation. To understand the behaviour of these materials, particularly with regard to thermal expansion it is necessary to examine the electrostrictive behaviour of the material.

Considering only the transverse strain (orthogonal to the poling axis) with the ∞m point group symmetry of a poled ceramic [19], the electrostriction can be expressed as [9]:

$$S_1 = Q_{12} P_3^2 \tag{7}$$

where S_1 is the strain, and P_3 is the polarisation in the direction of the poling axis. Q_{12} is the longitudinal electrostriction coefficient (Q_{12} is a fourth rank tensor but considerable simplifications are made using crystal symmetry and the averaging over crystallites in a ceramic). In the case of a ferroelectric below the Curie temperature, the polarisation and strain may be considered as comprising both spontaneous and field induced components:

$$S_r + S_i = Q(P_r + P_i)^2 \tag{8}$$

where the subscript r refers to the spontaneous component and i refers to the induced component. Expanding Eq. 8 gives

$$S_r + S_i = QP_r^2 + 2QP_rP_i + QP_i^2 \tag{9}$$

this equation can be used to evaluate the small signal piezo coefficient where P_r remains approximately constant and P_i is small, so we can neglect the term in P_i^2 and differentiate to give the piezoelectric strain-polarisation coefficient, e.

$$\frac{1}{e} = \frac{dS_i}{dP_i} = 2QP_r \tag{10}$$

this can be related to the piezoelectric strain coefficient d by

$$d = \frac{dS_i}{dE} = \frac{dS_i}{dP_i}\frac{dP_i}{dE} \tag{11}$$

$\frac{dP_i}{dE}$ is the permittivity ϵ so

$$d = 2QP_r\epsilon \tag{12}$$

In principle Eq. 7 can be used to estimate the thermal expansion due to polarisation changes from a knowledge of the electrostrictive coefficient. This can be added to the underlying lattice expansion to estimate the total thermal expansion of the ceramic from some simple measurements or a knowledge of a few material parameters. A knowledge of the thermal expansion of the ceramic allows the device engineer to estimate the performance of a real device, and to design a mechanism to make best use of the available strain over the widest possible temperature range. In practice, we measure values of the strain and the polarisation relative to a reference value. Equation 7 can be re-stated more generally as:

$$S_1(T, P) = S_0(T) + Q_{12}(T)(P_3 - P_0)^2 \tag{13}$$

where S_1 is the strain measured relative to a reference, S_0 is the strain maximum of the parabola relative to the same reference. Note that Q_{12} is usually negative reflecting the fact that the 'striction is negative in the transverse direction, so $S_1 < S_0$. S_0 and Q_{12} can therefore be found from a plot of S_1 against $P_3 - P_0$, but to do this

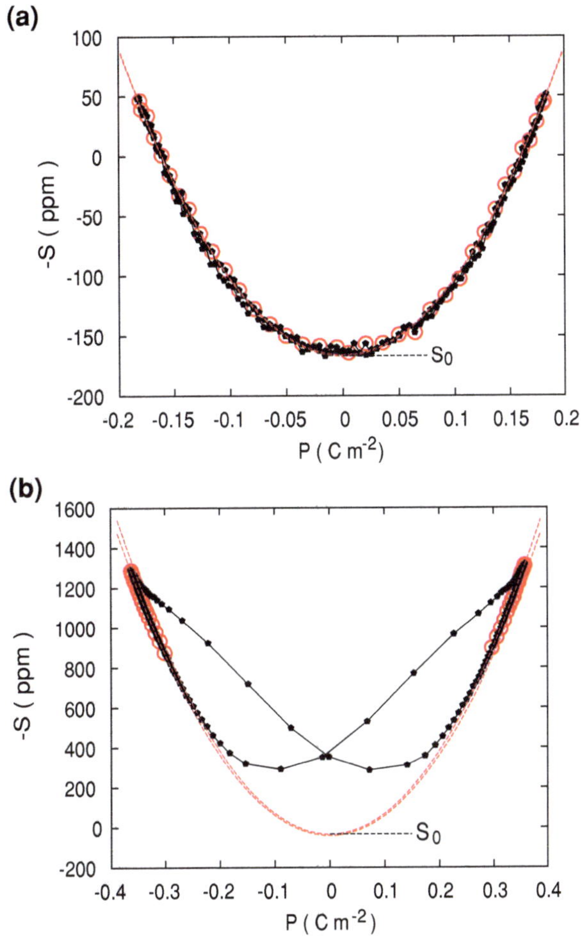

Fig. 7 SP loops at **a** $T > T_C$ (186 °C) and **b** $T < T_C$ (29 °C). *Red lines* shows the quadratic fit to the points marked as *red circles*. S_0 is the minimum of the parabola.

we need to know P_0. This is straightforward for measurements above the transition temperature. P_0 is simply the value of the polarisation at zero electrical field. This is not so obvious below the transition temperature, and not at all obvious for unipolar loops, and so requires clarification and is discussed further in Sect. 4.

Figure 2b shows the strain–polarisation loops. Above the transition temperature, these loops become very closely parabolic. This is shown more closely in Fig. 7a where the experimental data are compared with a parabolic line obtained by a least squares fit (see below for line fitting). The close parabolic relationship is only seen in the strain as a function of polarisation and is characteristic of electrostrictive behaviour [9]. The strain shows a much more complicated dependency on electrical field.

Having established a value for P_0 in Sect. 4, it is now possible to evaluate the coefficients of Eq. 13 by linear regression. An excellent fit to the experimental data

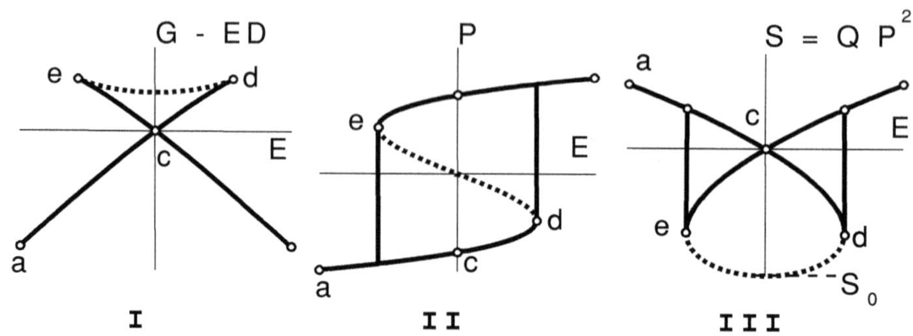

Fig. 8 Stability from free energy expansion. From *left* to *right* free energy $G_1 - ED$, polarisation, and strain as functions of electric field. **a–c** is stable, **c–d** is metastable

is obtained, as shown in Fig. 7a. The value of S_0 thus obtained corresponds to the temperature dependent strain of the non-polar centrosymmetric structure obtained at zero electrical field. The states obtained at non-zero electrical field are all stable and single valued for a given electrical field. This behaviour can be examined qualitatively by consideration of the free energy expansion in the presence of an applied field. The free energy expansion as a function of electric displacement, D, and the electrical field, E, are given in the usual way [2]:

$$G_1 = \frac{\alpha}{2}D^2 + \frac{\gamma}{4}D^4 + \frac{\delta}{6}D^6 \tag{14}$$

$$E = \left(\frac{\partial G}{\partial D}\right)_T = \alpha D + \gamma D^3 + \delta D^6 \tag{15}$$

where α varies with temperature, being positive above the Curie temperature, T_C, and negative below it. The coefficients γ and δ are assumed to be constant and positive corresponding to a second order phase transition. These assumptions are unlikely to be valid for the temperature range of interest here, but this does not affect the qualitative conclusions of this discussion. In the following discussion we use the approximation $D \approx P$ and treat the two terms interchangeably. From Eqs. 14 and 15 we can evaluate the free energy in the presence of an electrical field:

$$G = G_1 - ED \tag{16}$$

The function corresponding to $T > T_C$ is stable for all values of the electrical field in the sense that there are no states of lower free energy for any given value of the electrical field. In the plot of Fig. 7a we are therefore using only stable states to obtain the electrostrictive coefficient and S_0 corresponds to the strain of the centrosymmetric non-polar state. The function corresponding to $T < T_C$ is plotted in Fig. 8. The stable states are identified in the same way as for $T > T_C$, but now there are metastable

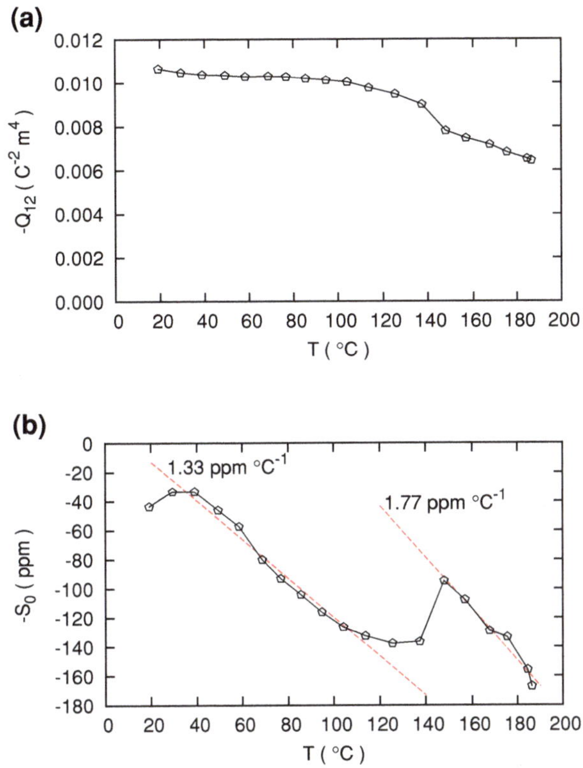

Fig. 9 Electrostriction coefficient (**a**) and strain zero (**b**) as functions of temperature (bipolar cycles). In (**b**) the *dashed lines* represent *straight line* fits to the approximately linear portions of the curves

states where a lower free energy state exists for the same electric field. An energy barrier separates the two states. Unstable states correspond to a free energy maximum with no energy barrier to the transition. Using the electrostrictive relationship (Eq. 7) we can calculate the strain as a function of electric field also shown in Fig. 8 for $T < T_C$. The familiar butterfly curve is obtained. This allows us to identify the regions of the strain curve corresponding to thermodynamically stable states i.e. the regions of reducing electric field between the maximum and zero electric field (between maximum and remanent strain). These are the points identified in Fig. 7b and used to calculate the electrostrictive coefficient and the strain zero from Eq. 13. The results thus obtained are shown in Fig. 9.

The electrostriction coefficient is approximately constant up to 100 °C with a value mostly between −0.0100 and −0.0105 C^{-2} m^4. This is slightly lower than the value of −0.0158 C^{-2} m^4 measured by Haun [10] for undoped PZT with x(Ti) = 0.52. Above 100 °C the coefficient drops rapidly until the transition temperature. Above the transition temperature the electrostriction coefficient (approximately 0.0075 C^{-2} m^4) reduces linearly with increasing temperature. It is thought that this reflects an increasing contribution from 180° domain wall motion which contributes to the polarisation changes, but not the strain, with increasing temperature. This may also explain why the electrostriction coefficient values reported here for soft PZT are somewhat lower that previously reported values. It is interesting to note that this process continues even above the transition temperature suggesting that domain processes are still contributing significantly to the dielectric and electromechanical response even well above the transition temperature.

Figure 9 shows the zero point strain variation with temperature. Up to 100 °C a fairly linear expansion is observed with an expansion coefficient of 1.33×10^{-6} °C^{-1}. Just below the transition temperature the expansion reduces to a very low value. A contraction is observed in the regions of the transition temperature, indicative of a diffuse first order crystallographic phase change. Above the transition, the expansion proceeds linearly with a coefficient of 1.77×10^{-6} °C^{-1}. These figures are in line with previous measurements of the expansion in PZT by Berlincourt [20] (reported in [21, p. 112]) of 2×10^{-6} °C^{-1}, falling to a very low value in the vicinity of the T_c and for unpoled ceramics [5, 6].

If we associate S_0 with intrinsic electrostriction [19], then S_0 describes a notional centrosymmetric state of the crystal (averaged over all the orientations of the ceramic crystallites). This can be interpreted as a measure of the underlying lattice expansion in the absence of extrinsic domain-related effects. The actual response is the sum of this lattice expansion and the electrostriction calculated through Eq. 13. For the transverse expansion measurements reported here, changes in the remanent polarisation couple electrostrictively to the strain causing an expansion on heating. This reinforces the normal thermal expansion of the lattice. If measured in the direction of polarisation, the changes in remanent polarisation would lead to a contraction on heating, which would oppose the underlying lattice expansion. For unpolarised samples, the electrostrictive contraction would be approximately zero, so a weak expansion would be observed. For poled samples, the electrostrictive contraction would dominate leading to negative expansion. Above the Curie temperature there is no remanent polarisation, so zero field strain in both directions reverts to a normal expansion on heating. The transverse measurements reported here therefore help to explain previous reports of negative thermal expansion in poled PZT ceramics [5, 6].

For unipolar cycles the polarisation zero can not be identified from a single loop measurement at a temperature below the transition temperature. Without this quantity we cannot evaluate the coefficients in Eq. 13. Figure 5 shows that for bipolar loops above the transition temperature the remanent polarisation approaches the same value as the loop mean polarisation. At 190 °C the remanent polarisation is within 0.002 C m^{-2} of the loop mean value. This is less than 1 % of the room temperature remanent polarisation. Because the material is non-polar at these temperatures, the high temperature remament polarisation for unipolar loops is expected to be the same as for bipolar loops. The value of unipolar remanent polarisation at 190 °C was taken as P_0 for the unipolar fit. This gave very similar values for the remanent polarisation at lower temperatures e.g. at 40 °C the zero field polarisation was 0.267 C m^{-2} unipolar and 0.274 C m^{-2} bipolar.

Figure 10 shows unipolar strain–field and strain-polarisation loops at 25 °C. The electric field is sufficiently high (1.3 kV mm^{-1}) to cause a significantly non-linear

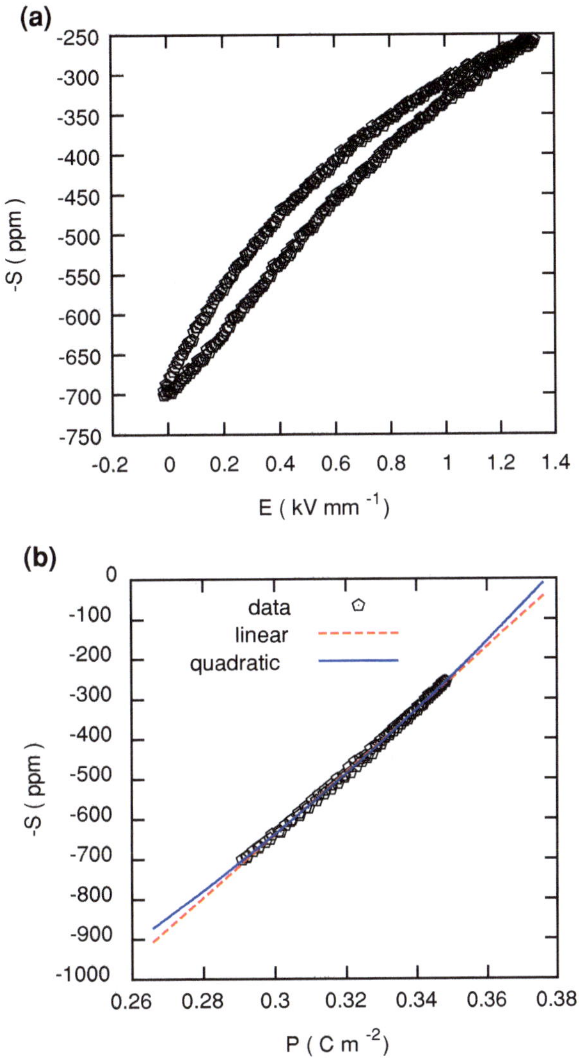

Fig. 10 Unipolar **a** strain–field and **b** strain–polarisation loops at 25 °C. In (**b**) the *dashed line* is a linear fit and the *solid line* is the electrostrictive curve

response with a large hysteresis. However, when strain is plotted as a function of polarisation, the hysteresis disappears completely, and the response is almost linear (Fig. 10b). This effect has been reported previously [22] where the linear response is used to improve control systems response by using charge control rather than voltage control. The linear fit to the data shows that the there is a very slight deviation from linearity. The electrostrictive fit, obtained by fitting the data using Eq. 13 and P_0 derived as described above, shows an excellent fit to the data. The electrostrictive relationship holds true to a high degree of accuracy right across the temperature range (Fig. 11), although the coefficients vary with temperature as discussed above. Figure 12 shows

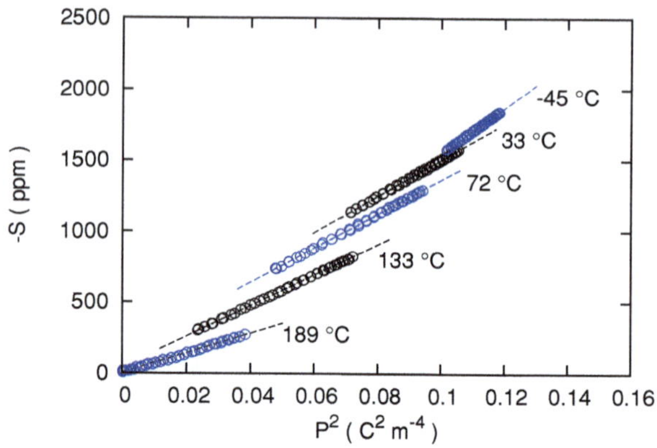

Fig. 11 Strain -v- P^2 for unipolar loops at different temperatures

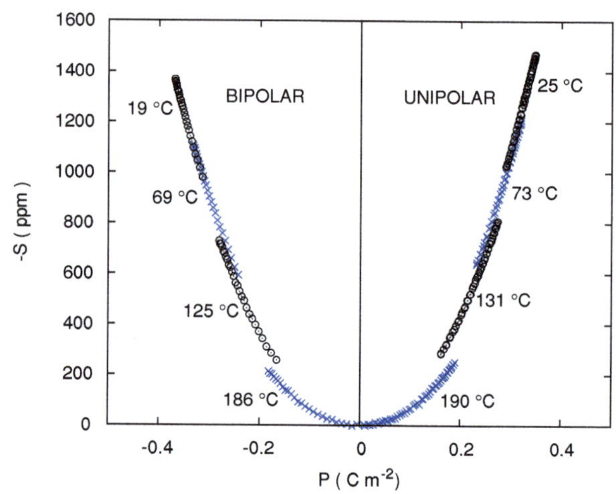

Fig. 12 Strain -v- Polarisation compensated for a thermal expansion of 0.8×10^{-6} °C^{-1}. The *left* half of the plot shows the stable regions of the bipolar cycles (negative half only shown). The *right* half of the plot shows unipolar cycles. Temperatures were (from *top* to *bottom*) 25 °C, 73 °C, 131 °C, 190 °C for unipolar cycles and 19 °C, 69 °C, 125 °C, and 186 °C

the measured strain-polarisation curves across the temperature range. As the temperature increases, the curvature becomes more pronounced. Approximate linearity is observed at lower temperatures because the high value of the remanent polarisation offsets the piezoelectric response to a low-curvature part of the quadratic function. The linear approximation becomes untenable at higher temperatures, but the quadratic electrostrictive description is accurate at all temperatures studied.

The strain–polarisation curves shown in Fig. 12 have been adjusted by applying a temperature dependent offset of 0.8×10^{-6} °C^{-1}. This is the value of the total expansion across the temperature range shown in Fig. 9b. The value is so low due to the contraction at the transition temperature. It is used here to compensate the curves to remove the average overall expansion (this figure is arbitrary as discussed below). We observe that all the curves nearly lie the same parabolic curve, indicating that there is an approximate underlying electrostriction that applies across the temperature range, and for unipolar and bipolar (stable region) cycles. However, the family of curves can not be made to lie on the same curve for any single pair of values of expansion and electrostriction coefficients. This is a reflection of the temperature dependence of these coefficients shown in Fig. 9.

These results confirm that the electrostrictive model provides a very accurate description of the device response to very high levels of induced polarisation and electric field, as long as the system is kept within the stable region of the free energy diagram. This is true even when there is significant non-linearity and hysteresis present in the strain–field response. This implies that the electrostrictive description of the piezoelectric behaviour applies not just to intrinsic crystallographic distortions, but also to the changes in strain and polarisation due to extrinsic domain wall motion. The fact that the relationship holds across the temperature range, and the electrostriction coefficient reduces with increasing temperature is a strong indication that both 180° and non-180° domain processes are accommodated within this model. However, the model does not apply in the region of metastability where the field is applied against the direction of the remanent polarisation.

7 Electrostriction and Domain Processes

An approximation to the butterfly strain loop can be obtained from the PE loop shown in Fig. 13a taking the square of the polarisation, to obtain the familiar butterfly loop shown in Fig. 13b. Note that there is some asymmetry in this loop which is exaggerated by the square power.

To make a more direct comparison with the measured strain loop (from the same experiment) the asymmetry is removed by averaging the polarisation amplitude, field amplitude and strain over positive and negative field values. To convert the P^2 values to strain a constant of proportionality is required. This is a measure of the electrostrictive coefficient. This was obtained by taking the strain amplitude from the zero field point to the maximum strain and dividing it by the P^2 amplitude between the same points. A value of $Q_{eff} = 0.01\,\mathrm{C^{-2}\,m^4}$ was obtained. For comparison a previous study [23] reported a value of $Q_{11} = 0.03\,\mathrm{C^{-2}\,m^4}$ for soft PZT ceramic. Q_{11} describes the strain in the direction of the applied field, which has approximately double the coefficient Q_{12} for the strain perpendicular to the field [10]. The value measured here is therefore slightly smaller than this previous report. Because the strain origin is arbitrary, an offset figure is also required. This was achieved by setting the strain maximum to the same value for the direct measurement and the value calculated from P^2. A comparison of the two butterfly loops is shown in Fig. 14.

The two loops match very closely for the region of decreasing electrical field from the maximum strain to zero field. This is the region that was used to obtain the amplitude for the calculation of the electrostrictive coefficient. The closeness of the fit over this region confirms that the strain very closely follows the electrostrictive quadratic function. However, as the field is reversed, the P^2 strain starts to deviate from the measured value, particularly around the coercive field. There is a whole region of strain amplitude predicted from the P^2 function that is not observed in

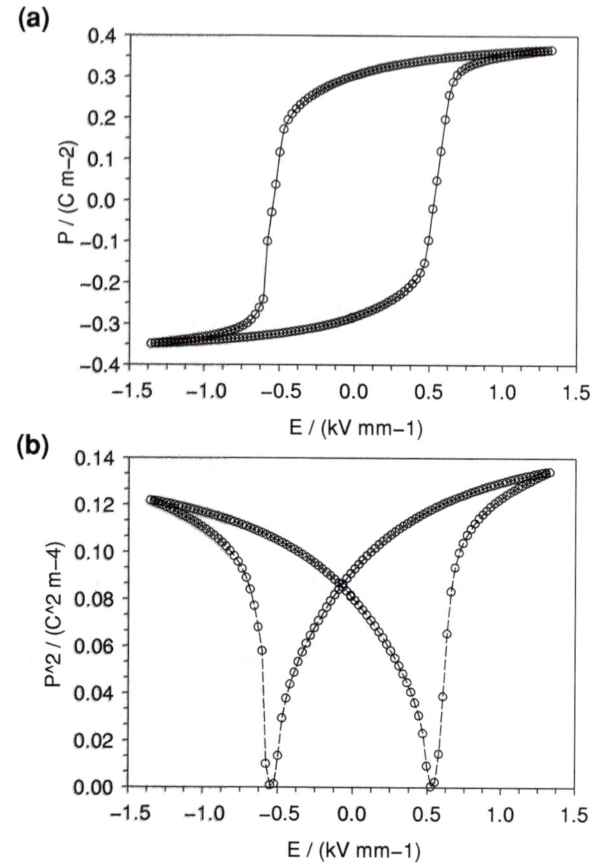

Fig. 13 (**a**) PE loop at 19 °C and (**b**) the same data plotted as P^2

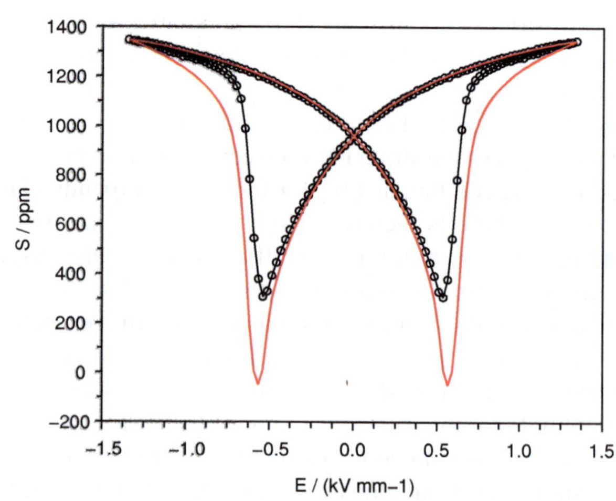

Fig. 14 Strain field loops directly measured (*circles*) and calculated from polarisation data (*red line*)

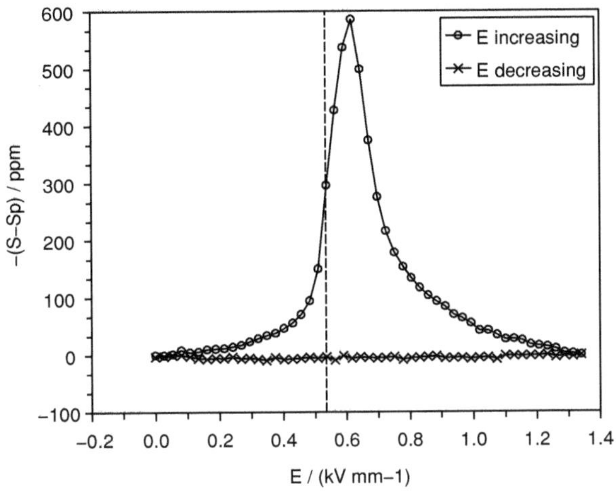

Fig. 15 Difference between the directly measured strain and the strain calculated from polarisation data. The *vertical line* represents the value of the coercive field

the strain measurement. To examine this more closely, the difference between the measured strain and the P^2 strain is plotted in Fig. 15.

The widely accepted view of the piezoelectric response of poled ceramics is [8] that the strain response is formed of two contributions—(i) distortion of the unit cells within a single domain referred to as 'intrinsic', and (ii) deformation due to the motion of non-180° wall motion. The region of proportionality between the measured strain and P^2 corresponds to one where a single value of Q_{eff} fits the curve i.e. Q_{eff} is not significantly affected by the electrical field. This was also shown to be true in PMN-PT [8] where it was ascribed to the absence of non-180° domain rotation i.e. purely intrinsic piezoelectric response. This cannot explain the observed electrostriction which is hysteretic with respect to the applied field. Unipolar loops over the same range of field are very accurately described by the same electrostrictive response whilst demonstrating significant non-linearity and hysteresis when viewed as a function of applied field. The electrostrictive model requires a reversal of the polarisation direction with a reversal in the field direction. The symmetry with respect to the electric field dictates that this cannot be considered as an intrinsic piezoelectric response. The coercive region shows an additional contribution to the strain causing deviation from electrostrictive behaviour. Figure 15 shows that there is a very significant strain effect for processes that become activated in the region of the coercive field. 180° domain wall movement does not contribute to the strain so cannot explain this phenomenon. If the part of the curve described by the electrostrictive model is due to 'intrinsic' piezoelectric distortion, then non-180° domain rotation could become activated in the region of the coercive field following the curve shown in Fig. 15. The role of 180° domain wall motion, in this model, is not clear.

An alternative explanation is as follows. For a ferroelectric below the Curie temperature, the strain response is gradually modified from the electrostrictive response by an increasing remanent polarisation and a coercive field. The dominant factor is still polarisation reversal, and a broadly electrostrictive response, but now the states

Fig. 16 Relative permittivity variation with temperature

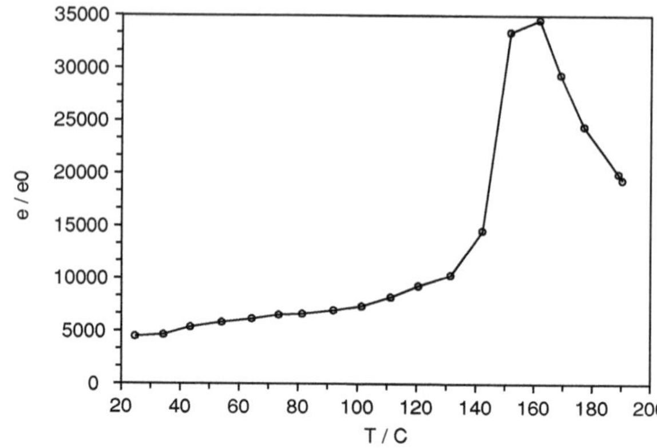

of lower strain at fields higher than the coercive field are no longer stable and can no longer be accessed. This explains the time dependency of states within the coercive region—a faster transition will show a broader curve with a lower minimum strain.

8 Piezo Coefficients and Permittivity

It was described in Sect. 6 how the low field piezoelectric coefficient can be obtained from the electrostriction coefficient, the remanent polarisation and the permittivity (Eq. 12). The experiments presented here did not include a direct measurement of permittivity, but an estimate can be obtained from the PE loops.

In principle the permittivity can be measured as the slope of the PE loop. However, with the large amplitude field used here, the permittivity measured in this way exhibits hysteresis and is multi-valued. The closest we can obtain from these experiments to a measure of the low field permittivity is the slope of the unipolar loop for increasing electric field. The variation of this measure of permittivity with temperature is shown in Fig. 16. The results display the classic ferroelectric temperature dependency of the permittivity with a peak at the transition temperature. The peak in the permittivity (the Curie-Weiss temperature) occurs at approximately 160 °C which is slightly higher than the temperature of the phase transition in the thermal expansion (the Curie temperature). This provides further confirmation of a first-order phase transition [9] in this material. The room temperature value of the relative permittivity measured by this technique was approximately 4,500. This compares reasonably well with the data sheet value for ϵ_{33} of 5500.

The piezoelectric coefficient calculated from Eq. 12 is shown in Fig. 17. The permittivity values used were those shown in Fig. 16. The room temperature value of the d_{31} is -260 pm V^{-1} which compares to the data sheet value of -330 pm V^{-1}. The size of the discrepancy in the piezoelectric coefficient is almost exactly the same as the discrepancy in the permittivity. The value of d_{31} increases slightly with increasing temperature, until it drops rapidly to zero around the transition temperature.

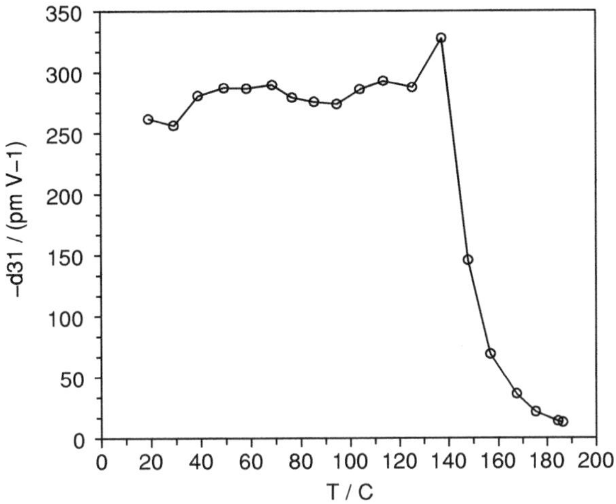

Fig. 17 Piezoelectric coefficient measured from electrostriction strain (fit parameter $E_{mar} = 0.3$)

9 Current Density and Strain Rate

According to Chong [24] the unit cell distortion is related to the spontaneous polarisation, strain rate and current density at the coercive point by the relationship:

$$\frac{3}{4}\frac{c_0 - a_0}{a_0 P_0} = \frac{\dot{e}_3}{j} \tag{17}$$

where a_0 and c_0 are the lattice constants, P_0 is the spontaneous polarisation and $\dot{e}_3$ is the strain rate in the direction of the applied field.

$$\dot{e}_3 = \frac{dS_3}{dt} \tag{18}$$

and j is the current density

$$j = \frac{dP}{dt} \tag{19}$$

$$\frac{\dot{e}_3}{j} = \frac{dS_3}{dt} / \frac{dP}{dt} = \frac{dS_3}{dP}. \tag{20}$$

From Eq. 17 we get

$$\frac{dS_3}{dP} = \frac{3}{4}\frac{c_0 - a_0}{a_0 P_0} \tag{21}$$

Room temperature values given are $a_0 = 0.4051$ nm, $c_0 = 0.4086$ nm, $P_0 = 0.32\,\mathrm{C\,m^{-2}}$, so using Eq. 21 we get a value of $0.0202\,\mathrm{C^{-1}\,m^2}$. Experimental results for a full bipolar loop at 19 °C are shown in Fig. 19 compared to the predicted value. To

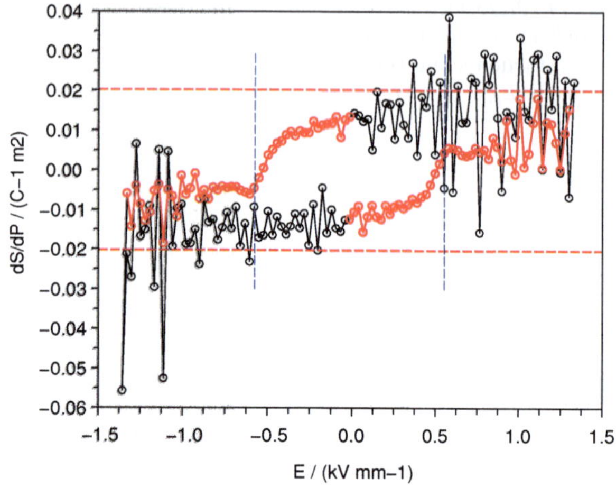

Fig. 18 $\frac{dS_3}{dP}$ as a function of electrical field at 19 °C. The *red markers* shows the data points in the coercive regions. The *vertical lines* are the positive and negative coercive field values. The *horizontal lines* represent the value calculated from the data of [24]

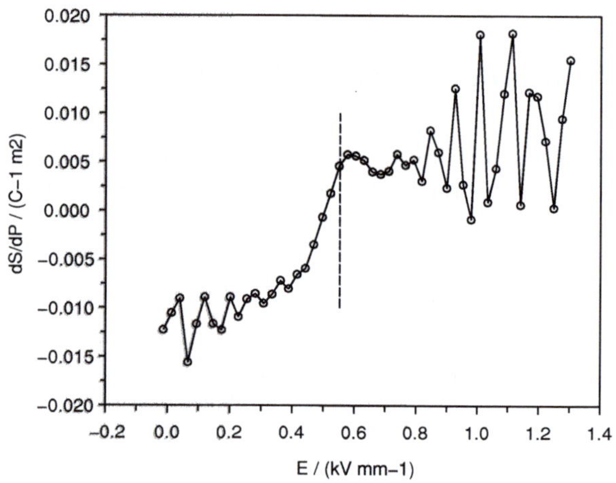

Fig. 19 $\frac{dS_3}{dP}$ as a function of electrical field at 19 °C for the positive going coercive region. The *vertical line* marks the coercive field value

estimate S_3 from the measured S_1 strain a factor of 2 was used i.e. $S_3(estimated) \approx 2 \times S_1$. Figure 19 shows the data points in the loop quarter that passes through the coercive region.

$$\alpha = \frac{c_0 - a_0}{a_0} = \frac{4}{3} P_0 \frac{\dot{e_3}}{j} \approx \frac{4}{3} P_0 \frac{2\dot{e_1}}{j} \tag{22}$$

Equation 22 shows how the lattice distortion parameter α can be obtained from polarisation–strain loops across the temperature range. Results are shown in Fig. 20. The value of the slope in Fig. 20 is -13.12 ppm °C^{-1}.

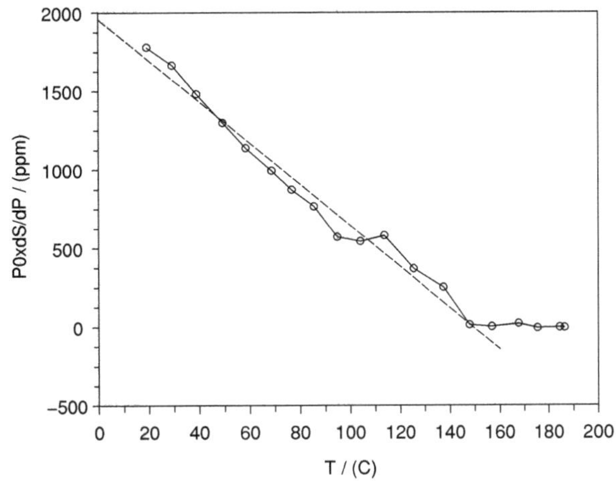

Fig. 20 Unit cell distortion parameter α as a function of temperature

10 Pyroelectricity and Polarisation

Polarisation was measured from the voltage across a sense capacitor (17.2 uF) in series with the piezoelectric sample (Fuji C91 25 mm × 5 mm electrode area × 0.3 mm thick). A high impedance instrumentation amplifier was used to monitor the sense capacitor with extremely low leakage ensuring no significant drift in recorded charge. The capacitance is approximately 1000x that of the sample so errors due to the voltage drop across the sense capacitor were negligible. Temperature was cycled from 20 °C to approximately 190 °C. Voltage loops were applied approximately every 10 °C . These were either unipolar 0 V to +400 V or bipolar 0 V to +400 V to −400 V to 0 V. The individual loops are not shown in the results presented here and have been filtered out from the continuous polarisation measurement. For ease of comparison between unipolar and bipolar results (positive and negative remanent polarisation respectively) measurements are multiplied by +1 or −1 so that polarisation is presented as decreasing with increasing temperature.

Figure 21 shows the change in polarisation with heating for the unipolar experiment. There is a steep transition of approximately 0.15 C m^{-2} at between 140 and 150 °C corresponding to the Curie temperature and the loss of spontaneous polarisation. Below this temperature the polarisation varies fairly linearly with temperature giving an effective pyroelectric coefficient of 1.3 mC m^{-2} K^{-1}. A similar result is seen for the experiment with bipolar loops (Fig. 22) with an effective pyroelectric coefficient of 1.4 mC m^{-2} K^{-1} .

The situation becomes more complicated on cooling. Cooling measurements were not available for the unipolar loop experiment, so we focus on the bipolar results shown in Fig. 23.

The heating curves vary linearly up to the Curie temperature above which the polarisation drops to zero. The cooling results follow the same general pattern, but below the Curie temperature the polarisation changes in steps corresponding to the

Fig. 21 Polarisation changes with temperature (unipolar loops)

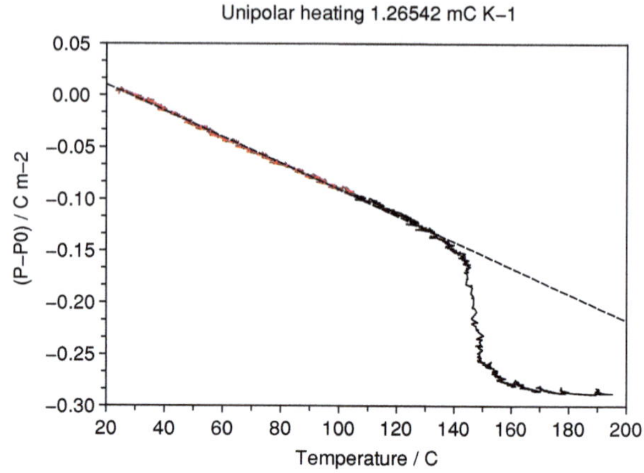

Fig. 22 Polarisation changes with temperature (bipolar loops)

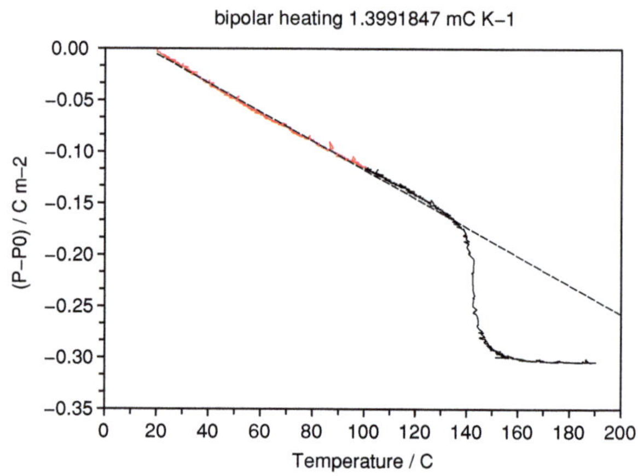

Fig. 23 Polarisation changes with heating (*red*) and cooling (*blue*) (bipolar loops)

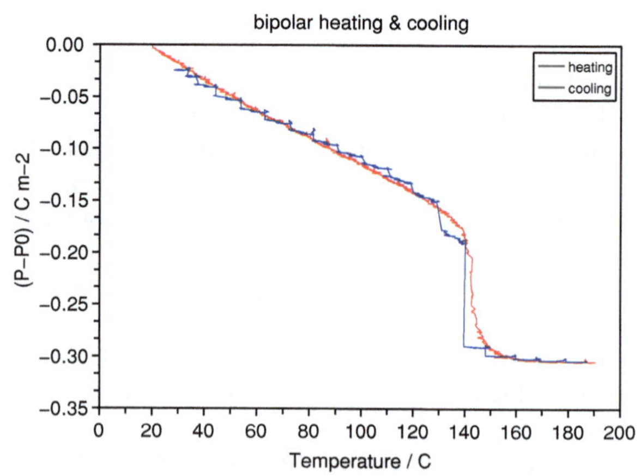

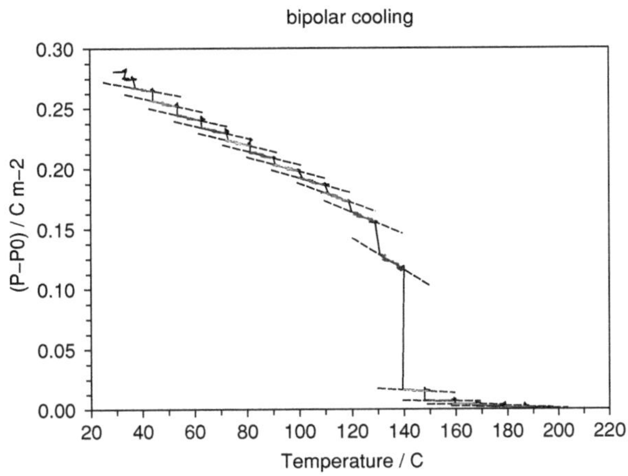

Fig. 24 Polarisation during cooling (bipolar loops) with linear fit to the sections between steps

points in time where the electrical cycles were applied. Between steps the polarisation increases slightly with cooling. The reason for the slope between loops could be the effect of changing lattice parameters coupling to the remanent polarisation. The non-centrosymmetric lattice distortion is known to increase with decreasing temperature, thus increasing the unit cell asymmetry and therefore the magnitude of the polarisation per unit cell. In this model there is no spontaneous increase in long range ordering as the temperature decreases. When the electrical cycle is applied this re-polarises the crystallites in the direction of the applied field, resulting in an increase in the net remanent polarisation. This is the same process as the change in remanent polarisation commonly observed between the first and second loops on a sample that hasn't been recently tested. In this model the changes in remanent polarisation during heating are irreversible i.e. they occur once during heating after which the smaller slope of the spontaneous cooling curve will be followed reversibly (requires experimental confirmation). It is therefore important to measure the value of the spontaneous cooling slope as this should more closely reflect the pyroelectric coefficient that would be obtained from small amplitude thermal cycling.

Figure 24 shows the same cooling curve as before, but divided up into sections between the steps. A straight line fit is applied to each section of the curve. The gradients of these lines are the effective pyroelectric coefficient for cooling. A similar analysis was done for the heating curve, and the results are shown in Fig. 25.

The size of the large peak at the Curie temperature is somewhat arbitrary because it depends on where the loops were taken with respect to the large change in polarisation at this temperature. Above the Curie temperature the pyroelectric coefficient is zero as expected. More detail is seen in the results plotted up to 140 °C shown in Fig. 26.

The coefficient for heating is approximately 1.4 mC m^{-2} K^{-1} as shown earlier, and is fairly constant with temperature. However, as discussed earlier this is likely to be a one off, and for repeated temperature cycles with no re-polarisation of the ceramic, this high figure would not be maintained. The lower figure from the cooling

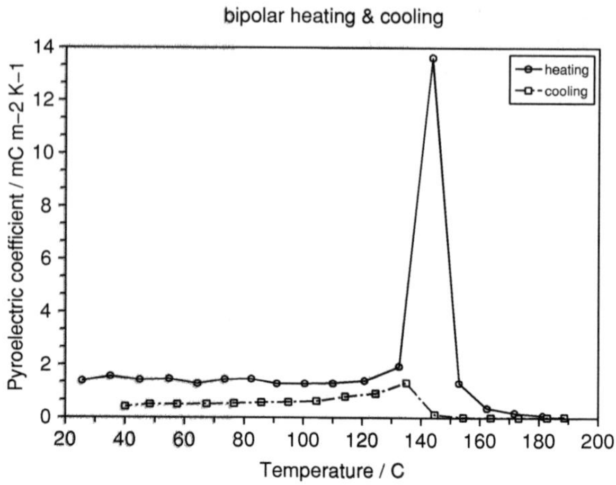

Fig. 25 Pyroelectric coefficient obtained from heating (*solid*) and cooling (*dashed*) curves 20 to 200 °C

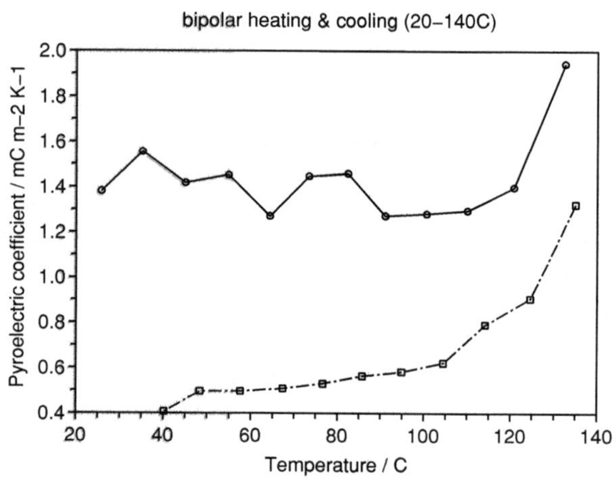

Fig. 26 Pyroelectric coefficient obtained from heating (*solid*) and cooling (*dashed*) curves 20 to 140 °C

curves is around 0.5 mC m^{-2} K^{-1} and this is likely to be maintained reversibly on subsequent heating and cooling cycles.

11 Actuators

11.1 Review of Roark's Equations for Bi-metal Strips

This section describes the application of formulae for the bending of a bimetal strip from Roark [25] to the case of a 2 layer unimorph actuator. Table 8.1 of Roark's provides formulae for the curvature of a beam subjected to a uniform temperature gradient from the top to the bottom surface. We are interested a beam that is fixed at one end and free at the other i.e. case 6a of Roark's table. We wish to calculate the tip deflection, y, of the free end of the beam ($a = 0$).

$$y = \frac{l^2}{2}\frac{\gamma(\Delta T)}{t} \tag{23}$$

where l is the length of the beam, γ is the coefficient of thermal expansion of the beam material, t is the thickness of the beam and ΔT is the temperature difference across the beam.

According to Roark we can apply this equation to the case of a bimetal strip formed from two materials, a and b, subject to a uniform temperature T by making the substitution:

$$\frac{\gamma(\Delta T)}{t} \equiv \frac{6(\gamma_b - \gamma_a)(T - T_0)(t_a + t_b)}{t_b^2 K_1} \tag{24}$$

so

$$y = \frac{3l^2(\gamma_b - \gamma_a)(T - T_0)(t_a + t_b)}{t_b^2 K_1} \tag{25}$$

where γ_a and γ_b are the coefficients of thermal expansion and t_a and t_b are the thicknesses of materials a and b respectively. T_0 is the temperature at which the beam is flat, and K_1 is given by

$$K_1 = 4 + 6\frac{t_a}{t_b} + 4\left(\frac{t_a}{t_b}\right)^2 + \frac{Y_a}{Y_b}\left(\frac{t_a}{t_b}\right)^3 + \frac{Y_b}{Y_a}\frac{t_b}{t_a} \tag{26}$$

where Y_a and Y_b are the Young's moduli of the two materials. Furthermore, the equivalent stiffness EI is given by

$$EI = \frac{w t_b^3 t_a Y_b Y_a}{12(t_a Y_a + t_b Y_b)} K_1 \tag{27}$$

The derivation of these equations is not given by Roark. Further detail, including calculation of the neutral axis, is to be found in the work of Timoshenko [26].

11.2 Application of Roark's Equations to Piezoelectric Actuator

Equation 25 can be rewritten as

$$y = \frac{3l^2(\gamma_b(T - T_0) - \gamma_a(T - T_0))(t_a + t_b)}{t_b^2 K_1} \tag{28}$$

$\gamma_a(T - T_0)$ and $\gamma_b(T - T_0)$ are simply the thermal strains that would occur in the individual components of the bimetal. This equation is still valid no matter what the source of the strain, so we can substitute for the thermal strains general strains that could be thermal or piezoelectric in origin.

$$y = \frac{3l^2(S_b - S_a)(t_a + t_b)}{t_b^2 K_1} \tag{29}$$

In the piezoelectric case this assumes that the piezoelectric material behaves elastically and the piezoelectric response is not significantly affected by the restraint imposed by the substrate. Restraint of the piezoelectric material is known to affect the piezoelectric response, and this may produce some error in the result. Within the level of approximation of this simple model this is unlikely to have a really significant effect, but should be taken into account if better accuracy is required.

For an actuator comprising a piezoelectric layer and an inactive substrate bonded together, we can allocate the piezoelectric material to a and the substrate to b. The strain in the piezoelectric material is $S_a(E)$ where E is the electric field applied across the thickness of the piezoelectric layer and S_a is the strain in the plane of the beam. The strain in the substrate layer S_b is zero:

$$\gamma_a(T - T_0) = S_a(E) \tag{30}$$

$$\gamma_b(T - T_0) = S_b(E) = 0 \tag{31}$$

$$y = -\frac{3l^2 S_a(E)(t_a + t_b)}{t_b^2 K_1} \tag{32}$$

The piezoelectric strain $S_a(E)$ will in general be a non-linear multi-valued function of the field E. The deflection of the actuator could be calculated directly from this strain derived either from experiment or theory. For an approximate calculation we can use the piezoelectric strain coefficient d_{31}

$$S_a(E) = d_{31}E \tag{33}$$

so the actuator deflection within the approximation of a linear piezoelectric response is

$$y = -\frac{3l^2 d_{31}E(t_a + t_b)}{t_b^2 K_1} \tag{34}$$

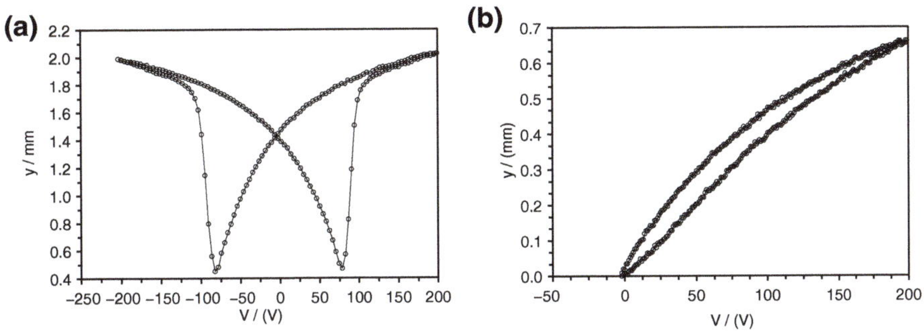

Fig. 27 Actuator beam model results at room temperature. **a** Bipolar deflection at 19 °C and **b** unipolar deflection at 25 °C

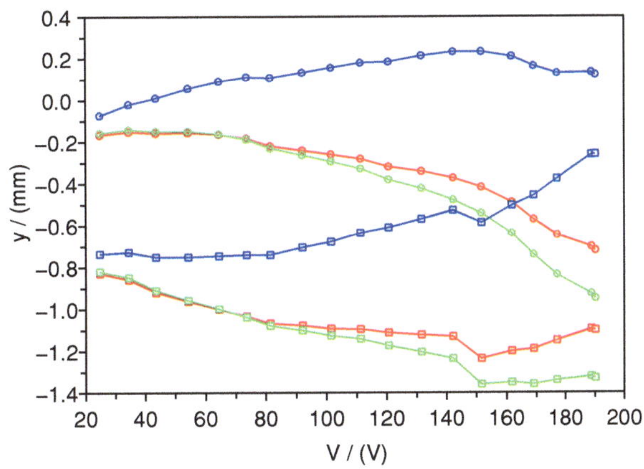

Fig. 28 Actuator beam model including temperature variation. Unipolar cycles ON and OFF positions for different expansion substrates

11.3 Actuator with Temperature Variation

In general and actuator will be subjected to both piezoelectric and thermal strains. The expansion coefficient of the substrate will vary with temperature so a more accurate model will use the directly measured strain $S_b(T)$. The error involved in assuming linear thermal expansion for the substrate will depend on the material used and the temperature range of interest.

In most piezoelectric materials the thermal and piezoelectric strains are coupled in a rather complicated way. This simple model allows analysis of how the complicated piezoelectric strain, $S_a(E, T)$ interacts with the substrate expansion to produce the temperature dependent actuator movement, and to estimate the effects of different substrate materials on the actuator performance. In this case direct measurements of $S_a(E, T)$ can be obtained from dilatometer measurements under different electrical and thermal cycling regimes and used to predict actuator performance by substituting directly into Eq. 29.

Model results based on ceramic expansion data at room temperature are shown in Fig. 27 for a 25 mm × 5 mm composite beam comprising 0.15 mm thickness of PZT, and 0.15 mm thickness of Kovar. for unipolar loops the predicted deflection is just under 0.7 mm. This is close to the values measured for actuators of these dimensions. These results demonstrate how ceramic based strain measurements can be used to obtain realistic predictions of real actuator behaviour.

References

1. Vopsaroiu, M., Blackburn, J., Muniz-Piniella, A., Cain, M.G.: Multiferroic magnetic recording read head technology for 1Tbitin.[sup 2] and beyond. J. Appl. Phys. **103**(7), 07F506 (2008)
2. Lines, E., Glass, A.M.: Principles and applications of ferroelectrics and related materials. In: International Series of Monographs on Physics. Oxford University Press, Oxford (1977)
3. Sabat, R.G., Mukherjee, B.K., Ren, W., Yang, G.: Temperature dependence of the complete material coefficients matrix of soft and hard doped piezoelectric lead zirconate titanate ceramics. J. Appl. Phys. **101**(6), 064111 (2007)
4. Maiwa, H., Kim, S.-H., Ichinose, N.: Temperature dependence of the electrical and electromechanical properties of lead zirconate titanate thin films. Appl. Phys. Lett. **83**(21), 4396 (2003)
5. Cook, W., Berlincourt, D., Scholz, F.: Thermal expansion and pyroelectricity in lead titanate zirconate and barium titanate. J. Appl. Phys. **34**(5), 1392–1398 (1963)
6. Seveyrat, L., Lemercier, M., Guiffard, B., Lebrun, L., Guyomar, D.: Temperature dependence of macroscopic and microscopic PZT properties studied by thermo-mechanical analysis, dielectric measurements and X-ray diffraction. Ceram. Int. **35**(1), 45–49 (2009)
7. Kungl, H., Hoffmann, M.J.: Method for the estimation of the total displacement of ferroelectric actuators under mixed thermal and electrical loading. Sens. Actuators A: Phys. **144**(2), 328–336 (2008)
8. Kholkin, A.L., Akdogan, E.K., Safari, A., Chauvy, P.F., Setter, N.: Characterization of the effective electrostriction coefficients in ferroelectric thin films. J. Appl. Phys. **89**(12), 8066 (2001)
9. Jona, F., Shirane, G.: Ferroelectric crystals. In: International Series of Monographs on Solid State Physics. Pergamon Press, New York (1962)
10. Haun, M.J., Zhuang, Z.Q., Furman, E., Jang, S.J., Cross, L.E.: Electrostrictive properties of the lead zirconate titanate solid solution system. J. Am. Ceram. Soc. **72**(7), 1140–1144 (1989)
11. Weaver, P.M., Cain, M.G., Stewart, M.: Temperature dependence of high field electromechanical coupling in ferroelectric ceramics. J. Phys. D: Appl. Phys. **43**, 165404 (2010)
12. Weaver, P.M., Cain, M.G., Stewart, M.: Temperature dependence of strain-polarization coupling in ferroelectric ceramics. Appl. Phys. Lett. **96**, 142905 (2010)
13. Weaver, P.M., Cain, M.G., Correia, T.M., Stewart, M.: Electromechanical coupling and temperature-dependent polarization reversal in piezoelectric ceramics. IEEE. Trans. Ultrason. Ferroelectr. Freq. Control **58**, 1730–1736 (2011)
14. Li, F., Jin, L., Xu, Z., Zhang, S.: Electrostrictive effect in ferroelectrics: An alternative approach to improve piezoelectricity. Appl. Phys. Rev. **1**, 011103 (2014)
15. Jaffe, B., Cook, W.R., Jaffe, H.L.: Piezoelectric ceramics. In: Non-metallic Solids. Academic Press, New York (1971)
16. Cross, L.E.: Relaxor ferroelectrics. Ferroelectrics **76**(1), 241–267 (1987)
17. Mason, W.P.: Electrostrictive effect in barium titanate ceramics. Phys. Rev. **74**(9), 1134 (1948)
18. Kay, H.: Electrostriction. Rep. Prog. Phys. **18**(1), 230 (1955)
19. Damjanovic, D.: Ferroelectric, dielectric and piezoelectric properties of ferroelectric thin films and ceramics. Rep. Prog. Phys. **61**, 1267 (1998)

20. Berlincourt, D.: Properties of lead titanate zirconate ceramics. Technical report no. 3 for 1 Feb 1956–31 Jan 1957 (1957)
21. Burfoot, J.C., Taylor, G.: Polar Dielectrics and Their Applications. University of California Press, Berkeley (1979)
22. Newcomb, C., Flinn, I.: Improving the linearity of piezoelectric ceramic actuators. Electron. Lett. **18**(11), 442–444 (1982)
23. Zhang, Q.M., Pan, W.Y., Jang, S., Cross, L.E.: Domain wall excitations and their contributions to the weaksignal response of doped lead zirconate titanate ceramics. J. Appl. Phys. **64**(11), 6445–6451 (1988)
24. Chong, K.B., Guiu, F., Reece, M.J.: Thermal activation of ferroelectric switching. J. Appl. Phys. **103**(1), 014101 (2008)
25. Roark, R.J., Young, W.C., Budynas, R.G.: Roark's Formulas for Stress and Strain. McGraw Hill, New York (2002)
26. Timoshenko, S.: Analysis of bi-metal thermostats. J. Opt. Soc. Am. **11**(3), 233–255 (1925)

Measurement and Modelling of Self-Heating in Piezoelectric Materials and Devices

Mark Stewart and Markys G. Cain

1 Introduction

There are many uses of piezoelectric ceramics where the desire for increased power output means increased drive levels, which subsequently can lead to thermal problems within the device. Applications such as:

- Ultrasonic Cleaning
- Ultrasonic Welding
- Sonar Transducers
- Diesel Injectors
- Ultrasonic Sewage Treatment

all use piezoelectric materials operated at high drive levels, where thermal loading on the device becomes an issue, and where potentially expensive cooling is needed to maintain device performance.

When piezoelectric materials are used as actuators they make use of the indirect piezoelectric effect, where the application of an electric field gives rise to an internal strain. In this solid-state energy transformation there will always be a balance between electrical energy input and work done by the device. The coupling coefficient, k, is used to describe this efficiency for an ideal case where there are no losses. Here, k is essentially the ratio of the open circuit compliance to the short circuit compliance. For most real piezoelectric materials this conversion process is also associated with losses—both mechanical and dielectric. These losses manifest themselves in the form of heat, causing a temperature rise in the device, which, depending on the thermal boundary conditions can be detrimental to device performance. This self-heating effect is most often encountered in resistive components and is termed "Joule

M. Stewart (✉) · M. G. Cain
National Physical Laboratory, Hampton Road, Teddington, Middlesex TW11 0LW, UK
e-mail: mark.stewart@npl.co.uk

M. G. Cain (ed.), *Characterisation of Ferroelectric Bulk Materials and Thin Films*, Springer Series in Measurement Science and Technology 2,
DOI: 10.1007/978-1-4020-9311-1_7,

Heating". However, it is also seen in non-ideal dielectric materials where the dielectric loss gives rise to internal heat generation. To a first approximation, piezoelectric actuators can be thought of as a non-ideal or lossy dielectric but, because the material is moving, additional mechanical terms are needed to model this behaviour. If the energy loss to the surroundings is greater than the internal power generation, then the sample will eventually reach an equilibrium temperature. If the sample losses are greater than those to the environment, or if the losses increase with increasing temperature, then the sample will heat up until some catastrophic event is reached—such as the soldered connections failing, softening of adhesives, or depolarisation of the material.

There are several factors that limit the high power operation of piezoceramics in dynamic applications, such as sonar or ultrasonic welding transducers [1].

1. The dynamic mechanical strength of the ceramic
2. Reduction of the efficiency due to dielectric losses
3. Reduction in efficiency due to mechanical losses
4. Depolarisation of the ceramic due to the applied electric field
5. Depolarisation of the ceramic due to temperature rise.

The first can be largely overcome by correct prestressing or biasing of the ceramic, in order to limit large tensile stresses. The fourth factor, electric field, can be ignored for dynamic operations, since by the time this has occurred the field would be sufficient to cause failure by factors 2, 3 or 5. The most common causes are factors 2, 3 and 5, and the dominant factor depends on the type of operation.

1.1 Where Does Self Heating Occur?

In many cases this self heating does not present a significant problem, for instance in quartz crystals used for timing in electronics, the material's intrinsically low loss and the very low drive levels means that self heating does not adversely affect timing. However, in an ultrasonic humidifying unit, if the water reservoir is allowed to evaporate completely, they will fail due to overheating and for this reason they will usually have a cut out to prevent operation when the water runs out. Typically most self-heating related failures are in resonant continuous wave (CW) operation, since these are likely to be the most aggressive in terms of power input. However, even in non resonant operation, for instance in multilayer stacks where soft compositions are used, self heating can still be a significant problem.

1.2 How Does It Fail?

The self-heating can be benign, or it can manifest itself in several ways in the operation of a device. At one end of the scale, the changes in material and device parameters caused by the temperature rise will mean that a carefully designed resonant device will not resonate at the desired frequency. At the other end of the scale, the temperature rise can be such that material itself fails through depoling, or through thermal stresses. Quite often, it is not necessarily the temperature rise in the ceramic that gives rise to failure, but the ancillary components that overheat and fail. For instance, solder may soften, insulation may breakdown, thermal expansion can lead to the release of prestress, and softening of adhesive leads to increased losses and failure. Tokin [2] state that "the main failures with multilayer piezoelectric actuators are deterioration of insulation resistance, short-circuit, and open-circuit". Although this need not be due to self-heating it can be a contributing factor.

1.3 Thermal Runaway

Much of the time, the phenomenon of self heating in piezoelectric materials is governed by linear processes and the temperature rise seen in a device reaches an equilibrium state, where the extraction of energy balances the internal energy generation through joule heating. It is possible, through positive feedback mechanisms, that the increase in temperature can lead to increased internal energy generation, leading to a rapid and uncontrolled temperature rise and subsequent failure. This phenomenon is also found in conventional capacitors [3] and is termed "thermal runaway", where the temperature dependence of properties can lead to positive feedback. Although thermal runaway is often associated with catastrophic failure, in Lead Zirconate Titanate (PZT) ceramic devices, failure is usually associated with the failure of ancillary components, such as soldered connections, adhesively bonded joints, etc rather than mechanical failure of the ceramic. If the temperature of the device approaches the Curie temperature, then the ceramic will become depoled, no longer piezoelectrically active, resulting in a failed transducer.

1.4 Examples

Figure 1 shows a typical failure in a high power ultrasonic transducer, which was probably caused by a failure in the insulating materials, as a result of the temperature rise, leading to dielectric breakdown and the catastrophic failure. The bunching of the wires together will exacerbate the overheating, due to decreased convection around the wires, and the pinning of the cables to the rear mass will also contribute to heating through mechanical strain.

Figure 2 shows a failure in a transducer from an ultrasonic cleaning bath. Evidence of thermal effects can be seen in the discolouration of the epoxy bonding the piezoelectric to the aluminium base plate. The cracking of the ceramic is very fragmented, indicating that the failure was not purely mechanical, but it is difficult to attribute the failure to purely a thermal cause without further investigations. The failure of high power devices involving epoxies is highlighted in 1–3 composites, where poor thermal transport away from the active ceramic causes temperature related failures in the epoxy resin [4].

Fig. 1 Typical failure in a high power ultrasonic transducer

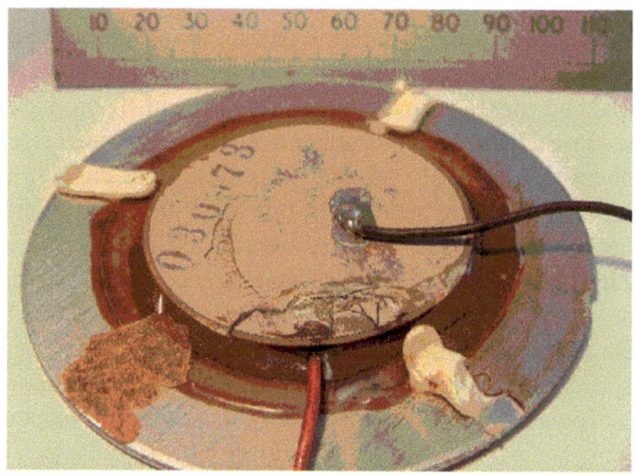

Fig. 2 Failure in a transducer from a laboratory ultrasonic cleaning bath

2 How to Minimise Self-Heating

This chapter is intended to help device designers to predict the effect of joule heating in their chosen device and its impact on device performance. If self-heating is a significant mechanism, then there are a number of ways that the problem can be lessened, most of which are common sense, although some may be less obvious.

Minimise Power input: The internal heat generation is dependent on the power input to the device and this can be limited by reducing the device operating conditions, such as driving frequency [5], field [6], or duty cycle. As many of the relevant material properties are non linear, a small decrease in power input may give a larger decrease in the heat dissipated in the device. Berlincourt [7] has suggested that for safe and efficient operation the dissipated power should be limited to less than 0.5 W/cm^3.

Maximise heat extraction: Obviously, the more energy that is removed through cooling processes, the less likely it is that self-heating will cause failure. This can be achieved by adding heatsinks, forced air or liquid cooling. Conduction processes are usually most efficient at heat removal and increasing the gauge of the power leads will increase conduction and reduce resistive losses in the cable. Sometimes, the orientation of the device might help the convection process remove energy more efficiently. Hu et al. [8] found that by orienting a long vibrating plate vertically, rather than horizontally, the increase in convective heat removal, lead to almost a halving of the temperature rise.

Maximise device size: Many thermal problems within the microelectronics industry come from device miniaturisation, and if size restrictions can be relaxed, benefits in terms of thermal performance can be achieved. Since the temperature rise in a device can be controlled to some extent by the heat extraction, for a given power level a larger device will usually have larger, conduction, convection and radiation heat transfer paths. Uchino et al. [9] suggest for temperature rise suppression, a device with a larger surface area is to be preferred, for example a tube rather than a rod. Another approach might be to use a device designed for higher power levels, but drive it a smaller fraction of the maximum.

Include thermal safety cut outs: A thermal cutout is advisable, however it is not always feasible to measure the temperature of the ceramic itself if high voltages are involved or the device is resonant. In this case the temperature of a nearby component, or perhaps the air or coolant media can be measured, and a cutout based on this reading can be used. A less expensive option might be to prevent operation under certain conditions, for example without a coolant present, or when the device has been on for more than a certain length of time.

Reduce internal heat energy production by choice of material: As discussed in more detail later, the major contribution to the internal heating is dielectric losses in the material, so choice of materials with lower dielectric losses help to minimise internal heat generation. At a simple level this can be through selecting a hard, rather than soft PZT composition, since the hard materials have lower dielectric losses and can withstand higher operating temperatures because of their higher Curie temperatures. The selection is complicated by the fact that the dielectric loss is dependent on many factors such as driving field, compressive stresses and temperature. Berlincourt et al. [7] give several comparisons of efficiency depending on criteria such as, maximum internal loss limited to 0.5 W/cm^3·kcps, and a maximum tan delta of 0.04.

Reduce Mechanical Losses in the system: Although dielectric losses are usually the dominant source of heat generation, mechanical losses can also add to the overall heat generating processes. As we are considering an actuator, any motion in the device, apart from rigid body motions, has the potential to cause a stress and therefore a mechanical loss. It is often seen where the active material joins the passive components, and after all, this is how ultrasonic welding works. In 1–3 composite actuators, the active ceramic is in intimate contact with a passive epoxy, which often has a high mechanical damping coefficient, contributing to heat generation [10]. Where the electrical leads are soldered onto the active ceramic, stresses are introduced into the solder, which again will have a high damping coefficient. Hu et al. [8] have seen increased temperatures around solder joints on actuators, although they have attributed this to contact resistance, which can also contribute to heating.

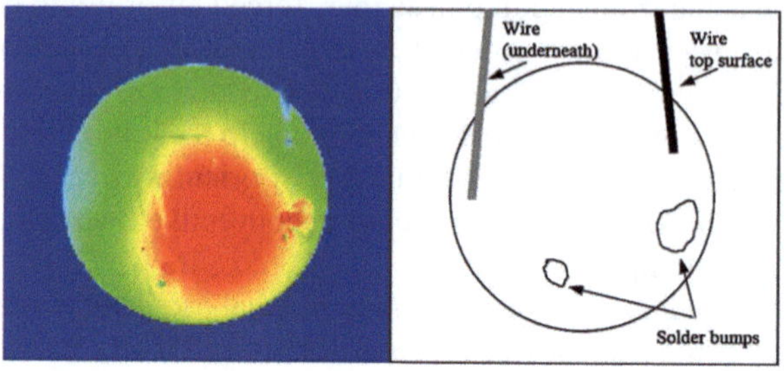

Fig. 3 Thermograph of self-heating in piezoceramic disc. Colour scale from *blue* (coolest) to *red* (hottest). Sample driven at thickness mode resonance of disc at 1.745 MHz. Coolest regions (*blue*) correspond to where power cables connect, and hottest points (*red*) to the two passive solder bumps

They suggest that making electrical contact at the nodal point, traditionally used to maintain a high Q in a device, may be counterproductive in terms of self heating, since this is where the strain transfer will be a maximum. Heating through mechanical losses can also be seen in the electrical contacts depending on how the wires are routed, often one end will be rigid and the other attached to the moving actuator.

Figure 3 shows the effects of solder connections on a disc resonating in the thickness mode at 1.745 MHz. The electrical power to the disc supplied via two soldered wires, one on the top surface, and one on the bottom. In addition to this, two extra dummy solder bumps that have been added to the top surface to examine the effect of these additions have on the temperature profile. The temperature of the device is non-uniform for several reasons. Although the sample is operating in thickness mode, non-uniform strains are still introduced through Poisson's ratio effects, which should lead to a temperature maximum at the centre of the disc. However because of the soldered connections, this maximum is offset towards the dummy solder bumps. The coolest parts of the device are where the power wires are connected; here the heat transfer through conduction in the wires leads to a smaller temperature rise. The two dummy solder bumps show increased temperature rise, probably because of the mechanical loss associated with the material. It is interesting to note that the solder joints associated with the power cables do not show the heating effect, since it is overshadowed by the increased thermal conduction in the cables.

In summary the contributions to the non-uniform self-heating in this device are in order:

1. Non-uniform strain in the soft piezoelectric material
2. Increased heat transfer through conduction in the power cables
3. Increased heat generation due to mechanical losses in the solder.

3 Measuring the Temperature of Piezoelectric Devices

When attempting to measure the temperature of piezoelectric materials used in actuators, there are several practical difficulties to be overcome. Firstly, the devices often have high voltages applied, which can have implications for safe and accurate temperature measurement. Secondly, if the device is resonant, it may not be practical to attach contact temperature probes, since this will interfere with the resonant nature of the device. Thirdly, as discussed previously, when a moving actuator comes into contact with a passive (not moving) temperature probe, there is a potential for frictional heating, giving rise to erroneous readings. These three factors point towards using non-contact temperature measurement methods such as optical pyrometry, or even thermally sensitive paints. Thermally sensitive paints undergo a colour change, sometimes permanent, on reaching a certain temperature. The accuracy is limited, but they are cheap and essentially non-contact. If greater accuracy and range is needed and the temperature reading needs to be interfaced to data collection systems, then infrared (IR) temperature sensors are a reasonably economical solution. Errors can arise in these systems if the emissivity of the sample is low, or not precisely known, and the temperature measured by the radiation thermometer will not represent the true temperature of the sample. Also, if the sample is small, it might not be fully filling the field-of-view of the radiation thermometer, and the temperature measured will be an average of the sample and its surroundings. Occasionally the sample can act as a mirror and the sensor measures the radiation of an object from behind the sensor.

Another non-contact method of determining the temperature of a piezoceramic device is to use the temperature dependence of the material properties as a temperature indicator. The simplest property to use is the capacitance of the device. For instance a soft PZT-5H composition changes permittivity by 33 % over the temperature range 0–40 °C [1], and assuming this change is linear over the range, a change of 1 % in the capacitance represents roughly a degree centigrade. The measurement of capacitance could be realized using the device drive wiring and electronics, however it may be difficult to achieve the required accuracy, particularly when measuring hard materials, where the temperature dependence of capacitance is much less. Lente et al. [11] have shown a one to one correspondence between sample temperature and the polarisation during fatiguing of PZT discs. As the device fatigues, the change in polarisation causes a change in the current drawn by the device, leading to a change in temperature. Ronkanen et al. [12] have also identified the link between temperature and current drawn by the device, and suggest that this could be used a mechanism for compensation of output, as the change in temperature causes a change in piezoelectric properties such as d_{33}.

If contact methods are preferred, then there are several work-arounds that can be used to mitigate some of the problems discussed before. In order to mitigate the high-voltage danger, then the thermocouple can be placed on the ground side of the device, or a thin insulating varnish or cyanoacrylate can be used to attach the sensor. Of course, should the insulation or piezoceramic breakdown then there is a potential for damage to sensitive measuring electronics and also potential for hazardous voltages to come into contact with users. For this reason, it is advisable to have the temperature measurement electronics electrically isolated from the outside world. From the point of view of interfering with a resonant device and also in order to minimize ultrasonic heating effects, it is best to have very small temperature sensors that act as point sensors. In this respect, thermocouples are readily available

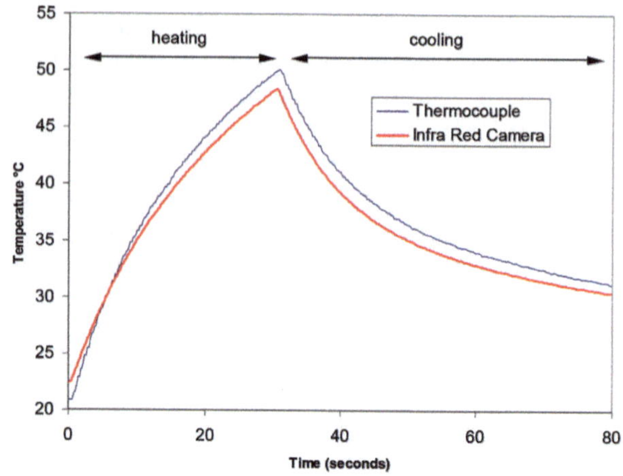

Fig. 4 Comparison of temperature measurements of a piezoceramic disc undergoing self-heating. Thermocouple is fixed to the disc with cyanoacrylate, IR measurements made with IR camera and temperature readings taken near thermocouple position

in thin wire gauges as small as 0.003″ (0.076 mm) diameter and smaller. Because of their low thermal mass, these small thermocouples are also advantageous in terms of their transient response and also their efficacy in measuring small samples reliably. Other contact mode sensors including Resistance Temperature Detectors (RTDs), thermistors and IC sensors, can all in principle be used, but the most commonly used sensors for these types of measurements are thermocouples and non-contact IR sensors. It should be pointed out that although no evidence of problems was seen in this work, there have been reports of thermocouples giving spurious readings in materials experiencing high power ultrasonic vibrations. Mignogna et al. [13] used copper constantan thermocouples bonded with epoxy to various resonant and non resonant bodies, and reported that the thermocouple would often record a temperature rise of as much as 100 °C, yet the sample was barely warm to the touch. Experiments to find the cause of these problems were inconclusive. However, they state "the results cast doubt on all previous measurements where thermocouples were used to measure heat generated during high-power ultrasonic insonation of metals". It should be stressed that the thermocouple measurements carried out in this work were used to record heating generated in the piezoceramic, and in general, temperature rise in the metallic parts was through heat conduction into the part, rather than internal heat generation in the metal.

Figure 4 shows a comparison of temperature measurements on a 30 mm diameter, 1 mm thick, disc of soft PZT, using a type K thermocouple and an infrared camera sensor. The thermocouple was attached near the centre of the disc using cyanoacrylate and the camera used gave a full field image of the device, but only results near the thermocouple position were used in the comparison. The IR system was corrected for emissivity of the target surface by placing the sample on a hot plate and comparing with the thermocouple readings. The calibration curve was highly non-linear, with a 4th order polynomial used to represent the readings, but this is probably a due to the camera sensitivity rather than the temperature dependence of the surface emissivity.

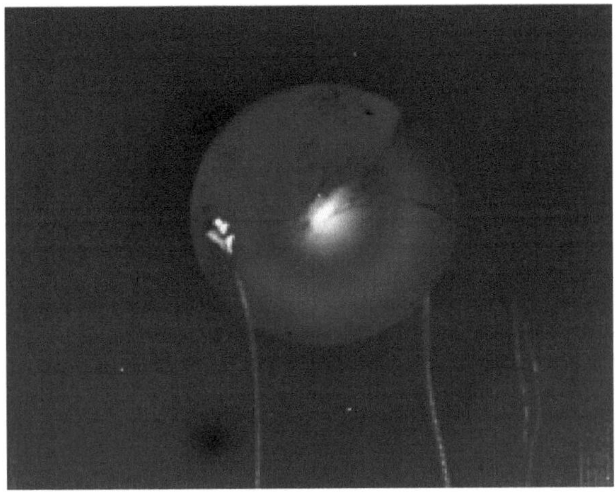

Fig. 5 Greyscale thermograph of PC 5H disc resonating in radial mode. Sample has been covered with graphite to give uniform emissivity, but then some of this coating has been removed to illustrate the effect of emissivity on the image. *Left hand* solder connection appears to be as hot as the *centre* of the disc. The *circular* pattern due to nodal strain heating effects is only visible on the graphite coated sector

The readings are a maximum of 2 °C apart and on average less than a degree apart, which is the order of the accuracy of the two systems. The differences could be because of the slightly different spatial positions of the sensors, (the disc was operating in a resonant mode so the temperature was spatially dependent), or because of the different time constants of the sensors. The fact that the two temperature traces cross on the heating cycle, but not on the cooling cycle could again be due to sensor time constants, or it could be due to electrical pickup. The only difficulty encountered in this work relating to thermocouple measurements came through electrical interference of a high power transducer running at around 1kW. In this case, when the power was applied readings shot to over 200 °C but instantaneously returned to ambient when the power was removed. The overall conclusion is, that thermocouple and IR sensors give identical results, except when substantial electrical interference is present, where these effects are easily identified.

The temperature sensors discussed so far have essentially been single point sensors. However, much information regarding the thermal performance of a piezoelectric device can be gained by using IR cameras. Calibration of these sensors is more difficult since the performance of each camera pixel could be different, however the information is usually used to detect differences point to point in an image and how this progresses in time. In this respect these types of images can often be misleading since the images will often contain areas with different emissivity that will be difficult to account for. Figure 5 shows a raw (uncorrected for emissivity) thermal image of a thin piezoelectric disc under resonant drive in the radial mode. To highlight the emissivity issue, one half of the discs top surface has a graphite coating, whilst this has been removed from the other half. The graphite gives an increased signal so it is possible to see the ring pattern produced by non-uniform self-heating, whilst this pattern is absent in the uncoated sector (uppermost). In the uncoated region the hottest part appears to be the soldered power connection, which is similar to the centre of the disc. In fact this is an artefact, the solder is acting as a mirror reflecting some other part of the scene, giving a false impression of temperature. To overcome this the sample would normally be entirely coated with graphite.

4 Thermal Modelling

The aim of the modelling described here is to be able to predict the temperature rise in any piezoelectric device, given the sample type, applied voltage, frequency, and some environmental conditions.

4.1 Energy Balance

The temperature rise in a piezoelectric is a result of a change in its internal energy, and that internal energy can be found by an energy balance criteria, such that:

Change in internal energy = energy generated in device − energy lost to surroundings

Each of these energy transfer processes can be a function of many different processes:

Energy generated in device = f (loss, frequency, capacitance, voltage)
Energy lost to surroundings = f (conduction, convection, radiation).

In order to model piezoelectric device behaviour as far as thermal conditions is concerned, there is a need to understand both the heat generation, and the heat transfer. The heat transferis a general problem that is covered by many textbooks and software solutions. There are many practical heat transfer problems that include internal energy generation, such as chemical reactions, nuclear radiation, resistive heating, where the solutions are common and the exact nature of energy generation process is not important. The mechanisms of heat generation in piezoelectrics, although similar to those in dielectric materials, are further complicated by the piezoelectric coupling. We will assume that the internal energy generation is the largest unknown in these problems and that the heat transfer constants are largely constant over the temperature range of interest for most cases.

4.2 Heat Transfer Processes

The three mechanisms of heat transfer; conduction, convection and radiation can be described by their individual rate law [14].

Conduction, the diffusion of energy by random molecular motion can be described by the rate equation, Fourier's Law.

$$q''_x(\mathrm{W/m^2}) = -\kappa \frac{dT}{dx} \tag{1}$$

where κ is the heat transfer coefficient for conduction, and has units of W/mK, T is temperature and q''_x is the heat flux in the x direction.

Convection, which is a combination of conduction and energy transfer due to mass motion, advection, can be described by the following rate equation, sometimes known as Newton's law of cooling.

$$q_x(\mathrm{W/m^2}) = -h(T_s - T_\infty) \tag{2}$$

where h is the heat transfer coefficient for convection, and has units of W/m^{2}K. Values for h can range from 2 for free convection in a gas, to 20 for forced convection, to many thousands for convection with a phase change.

Radiation, the transfer of energy by electromagnetic radiation, can be described by the following rate equation,

$$q_x(\mathrm{W/m^2}) = \varepsilon_m \sigma (T_s^4 - T_{sur}^4) \tag{3}$$

where ε_m is the emissivity, and σ is the Stefan-Boltzmann constant ($5.67 \times 10^{-8}\,\mathrm{W/m^2K^4}$).

In general the most important heat transfer processes for piezoelectric devices are conduction and convection. Since the devices are solid state the heat transfer within the device will be through conduction, and much of the energy transfer to the surrounding media will be through convection into air, water or other fluids.

4.3 *Piezoelectric Device Operation*

Piezoelectric devices are used in an ever-expanding range of applications that cover a wide range of operating regimes of frequency, power levels etc. However, as far as internal heat generation is concerned, there are three key types of driving conditions, resonant drive, non-resonant drive and non-CW drive.

Off Resonant Drive

In general, the temperature of the sample/ device is uniform throughout the sample. This is because strain throughout the sample is uniform, so each part of the sample volume is identical, apart from those that see a different heat transfer rate, such as sample surfaces or regions next to internal electrodes. It is usually accepted that under these conditions the dielectric losses contribute to self-heating. Several theoretical analyses have been developed to predict the temperature rise of devices under off resonant conditions. Most of these models assume that the thermal conductivity and the heat generation mechanism are independent of temperature. Over the operating temperatures of most piezoelectric devices it is likely that the thermal conductivity is temperature independent, however the assumption that the internal heat generation is temperature independent may lead to underestimation of internal temperature profiles.

Resonant Drive

Here, standing waves are set up within the sample, which introduces a non-uniform strain and thus potentially non-uniform temperature. Non-uniform temperatures have been observed at nodal points in resonating devices [9, 15]. It is not clear if these non-uniform temperatures play a part in thermal runaway type behaviour. For example, a small part of the sample may become locally hotter and, through heat transfer, will heat up the rest of the sample. Here, the question arises: does failure come about from a small part reaching a critical temperature, or the whole sample reaching this temperature?

The critical factor under resonant drive conditions is that the mechanical loss is increased, and may contribute significantly to the internal heat generation, and also that this heat generation process is spatially dependent. Little work has been done on predicting temperature rise under resonant conditions, and the non-uniformity of temperature profiles.

Non CW Operation, Pulse Drive and Low Duty Cycle Operation

Non continuous wave (CW) drive waveforms of piezoceramics are often used, either because it is necessary for the particular operating pattern, or to overcome some of the overheating problems associated with CW operation. A diesel injector valve is an example of pulse drive, where the duty cycle is actually very small, but the operating cycle is very high power. In contrast, a high power ultrasonic cleaning bath may be a resonant device, but in order to restrain the temperature rise, the operation may be limited to duty cycles of 10 % or less.

For the ultrasonic bath example, the low duty cycle operation only slightly complicates the heat transfer problem, in that the heat generation is modulated in the time domain only. For the diesel injector example not only is the operation non continuous, the drive signal will likely be square wave, meaning that the internal heat generation will have many additional components in the frequency domain.

4.4 Modelling of Heat Generation

Dielectric Heating

In its simplest form a piezoelectric device can be thought of as a capacitive load, whose energy at any point can be modelled as the energy in a capacitor. If the capacitor is not ideal then some of this energy will be lost, dependent on the dielectric loss tangent of the material. If we assume that all this lost energy is converted to heat in the capacitor through dielectric heating then the power dissipated can be given by Jordan and Quanies [16]

$$Power = 2\pi f C tan\delta V^2 \tag{4}$$

where f is the frequency, C the capacitance of the device, V the applied voltage and tan δ the dielectric loss. Dielectric heating is the mechanism for heating in a domestic microwave oven and the temperature rise of the food in an oven can be predicted similarly using (4), coupled with the density, specific heat capacity and the thermal boundary conditions.

Energy dissipation in piezoelectric materials is further complicated by the materials intrinsic non-linearity when driven under high field, such that the permittivity and dielectric loss become field dependent [17]. Coupled with this, the permittivity and dielectric loss will also be temperature dependent [18, 19].

The *dielectric heating model* has been used by many workers [20, 21] as a basis for heat generation in piezoelectrics driven at low frequency, under off resonance conditions. Uchino [9] used a modification to this model to predict temperature rise in multilayer piezoelectrics under off resonance drive. Uchino [9] expresses the rate of heat generation in the multilayer q as:

$$q = u f V_{eff} \tag{5}$$

where u is the loss of the sample per driving cycle per unit volume, f is the frequency, and V_{eff} is the effective volume of active ceramic material. The effective volume is dependent on the amount of active ceramic in a multilayer. The loss u is given by:

$$u = \pi \varepsilon^x \varepsilon_0 E_0^2 tan\delta_e \tag{6}$$

which is essentially a geometry independent version of the power loss in a capacitor seen previously [4].

Strain Heating

Under adiabatic conditions, a material undergoing a change in stress state can undergo a temperature rise in the same way that an ideal gas does. This temperature rise as a result of a volume change is termed the *thermoelastic effect* and can be described by the following equation:

$$\Delta T = \frac{\alpha_1}{\rho C_p} T(\sigma_1 + \sigma_2 + \sigma_3) \tag{7}$$

where α_i is the thermal expansion coefficient, and σ_i are the changes in the principal stresses. This temperature rise is fully reversible and results in cooling on expansion and heating on contraction. In practice these changes are of the order 0.1 K or less for fully elastic conditions in most metals. When the material is no longer perfectly elastic the material can undergo heating effects that are sometimes described as thermoplastic heating, and are attributed variously to grain boundary motion, dislocation movements, bond rotation in polymers and others. Here we will use the term, *strain heating* to describe the process of temperature rise due to mechanical motion in piezoelectric materials.

Under resonance drive conditions the dominant mechanism for heat generation is thought to be due to *strain heating*, rather than *dielectric heating*. This phenomenon is seen in many materials undergoing high power ultrasonic vibrations [13], and is mechanically analogous to dielectric heating. In dielectric heating there is a lag between the applied voltage and current, which corresponds to an energy loss, which is converted internally into heat. In strain heating there is a lag between stress and strain, and the corresponding energy loss is converted into heat. For dielectric heating the energy loss is proportional to the dielectric tan δ_e times the square of the field, whereas for strain heating, the energy loss is proportional to the mechanical loss, tan δ_m, times the square of the strain.

Ando [22] has used the following relationship to determine the mechanical loss for a given volume:

$$q_v = \frac{\alpha_v S^2}{2} \tag{8}$$

where α_v is the damping coefficient, and S the magnitude of the vibratory strain.

Other workers have used similar expressions using the applied stress, to describe the losses due to mechanical vibrations, [9, 23]. Uchino [9] determines the hysteresis loss in a full cycle using the following expression;

$$w_m = \pi s^E X_0^2 tan\delta_m \tag{9}$$

where s^E is the compliance and X_0 is the amplitude of the stress. Blotmann et al. [23] use a very similar expression, but divided by 2π, to define the mechanical dissipation, where the discrepancies with Uchino are due to imprecise definitions of the cycle. Lu and Hanagud [23], use irreversible thermodynamics to develop a model for self heating, but similarly show the strain heating is proportional to the square of the strain multiplied by various viscous damping coefficients.

The relationship between strain and heat generation implies that, under non-uniform strain conditions, for example at resonance or antiresonance, the heat generation will be spatially dependent. This is the key difference between strain heating and dielectric heating where, in the latter, the internal heat generation is assumed uniform throughout the volume.

Sherrit [24] has shown that for a plate sample of thickness, L and area A, with x=0 at the centre, the power distribution as a function of distance x is given by

$$P(x) = 2P_m \cos^2\left[\frac{\pi x}{L}\right] \quad (10)$$

where P_m is the mean power level. The cosine squared variation of temperature in a resonating piezoelectric device has been confirmed by Tashiro [15], although the form of the temperature variation at the fundamental resonance is not very different from a case with uniform heat generation. Several workers [25, 26] have modelled and measured temperature profiles of piezoelectric transformers driven at resonance, and shown that hot spots coincide with regions of high strain.

Dielectric or Strain Heating

In a capacitor the only internal heating mechanism is dielectric heating. However, for a piezoelectric material whenever an applied field leads to dielectric heating there will also be some associated strain heating. In general, most workers have assumed that under off resonance conditions dielectric heating is dominant, and ignored any strain heating contribution. Under non-resonant operation the induced strain will most probably be uniform throughout the sample, and because of this the errors associated with this will be equivalent to an error in the tan δ. Conversely, under resonant conditions the dominant mechanism is assumed to be mechanical heating, and dielectric heating is ignored [15, 24]. However, as before, where there is an applied voltage, there will be associated dielectric heating, but this will be spatially independent. Hu [25] has derived an expression for the internal loss per unit volume in a piezoelectric transformer as:

$$p(x) = A_0 \exp^{-\alpha x} \cos^2 kx + p_e \quad (11)$$

where the first term is the spatially dependent mechanical dissipation, and the second p_e is the dielectric heating contribution. As can be seen, neglecting this contribution will result in an offset in the temperature rise predicted, depending on the relative contribution of this second term. Determining the mechanical and dielectric loss for a given situation is difficult, and there are various approaches. Hu [25], uses the phase difference between the input voltage and current to determine the overall level of energy loss, and uses a value of 7:3 for the ratio of mechanical to dielectric loss, based on previous experimental evidence. Other workers [22, 23] use the low field values of mechanical and dielectric loss and Eqs. (6) and (9) to determine the individual contributions in FEA simulations.

4.5 Modelling of Heat Transfer

Steady State Heat Transfer

In the preceding discussion on heat generation, the power dissipatedwhen a certain sinusoidal field level is applied to a device can be determined. However, this is not a means by itself to calculate the rise in temperature of a device. To achieve this, there must be some understanding of the material's thermal properties and the thermal boundary conditions of the device. At the simplest level we can imagine a volume of material that loses power P to ambient air, with a heat transfer coefficient h, (surface conductance into air), where the heat loss is essentially dependent on the surface area to volume ratio.

$$P = hT_m Area \tag{12}$$

Here P is the dissipated power and T_m is the temperature rise of the sample.

For a simple disc shaped device the surface temperature rise is given by [27]

$$T_m = \frac{P_v}{h}\left[\frac{2}{t} + \frac{4}{d}\right]^{-1} \tag{13}$$

where d and t are the diameter and thickness of the device, and P_v is the dissipated power per unit volume.

Equation (13) describes the surface temperature rise of a disc that has a uniform power generation P_v, and is losing energy to the surroundings via convection only. This assumes the temperature of the whole surface is the same, and the internal temperature is constant, i.e. the volume is small enough for internal heat conduction to be ignored.

The internal temperature, and it's spatial variation, of a plate, thickness L and area A, where the external temperatures are maintained at T_0, with a uniform internal power generation, P, is given by (Fig. 6).

$$T_x = \frac{P}{2kAL}\left[\frac{L^2}{4} - x^2\right] + T_0 \tag{14}$$

The external temperature T_0 can be determined by defining the control volume and the mechanisms of heat transfer in the media. However, to a first approximation T_0 is equal to the ambient temperature.

Equation (14) is valid for uniform power generation, and so would apply to off resonance drive, with dielectric heating. Sherritt et al. [24] have shown that if the power generation in the slab is not uniform, but proportional to the square of the strain, then the temperature distribution is given by:

$$T_x = \frac{PL}{2kA}\left[\frac{\cos^2(\pi x/L)}{\pi^2} - \frac{x^2}{L^2} + \frac{1}{4}\right] + T_0 \tag{15}$$

This distribution can be used to predict the temperature of a device driven at resonance, and so measurements of temperature distribution of resonating devices should be able to discriminate between dielectric heating and strain dependent heating. Figure 7 shows the predicted temperature distributions for uniform and strain squared heating, showing the different form of the behaviour. Unfortunately, if the

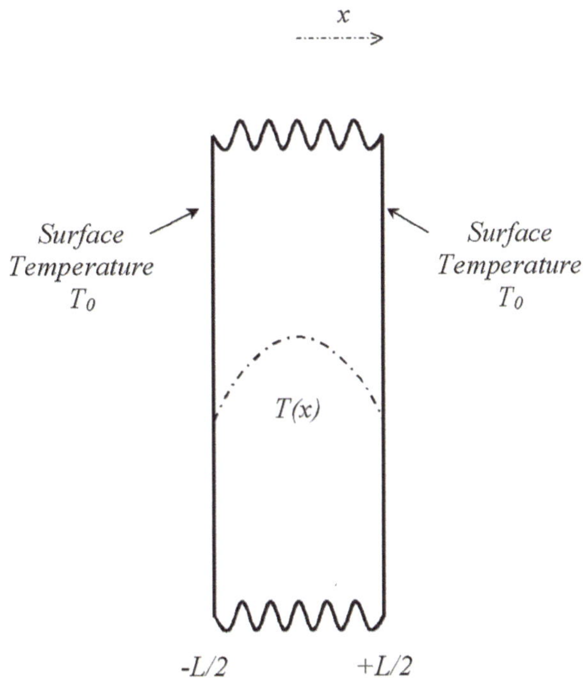

Fig. 6 Boundary conditions for analysis in Eqs. (13) and (14)

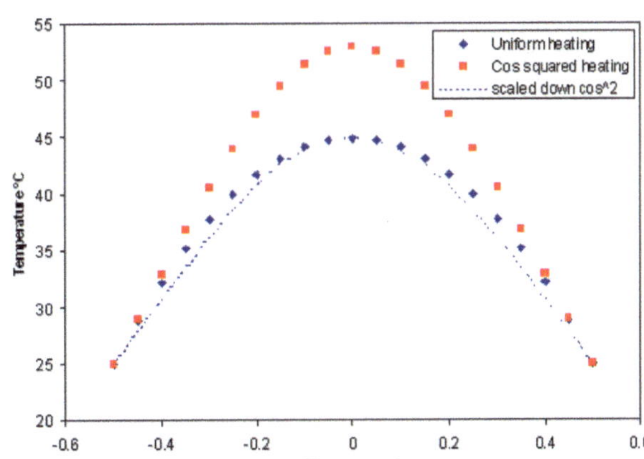

Fig. 7 Temperature profile of an infinite slab subject to internal heat generation. Comparing uniform heat generation with strain dependent heating

temperature scale of the cosine squared distribution is scaled down the behaviour is very similar to that in the uniform heating case, which means that very accurate experimental measurements are needed to definitively differentiate uniform and spatially dependent heating. As will be seen later the spatial dependent effects are more readily observed at the overtones than at the fundamental resonance of the device.

The preceding models only account for conduction in the sample. In order to account for convection at the surfaces the following model gives the temperature rise at a distance x from an infinite slab of thickness L, with a thermal conductivity k, and heat transfer coefficient for convection, h:

$$T_x = \frac{P}{h}\left[\frac{L}{2} + \frac{h}{k}\left[\frac{L^2}{4} - x^2\right]\right] + T_0 \tag{16}$$

The maximum temperature is at the centre of the slab, x=0, and the minimum at the surface $x = L/2$.

Transient Heat Transfer

The models so far have covered steady state heat transfer, that is the predicted temperatures and temperature distributions are of a system in an equilibrium state, and it tells us nothing about the temperature change with time. In many applications, this equilibrium state is the most important, since this defines the steady state operating conditions. However, the knowledge of transient behaviour can be useful for low duty cycle behaviour, or for the prediction of thermal runaway.

The analytical solutions for transient heat transfer are much more involved than the steady state solution, and as a consequence the solutions are often numerical, based on finite differences or finite elements.

As an example the analytical solution for the temperature distribution in a semi-infinite solid, with initial temperature T_0, with heat produced at a constant rate P, per unit time per unit volume is given by Carslaw and Jaeger [28]:

$$T_{x,t} = \left[T_0 + \frac{\kappa t P}{k} + \frac{Px^2}{2k}\right] erf\left[\frac{x}{2\sqrt{\kappa t}}\right] + \frac{Px}{k}\left[\frac{\kappa t}{\pi}\right]^{1/2} \exp^{-x^2}/4\kappa t - \frac{Px^2}{2k} \tag{17}$$

where the surface is maintained at zero temperature, and κ is the thermal diffusivity.

Temperature Dependent Heat Generation

In all the previously described models the thermal material parameters and the heat generation processes have been considered temperature independent, with a consequence that all the solutions will be stable. From experimental results, it appears that the thermal diffusivity is relatively temperature independent and the thermal conductivity and specific heat capacity increase approximately 20 % over the common operating temperature regime. However the heat generation process is likely to be temperature dependent, as many of the dependent parameters such as dielectric constant and loss are also temperature dependent. Inclusion of temperature dependent heat generation can lead to unstable solutions, which is a possible mechanism for thermal runaway.

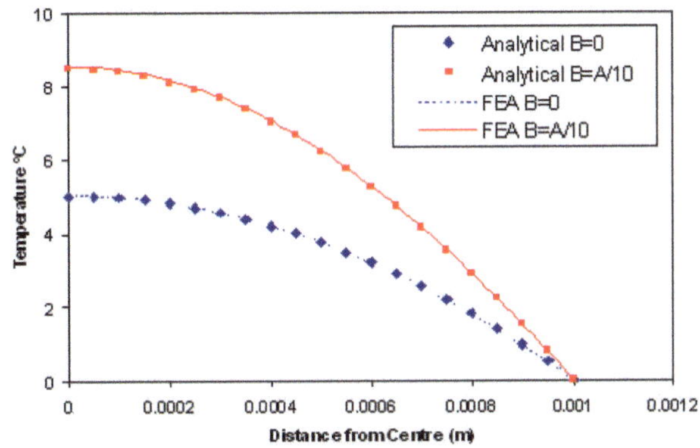

Fig. 8 Temperature dependent heat generation in an infinite slab, where B is the temperature dependent factor. Comparison of the analytical results (points) and a 1D finite element simulation (*lines*)

Adding temperature dependent heat generation to the models further complicates the solution and makes the analytical solutions even more complex. The following is the temperature profile of a slab length, l, with no flow of heat at $x=0$, and initial temperature of 0 K. The heat generation is defined by $k(A+B.T)$, where k is the thermal conductivity, $A$and B are constants where B is the temperature dependent part of the heat generation.

$$T(x,t) = \frac{A}{B}\left[\frac{\cos xB^{1/2}}{\cos lB^{1/2}} - 1\right] + \frac{16Al^2}{\pi}\sum_{n=0}^{\infty}\frac{(-1)^2\exp\left[\left(-(2n+1)^2\pi^2 + 4B^2l^2\right)kt/3l^2\right]\cos(2n+1)\pi x/2l}{(4Bl^2 - (2n+1)^2\pi^2)(2n+1)} \tag{18}$$

In order for this solution to be stable, the temperature dependent part of the heat generation, B, must be such that the following inequality holds:

$$B < \frac{\pi^2}{4l^2} \tag{19}$$

If the equality holds then there is a steady state solution to the transient problem, given by:

$$T(x) = \frac{A}{B}\left[\frac{\cos xB^{1/2}}{\cos lB^{1/2}} - 1\right] \tag{20}$$

The temperature profile determined using Eq. (18) is illustrated in Fig. 8, calculated at long times, so it is equivalent to Eq. (20), and shows the effect of increasing the temperature dependent heat generation factor, B. Obviously the larger B becomes the higher the temperature at the centre will become, the outside temperature being held at zero in the model. To add other, more realistic, boundary conditions to the model, the analytical solution increases in complexity, which is where FEA solutions come into their own. Also included on the figure are FEA solutions for the same problem, showing very similar results.

Finite Element Solutions

Finite element software such as ANSYS or PAFEC allows both transient and steady state solutions to heat transfer problems for a variety of boundary conditions, including internal power generation, temperature dependent thermal conductivity, and also some coupled field solutions, such as magneto thermal, i.e. the heat generated through joule heating in electromagnet coils. However, to model the self-heating problem observed in piezoelectrics using ANSYS, the temperature and piezomechanical solution would have to be solved sequentially. The thermal model would be treated as a system with internal power generation, and coupling between the piezomechanical and thermal solutions could be achieved through an iterative process.

Early work [29], using Finite Elements to model behaviour in an ultrasonic power transducer, used the dielectric heating model to determine the internal power generation, which was assumed uniform throughout the device. The dissipated power was determined by fitting results to experimental temperature profiles, rather than using the material dielectric loss and applied voltage to calculate the power.

Shankar and Hom [30], used dielectric heating alone to predict temperature related phenomena in an electrostrictive PMN sonar transducer. Contrary to PZT, dielectric loss in PMN decreases with increasing temperature [31], and so is not beset by the thermal failures seen in PZT. The internal power generation was calculated using the loss tangent and applied voltage, and this power distributed evenly through the device. In order to simulate pulse drive and reduced duty cycle, this power input was distributed temporally. For example, a 33 % duty cycle was simulated by full power on for 1 s, then off for 2 s.

Ando [22] carried out a comprehensive simulation of transient thermal behaviour in an ultrasonic transducer, using an iterative procedure as follows:

1. Determine the static stress and strain in the device
2. Calculate resonant frequency, stress, and strain etc using room temperature material parameters.
3. Determine the mechanical and dielectric heat dissipation.
4. Calculate heat diffusion and temperature distribution.
5. Calculate thermal stress, strain and thermal expansion.
6. Change material parameters due to new temperature distribution and repeat cycle.

This routine was continued until the required time was reached. He accurately predicted, not only the temperature rise of the device, but also its effect on the resonant frequency of the transducer.

Abboud et al. [32] have used the PZFlex finite element package to deal with heat generation in 1–3 piezocomposites. They point out that a coupled solution to the problem is difficult due to the different characteristic timescales of the different processes; the piezoelectric vibration at 150 kHz, and the thermal conduction that occurs over the order of seconds. Their solution to this problem takes the following form:

1. Model the piezoelectric effect for one cycle and then calculate the power generation.
2. Solve for the transient thermal problem to determine the temperature distribution.
3. Apply any piezoelectric, mechanical, thermal property changes due to temperature rise/fall. Repeat steps 1–3.

The solution is reached at 2 when the temperature dependence of the material properties is negligible for the heating levels involved.

More recently Blottmann et al. [23], have performed a full 3D FEA simulation of model sonar transducers using ATILA. Mechanical and dielectric losses were included in the model, and although the mechanical loss was constant and the low field value, the dielectric loss was both field and temperature dependent. Again an iterative routine was used to modify material parameters dependent on the temperature rise from the internal losses.

Although fully coupled piezo-thermal solutions are not yet realised in FEA packages, another approach to the problem could be to use thermally dependent internal heat generation. In the ANSYS package, by using point elements, heat generation rates can be modelled using a polynomial of the form:

$$Q = A_1 + A_2T + A_3T^{A_4} + A_5T^{A_6}$$

where the constants A_i can be input as "material parameters". Thus if it is possible to model the temperature dependence of the heat generation rate of the piezoelectric, this can be included in the thermal solution.

Example FEA Solution

As shown in Fig. 8, 1D FEA simulations agree well with analytical solutions but come into their own when the models become more complex.

Figure 9 shows a simple example of a 2D steady state FEA solution for a sample with uniform heat generation, and boundary conditions of 0° on the long sides, and held at 5° on the short sides. The temperature profile across the centre of the sample is similar to that in Fig. 8, however the core temperature is slightly higher because of the increased temperature of the short sides. This is relatively trivial example for FEA, however the 2D analytical solution would be much more complex.

5 Case Studies

In order to demonstrate some of the predictive modelling described in this document, a number of case studies will be discussed, where the models are compared to experimental measurements. Each case study is intended to highlight a different problem faced in modelling piezoelectric thermal behaviour.

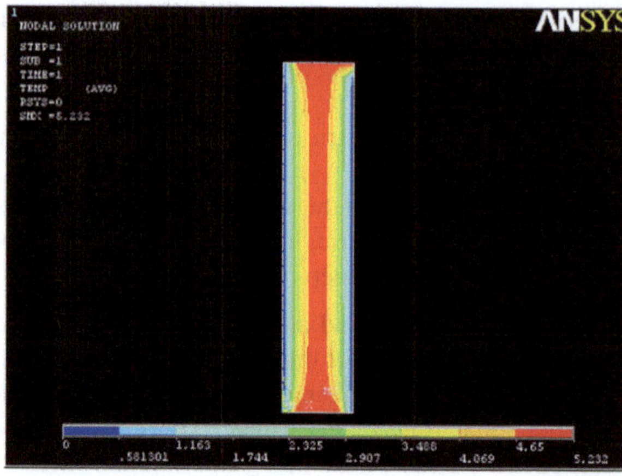

Fig. 9 2D steady state FEA solution for a sample with uniform heat generation. *Line* profiles are the same as Fig. 8, but the end effects are observed because the slab is not infinite

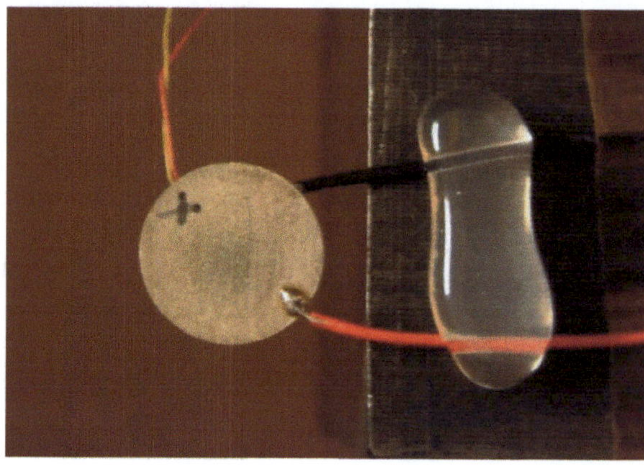

Fig. 10 PZT disc sample, suspended in air, showing power cables (*thick*) wires and thermocouple (*thin wire*)

5.1 Case 1: Off Resonant Drive of a Simple Monolithic Ceramic Disc Suspended in Air

Rationale: This is almost the simplest case encountered; the sample is thin, so there is no significant temperature profile through the sample. A balance between internal heat generation, and heat removal via convection from the surface, governs the sample surface temperature and there is no thermal conduction. The sample is operating off resonance, so the internal heat generation is assumed uniform throughout the sample.

The sample is a 0.2 mm thick, 10 mm diameter disc of soft PC 5H composition, mounted as shown in Fig. 10. The sample was placed in a large enclosure, approximately 0.03 m^3 (1 cubic foot), to control the air movement around the sample. The temperature of the sample was measured with a thermocouple, and comparisons were made using an IR sensor, but little difference was seen. The sample was driven bipolar at frequencies of 100 Hz–5kHz and the current and voltage measured during

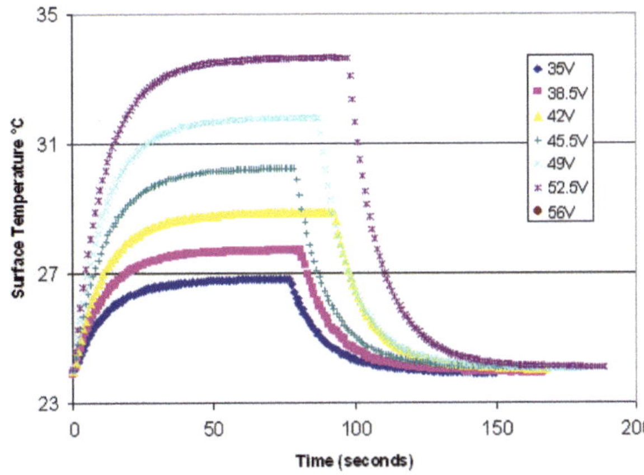

Fig. 11 Example heating/cooling curves for a thin piezoelectric disc driven at 5 kHz. Power is automatically cut off when sample temperature stabilises, resulting in cooling curves

each run. The data collection routine had a thermal cut-off point, should the sample surface temperature reach 100 °C, in order to prevent irreversible changes in the material. Otherwise the applied power would be automatically removed once a stable temperature was reached.

Figure 11 shows an example of heating/cooling curves for these experiments, for applied voltages at 5 kHz, where generally the self-heating leads to an equilibrium temperature. The equilibrium temperature depends on a number of variables including the geometry, convection coefficient as well as voltage and frequency, but as stated previously is essentially a balance of the input energy against the energy removal processes. This balance can be described by the following equation:

$$E_{st} = E_g - E_{out} \tag{21}$$

where;
E_{st}= Energy Stored in Sample with temperature rise via Cp (equation (A1))
E_g= Internal Energy Generation via dielectric Heating (Eq. 4)
E_{out}= Energy Removal via convection (Eq. 2) or radiation (Eq. 3).
As the sample is operating off resonance, the internal energy generation can be calculated using the dielectric heating model, Eq. (4). The thermal energy stored in any arbitrary block of material can be determined from Eq. (28), and assuming there is no thermal conduction, the energy removal can be determined from Eqs. (2) and (3).

Assuming that there are negligible heat losses from the edge of the disc, the instantaneous temperature rise is given by

$$\Delta T = \frac{2\pi f \varepsilon \varepsilon_0 V^2 \tan \delta / t - 2h(T - T_{amb}) - 2\varepsilon_m \sigma (T^4 - T_{sur}^4)}{\rho t C_p} \tag{22}$$

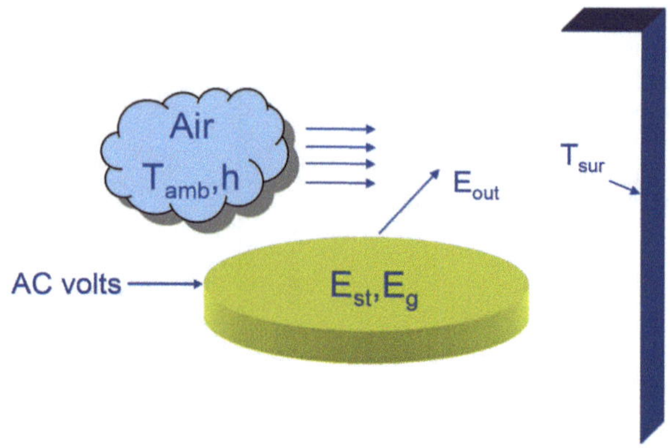

Fig. 12 Schematic of energy balance in self heating of piezoelectric disc

where t is the disc thickness. The terms in Eq. (22) are a mixture of material parameters, fundamental constants and experimental conditions, where the least accessible value is the convection coefficient, h. The convection coefficient is very dependent on experimental conditions and although it can be estimated it can also easily be determined experimentally from the cooling curves.

Newton's Law of Cooling

Newton's law of cooling simply states that the rate of change of the temperature is proportional to the difference between its own temperature and ambient, where the proportionality constant is the heat transfer coefficient, h.

$$\frac{dT}{dt} = -h(T - T_0) \tag{23}$$

Solutions to this equation take the form:

$$T(t) = T_{amb} + (T_0 - T_{amb}) \exp^{-ht} \tag{24}$$

This simple equation can be used to model many thermal problems, such as cooling cups of coffee, time of death calculations, and also thermal behaviour of piezoelectric devices. Several workers [33–35] have used this model to fit to experimental data of surface temperature changes of self-heated piezoelectric devices, and although the heat transfer coefficients obtained are dependent on specific device geometry and boundary conditions, it does permit limited predictive capabilities.

A value of $33\,\mathrm{W/m^2K}$ for the heat transfer coefficient gives a good fit to the experimental cooling curve, Fig. 13. Strictly speaking, this constant applies for the cooling curve, and it is possible that it may not necessarily be equivalent to the heating process. Zheng et al. [33] has attributed variations in heat transfer coefficients in active piezoelectric devices, to the increased convection due to the vibrating surface.

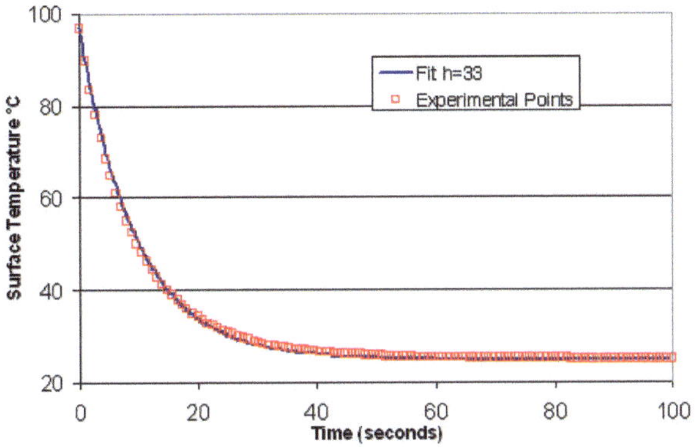

Fig. 13 Cooling curve of thin piezoelectric disc with experimental fit based on Newton's law of cooling

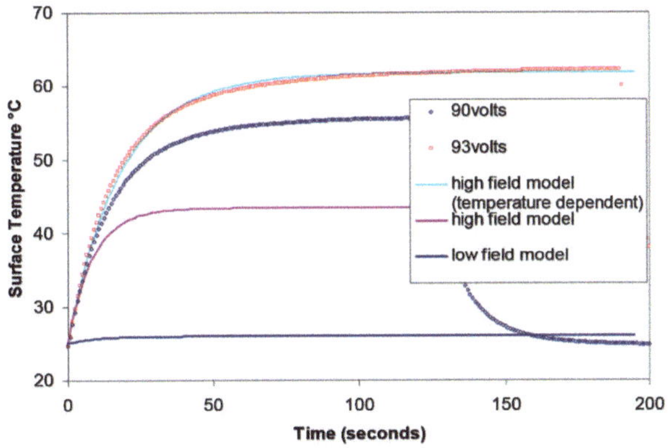

Fig. 14 Experimental results of self heating of thin piezoelectric disc driven at 500 Hz, compared with "low field" and "high field" models based on Eq. (22)

Model Assumptions

The model described by Eq. (22) has a number of assumptions including; negligible heat losses from the disc edge, uniform temperature throughout disc, radiation occurs between a small sample and large enclosure, and most importantly constant material properties, i.e. temperature invariant.

As an example the model is tested on heating curves when the sample is driven at 5 kHz, with an applied voltage of 93 V. If the model is used with the typical material constants, in particular the "small signal" or "low field" permittivity and loss values, this gives us a very poor fit to the experimental results. In the "low field" model the values used for permittivity and loss are 2600 and 0.02 respectively, leading to an overall temperature rise of 1 °C, Fig. 14. It is perhaps unrealistic to use the "small signal" dielectric constant values, since small signal is generally measured at fields of around 1 V/mm. If the dielectric constants measured at a "high field" of 94 V are used, permittivity = 3300 and loss = 0.3, then a stable temperature rise of 18.5 °C is predicted. The "high field" model is improved compared with the "low field" prediction, however the temperature rise is roughly half that measured.

The assumption in the model so far has been that all the material constants are temperature independent, however this may not be a valid assumption. The dielectric properties, particularly for soft materials are known to be temperature dependent, and measurements of the voltage and current during the experiment have shown this to be the case. The temperature dependence can be included in the model by assuming a linear variation with temperature for the permittivity and loss such that;

$$\varepsilon = 3300 + 70 * (T - T_{amb}) \tag{25}$$

$$\tan\delta = 0.3 + 0.001 * (T - T_{amb}) \tag{26}$$

A much improved fit is obtained when the temperature dependence of the dielectric properties is included in the model, Fig. 14.

Thermal Runaway

As discussed previously, thermal runaway occurs in systems when the internal energy generation is greater than the energy removal rate and the system becomes unstable, usually ending in system failure. It is common in chemical processes, charging of batteries and capacitors [3], many semiconductor devices, as well as piezoelectric devices. The key feature of this phenomenon is some kind of positive feedback between the internal energy generation and the increasing temperature.

Initial internal heat generation rate
⇓
Increases temperature
⇓
Increases heat generation rate
⇑

From the modelling of the thin disc it is clear that the temperature dependence of the dielectric properties plays a key part in temperature rise and is a possible mechanism for this positive feedback.

When the disc is driven with high fields at a higher frequency of 5 kHz, the sample starts to exhibit thermal runaway, Fig. 15a. This can be seen more clearly as a change in the temporal temperature gradient from negative to positive, Fig. 15b. In this case at a voltage of approximately 84 V the behaviour changes from stable equilibrium temperature to thermal runaway. This behaviour is chaotic and the exact voltage that this change occurs is highly dependent on slight changes in the experimental conditions such as the heat removal rate. Figure 15a also shows the modelled temperature rise (solid lines), based on the temperature dependent high field model. The fit for the higher voltages, cf. 91 V, is not as good as at 77 V and reflects the chaotic nature of the behaviour near the runaway point. In order to get a better fit at 91 V it is necessary to raise the modelled voltage to 100 V.

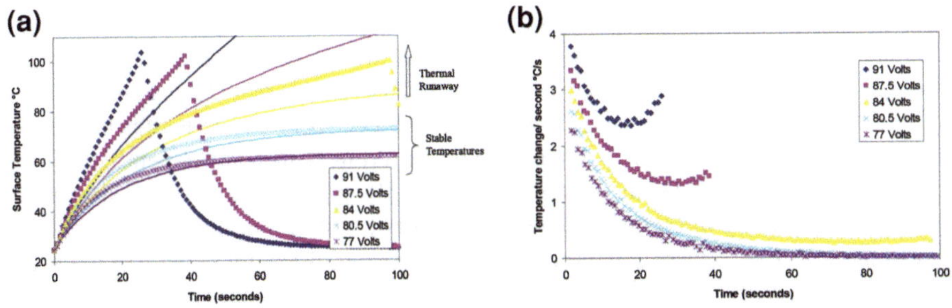

Fig. 15 Self heating in a thin PZT disc driven at 5 kHz exhibiting thermal runaway. **a** Experimental temperature rise (points) and modelled temperature rise (*solid lines*) based on a temperature dependent high field model. **b** Rate of temperature change of disc showing clearly the onset of thermal runaway. Power is automatically cut off when sample temperature stabilises, or temperature reaches 100 °C, resulting in cooling curves

Conclusion

A simple model based on the energy balance in a volume, where only energy generation, convection and radiation are considered has been developed that can accurately predict the temperature rise in thin piezoelectric discs. It has been shown that this model is vital if the internal heat generation mechanism is temperature dependent, and it can even extend into prediction of thermal runaway, although here the chaotic behaviour make accurate predictions difficult. The model is solved iteratively, and in principle this method can be incorporated into FEA solutions where complex geometries and thermal conduction can be included.

5.2 *Case 2: Resonant Drive of Simple Monolithic Ceramic Suspended in Air*

Rationale: This device was chosen to illustrate the effects of the spatially dependent heating of a piezoelectric, depending on the strain profile at resonance. The device has a large surface area to fill the field of view of the thermal camera, and is driven in the d_{31} mode, so the direction of electric field is perpendicular to the expected spatial thermal anomalies.

Experimental

A bar of hard PZT 4D material, 70 mm long, 10 mm wide and 2 mm thick is driven at longitudinal resonance and its overtones. The bar is poled across the width and the field is also applied in this direction, resulting in expansion along the length. The sample is driven at the resonance frequency, 24 kHz and the first and second overtones, 71.2 and 111.1 kHz. The sample is suspended in air on two wooden cocktail sticks, and thermal images recorded using an infrared camera. The temperature readings from the camera were corrected for emissivity by placing the sample on a hot plate and comparing the camera results with those from a type K thermocouple.

The thermal images, Fig. 16, clearly show different behaviour dependent on the resonant mode. At 71.2 kHz the first overtone of the length resonance, there are clearly three hot spots along the length, roughly coincident with the nodes where the strain is at a maximum. Similarly at the second overtone, 111.1 kHz, there are five hot spots and seven cool areas. The power cables also appear to heat up, being hotter where the wire makes contact with the bar. Some of this could be due to contact resistance as the silver electrode did not make good mechanical contact and was sometimes pulled off with the wire. Ideally the centre line scans should be perfectly symmetrical, however there is some evidence that the right hand side of all the images are nominally hotter. There are several possible reasons for this, including misalignment of the sample, uneven illumination of the sample and non-uniform pixel array sensitivity.

Modelling

The modelling of the resonator was performed using a 3D thermal FEA model, consisting of brick elements with eight nodes, with a single degree of freedom, temperature, at each node. This element was used for both, steady-state and transient 3-D thermal analysis. The heat loss from the resonator was through a uniform convection coefficient on all the surfaces, conduction through the wires and supports and radiation loss was ignored. The convection coefficient was determined by matching the predicted cooling curves to the measured temperature and a value of 10 W/m^{2}K was used.

The heat generation was modelled by distributing the dissipated power according to the square of the strain, since both tensile and compressive strains lead to a temperature rise. In the d_{31} mode the largest strains occur along the length, x direction, and the strain in the other directions was assumed to be uniform. Therefore the dissipated power was distributed according to the following equation:

$$P(x) = P \sin^2 \left[\frac{n\pi x}{L} \right] \qquad (27)$$

where P is the power per unit volume, L is the sample length, and n depends on the resonant mode, odd for resonance and even for antiresonant modes.

Figure 17 shows the measured centreline temperature profile after 50 and 100 s and the transient modelled profiles using n=3 and a power level of 0.2 W/cm^3. The model shows the same peaks and troughs as the real behaviour, and this holds for different times, however the match is not perfect. The fact that the measured profiles are hotter towards the right hand side has already been discussed, however the central point is consistently the hottest part on the measured device, a feature that is not mirrored in the model behaviour. It is possible that the electrical contacts, which

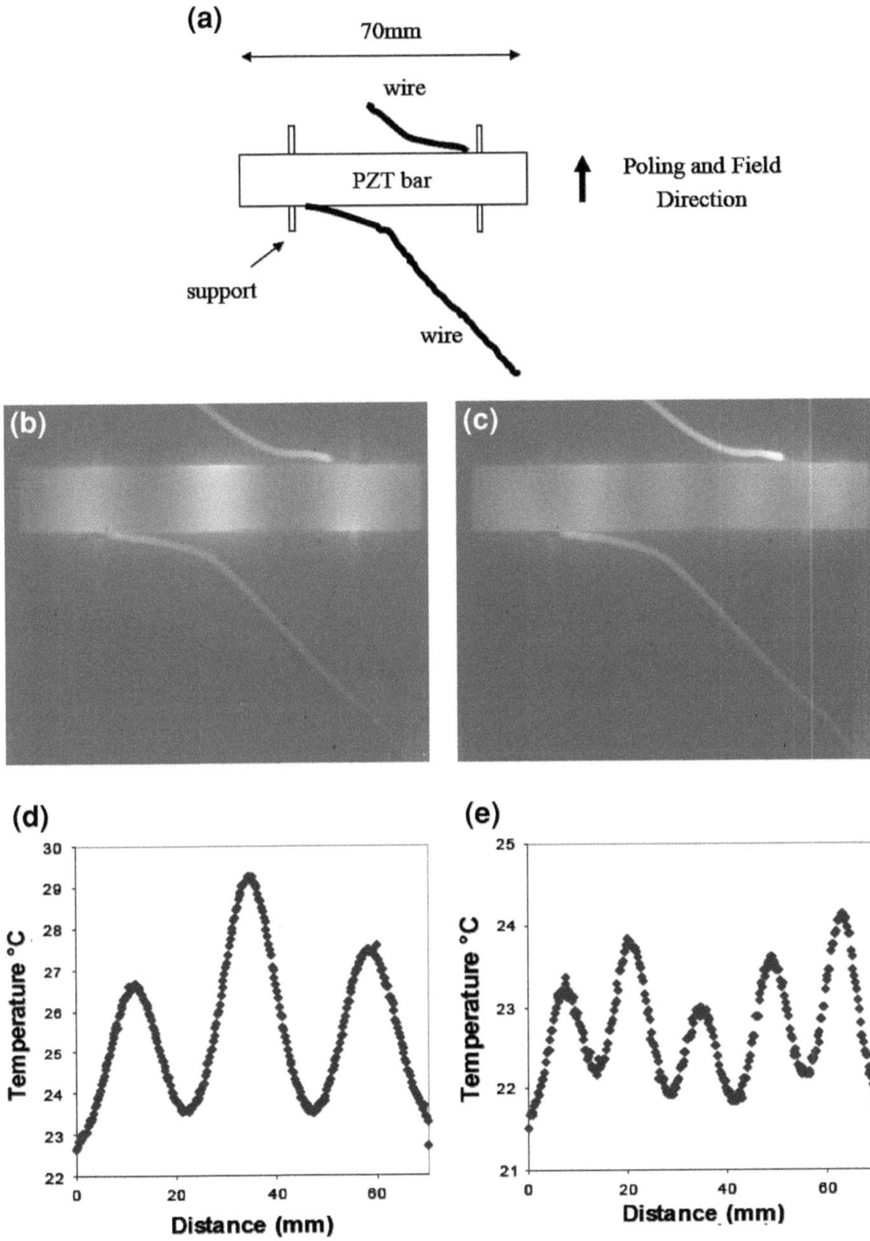

Fig. 16 **a** sample setup, **b**, **c** Thermal images of piezoelectric bar driven at resonance. **d** Thermal image and centreline temperature profile of first overtone, 71.2 kHz. **e** Thermal image and centreline temperature profile of second overtone, 111.1 kHz

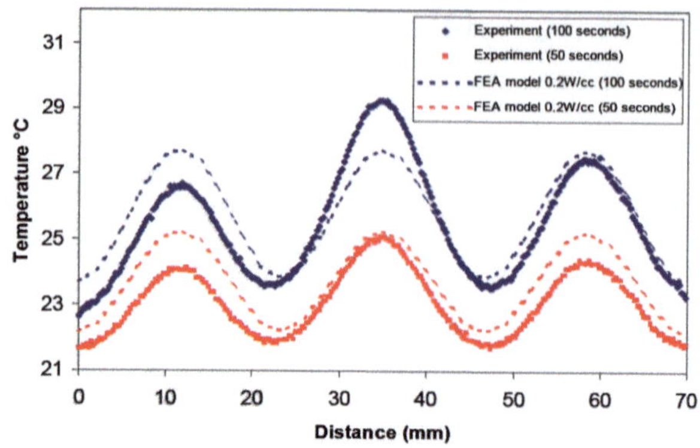

Fig. 17 Experimental and modelled temperature profiles for a bar driven at the first overtone 70.2 kHz

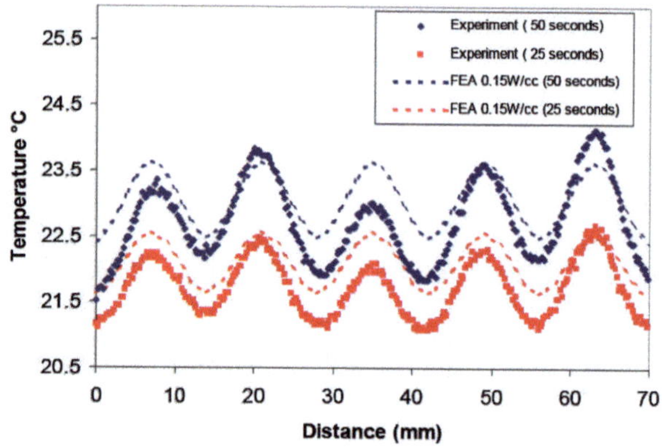

Fig. 18 Experimental and modelled temperature profiles for a bar driven at the second overtone 111.1 kHz

are coincidentally near the side lobes are the cause of increased heat loss through conduction.

The modelled and measured temperatures for the second overtone, $n = 5$, are shown in Fig. 18. As can be seen, the temperature difference is smaller for this mode, because the strain variations are closer together. The model still gives a reasonable representation of real behaviour, but now because of the small temperature rise, the errors in the measurements are more apparent.

So far, the results of the model have shown the temperature profile of the centreline of the top surface only. The FEA model is 3D, so it predicts the temperature of the entire sample, and the surface temperature is shown in Fig. 19. From this it can be seen that the predicted temperature is not uniform along the width direction and the contours around the hot spots are convex, whereas around the cool areas they are concave. This is due to the thermal edge effects where more heat is lost near the edges because the material in this region is close to two surfaces. In fact, the experimental behaviour shows the opposite, with the hot spots being concave, and the cool areas convex. In the model we have assumed that the spatial dependent effects are entirely in the x, length direction and uniform along the width, y direction. In fact, because

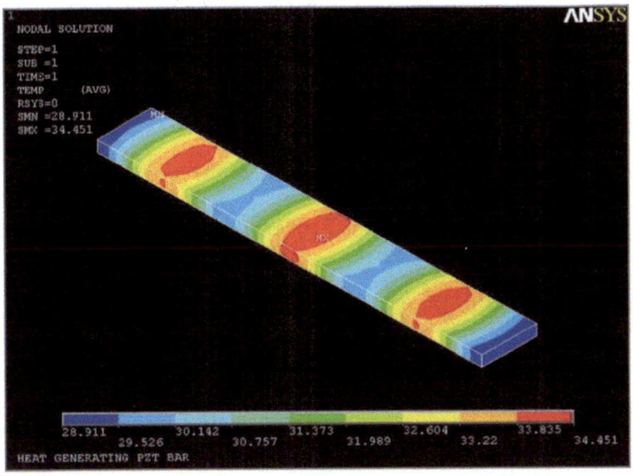

Fig. 19 Full 3D FEA simulation of temperature map of bar where n=3, first overtone

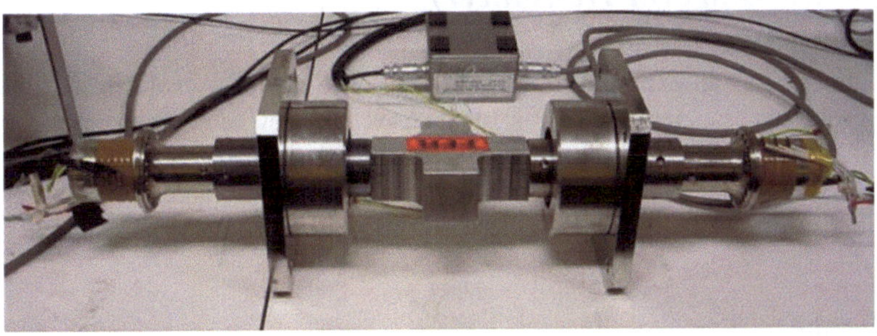

Fig. 20 Photo of the back to back high power transducer system

of mechanical edge effects, the strain levels near the edge decrease, which would amplify the thermal edge effects, thus increasing the curve of the contours rather than reversing them. Although the predicted contours take the wrong form, in practice the errors in actual temperature levels are less than 0.5 °C.

Conclusion

The case study clearly illustrates the spatial dependence of heat generation in a piezoelectric device driven at resonance. This behaviour can be modelled by assuming a sine-squared distribution of the power dissipation rather than a uniform one. The effects here are dramatic because the resonator is mechanically and thermally well insulated. In many practical situations the resonator will be attached to another medium and heat transfer through conduction will override the spatial dependent effects seen in this case.

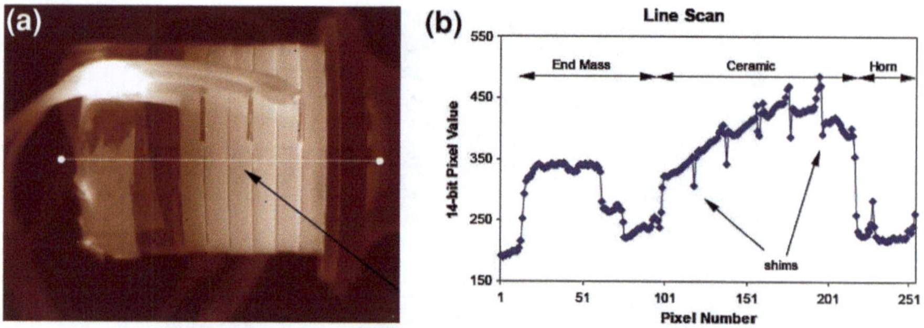

Fig. 21 Thermal image of high power transducer driven at an input power of 400 W for 400 s, with associated line scan pixel values

5.3 Case 3: Resonant Drive of a Clamped Ceramic (Bolted Langevin Transducer)

Rationale: This is an example of an industrial high power transducer and includes all the problems associated with non-ideal cases. The device geometry is more complicated, there are many different materials present, and the drive electronics are proprietary. Because of these complexities, it is no longer viable to solve this using analytical solutions and finite element methods must be used.

Experiment

The transducer is a commercial high power device, and in order to drive the system safely at high powers it is necessary to extract power from the system. This is achieved by mounting two transducers back to back in the form of a 1:1 piezotransformer, where the output is terminated into a resistive load consisting of several halogen lights. Figure 20 shows the system, which is held horizontally at the nodal points by two large steel plates. The transducers have a resonant frequency of 20 kHz, and the active piezoelectric consists of 6 rings of PZT, 5 mm thick and 50 mm OD, stacked sequentially but wired electrically in parallel. The ceramic is prestressed by a stainless steel back mass using a high tensile steel bolt which is attached to the aluminium front horn. The ceramics have silver electrodes, but in order to make a mechanically robust electrical contact to these a thin metal shim is placed between each PZT disc. These shims have a small tab to enable the electrical contact to be made.

The temperature of the transducer was measured with an Indigo Merlin InSb mid range infrared camera with a 320 by 256 pixel array. Figure 21 shows a greyscale thermal image of the device after being driven for 400 s at an input power of 400 W. The image shows hot spots associated with the three power cables, and also close to the metal shims. The hot spot where the cables are bunched together is real, i.e. this is the hottest part of the cables, but not necessarily the hottest area in the image.

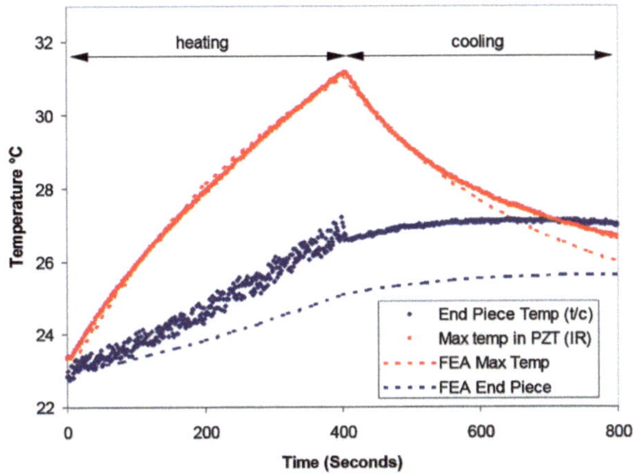

Fig. 22 Measured temperatures during application of 400 W electrical power for 400 s. *Red*, maximum temperature measured using IR camera, *blue* temperature measured using thermocouple attached to back mass of transducer. *Dashed lines* show FEA predictions based on 5 W power dissipated

The hot spot close to the left hand side of each shim is an artefact, where a small raised portion of the shim is reflecting radiation from some other part of the scene. This illustrates the necessity for emissivity correction. The camera gives a 14-bit number associated with each pixel, and this must be corrected for the emissivity of the object in order to convert this number to a temperature. The emissivity correction was carried out for the PZT only, by heating a similar piece of PZT on a hotplate, attaching a thermocouple to the surface and comparing the results. Because of the high voltages present it was not possible to attach a thermocouple to the active part of the transducer, but a thermocouple was attached to the steel back mass, roughly half way along the length. For ease of data capture only a line scan was captured every 200 ms, and the data for the line is also shown in Fig. 21. In order to get an idea of the transient behaviour, the maximum temperature in the ceramic and the temperature from the thermocouple are plotted in Fig. 22. A power of 400 W was applied for 400 s and then turned off, whilst still measuring the temperature. The thermocouple shows the effects of electrical pickup whilst the power is applied, but this noise is still only around 0.5 °C. It is interesting to note that temperature of the end mass still continues to rise after the power has been removed before eventually cooling down.

FEA Modelling

The modelling of the transducer was carried out by simplifying the device to an axisymmetric system, consisting of three thermal elements, the steel back mass, aluminium horn, and the active ceramic ring. The values used for the thermal properties for the elements are listed in Table 1. The heat loss from the device was assumed to be through convection only, since at these temperatures energy dissipation through radiation is small. All the external surfaces were subject to a uniform heat transfer coefficient, with the air temperature maintained at ambient. The value of this coeffi-

Table 1 Thermal properties used for FEA model of transducer

	Density (kg/m^3)	Thermal conductivity (W/m K)	Specific heat capacity (J/kg K)
Steel	7,800	75	640
PZT	7,500	1.2	320
Aluminium	2,700	200	900

cient was estimated by matching the predicted cooling curves to the measured and a value of 3 W/m^{2}K was used. The convective heat transfer coefficient is probably the biggest source of error in the analysis, since it is unlikely to be completely uniform due to the complex geometry of the system. However even with this assumption the model shows all the features found experimentally.

If we assume that the internal heat generation can be modelled by the power dissipated through dielectric loss then for an applied voltage of 1000 V and using a relative permittivity of 1300 for the material, this leads to a power of around 10.9 W for a tan delta of 0.2. If this power is uniformly distributed over the volume, then this leads to a power/unit volume of around 220, 200 W/m^3, or 0.22 W/cm^3, which is within the safe rule of thumb value of 0.5 W/cm^3 given by Berlincourt [7]. Figure 23 shows the predicted temperature rise of the model after 400 s at a power level of 10 W, where the maximum temperature is at the centre of the PZT. The model matches the transient behaviour reasonably well for different power levels, Fig. 24. Halving the power input in the model to 5 W agrees well with the experimental measurements. The model also predicts the temperature rise in the rest of the transducer, but the matching between the measurements taken by the thermocouple on the back mass and the modelled behaviour is not as good, Fig. 22. The form of the predicted behaviour is very similar, it predicts the slight temperature increase after the power is cut, but the measured temperature rise is roughly 2 °C greater than predicted. This is probably because the thermocouple was attached to the transducer using kapton tape, which had the effect of insulating the thermocouple from the surrounding air.

In order not to overheat the transducers the maximum operating time was limited to 800 s, however the thermal model predicts that steady state will be reached in around 12 h, where the maximum temperature for the 5 W dissipated power would be 100 °C and for 10 W, 180 °C.

The transducer is a resonant device; therefore it is perhaps unreasonable to assume that the internal heat generation is uniform throughout the device. At the fundamental resonance the strain in the transducer is distributed such that there is maximum strain in the ceramic directly next to the titanium horn, and it decreases sinusoidally to zero at the end of the device. If is assumed that the power distribution of 10.9 W is distributed according to the square of this strain, rather than a uniform distribution, then this should lead to a more accurate prediction of the temperature. The predicted temperature for uniform and strain heating are shown in Fig. 25 along with the measured temperature in the ceramic. As can be seen the strain dependent temperature profile is shifted slightly towards the point of maximum strain, but in fact there is very

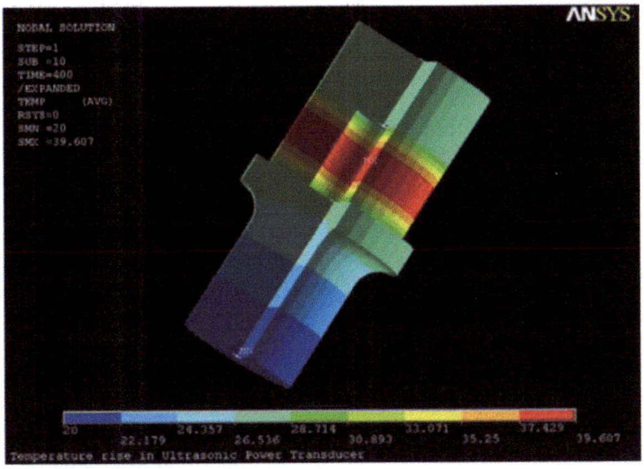

Fig. 23 Thermal FEA model of high power transducer

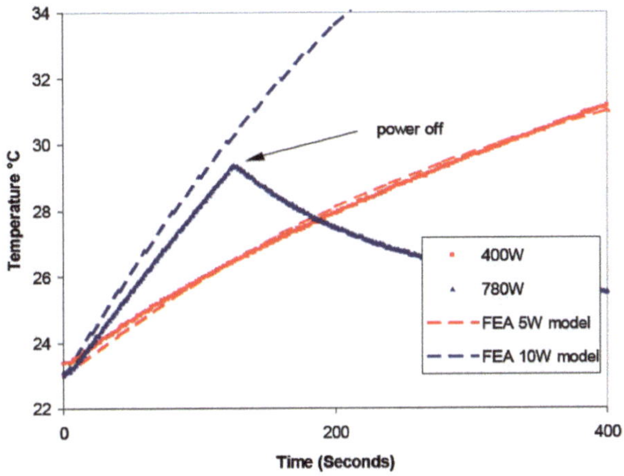

Fig. 24 Experimental maximum device temperature and modelled behaviour based on 5 and 10 W dissipated power

little difference between the two curves since the thermal behaviour is dominated by heat removal rather than heat generation. The form of the measured temperature rise in Fig. 25 is skewed towards the horn side, whereas the modelled temperature profile is roughly symmetric. These are transient curves, after 400 s operation.

Under steady state conditions the modelled temperature profile skews towards the end mass because of the reduced energy removal from this component, thus increasing the discrepancy between measured and modelled behaviour. It is clear that although the model may be used to predict approximate temperatures there are still discrepancies between modelled and measured behaviour. In order for the temperature profile to skew as seen experimentally, there must be less heat removal from the horn end. This could be because of inaccurate thermal properties for this material, or more likely due to thermal contact resistance between the ceramic and the horn. The initial FEA model has assumed that the six PZT pieces are one solid piece, and that there is perfect contact between the ceramic and the end mass, and the ceramic

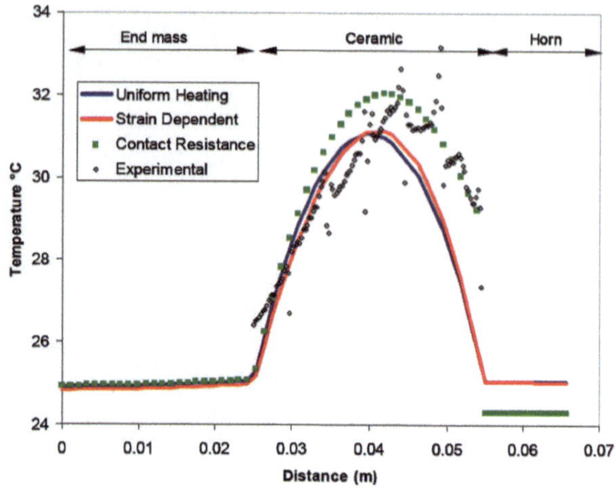

Fig. 25 Experimental temperature profile compared to FEA models based on dielectric heating, *blue line*, and strain heating model, *red line*, with added contact resistance, *green*

and the horn. If a thermal contact resistance element is placed between the ceramic and the horn, then a skew of the temperature profile similar to that seen experimentally is observed. The contact resistance is characterised by a thermal contact conductance (TCC) such that the heat flux per unit area across the contact is TCC multiplied by the temperature difference across the contact. A value of 250 W/m^2K for the thermal contact conductance gave the green trace in Fig. 25, where a value of 5000 W/m^2K is a 'perfect' contact. The model could be improved by adding the various extra interfaces, however it does mean that accurate values for these contact resistances would be needed.

Conclusion

A finite element model enables more complex geometries to be considered, and realistic results can be obtained. Using the heat generation values based on the dielectric heating model, obtained using input parameters of applied voltage, material permittivity and tan δ_e, gives acceptable results. The added complexity of space dependent heat generation does not give significantly different results, since unbalanced heat removal rather than heat generation dominates the behaviour in this system. For this particular transducer it appears that contact resistance at the component interfaces plays a significant part in controlling the temperature rise. Practically, the system cannot be run at high power levels without adequate convective cooling, however the temperature rise in the current configuration might be reduced by careful selection and manufacturing of the interfaces between the various components.

6 Concluding Comments

Failures due to thermal issues are common in high power piezoelectric devices. This chapter gives engineers an understanding of the problems, where they occur and how to avoid them. The work covers methods used to predict the temperature rise seen, based on simple analytical models and the use of Finite Element Models.

The case studies illustrate the use of these methods in examples that embody most of the conditions seen in real systems. The input data required in these simulations were measured using the methods describe in the chapter annexes.

Appendix A: Thermal Property Data of Piezoelectric Materials

In order to be able to make any predictions of how a piezoelectric device will perform in terms of self induced temperature rise, some knowledge of the thermal properties are needed, typically the thermal conductivity k, but also the thermal diffusivity α, and the specific heat capacity Cp. Data for these properties is scarce in the literature, with the most often quoted source coming from Berlincourt et al. [36], where a single value for thermal conductivity and specific heat capacity is given for all PZT materials with no information regarding the temperature dependence.

A.1 Specific Heat Capacity, C_p (J/kg K)

The specific heat capacity, C_p, is the amount of energy needed to raise the temperature of 1kg of the material by one degree. This can be used to determine the temperature rise of a volume of a material since the change in energy stored is proportional to the specific heat through the following equation;

$$\dot{E}_{st} = \frac{d}{dt}(\rho V C_p T) \tag{28}$$

where ρ and V are the density and volume of the material. If the internal power generation rate of the material is known, then the resultant temperature rise can be determined.

As discussed the most quoted value for the specific heat capacity of PZT is 420 J/kgK [36], but Table 2 gives a summary of other sources that give values for the room temperature value ranging from 350 to 491 J/kg K. Yarlagadda et al. [37] made a comprehensive set of measurements of heat capacity and thermal conductivity over the temperature range 20–300 K and show that the specific heat capacity varies greatly over this range, and is different for hard and soft materials. The soft material PZT-5H increases 15 fold over the temperature range, whilst the hard composition only increases by around 4 times. These values are not necessarily inconsistent with

Table 2 Specific Heat Capacity values for PZT type piezoelectric materials

Material	Temperature (K)	Specific heat capacity (J/kg K)	References
PZT 5H	22	23.3	[37]
PZT 5H	155	348	[37]
PZT 4S	22	42.6	[37]
PZT 4S	155	159	[37]
PZT(TRS-600)	Room	413*	[38]
PMN-PT-BT	Room	350	[30]
PZT	Room	420	[36]
Modified barium titanate	Room	500	[36]
PZT [MT-18]	Room	491	[22]
PZT	Room	350	[39]
PZT	Room	320**	[40]

*Value quoted as volume specific heat capacity, 3.26 × 106 J/m3 °C, using a density of 7900 kg/m^3 to convert to specific heat capacity
** Value quoted as volume specific heat capacity, 2.5 × 106 J/m^3 °C, using a density of 7900 kg/m^3 to convert to specific heat capacity

the room temperature values reported in the literature as some of the temperature dependence may be due to phase changes, indeed the authors attribute deviations at around 50–80 K to transition type behaviour.

The most common operating regime for piezoelectric actuators is from ambient temperature to around 300 °C and there is little published information on the change in specific heat over this temperature range. Measurements of the specific heat capacity of a soft and a hard PZT composition were made at NPL, from room temperature to 200 °C, using a Perkin Elmer DSC 7 Differential Scanning Calorimeter. The basis of DSC measurements is that a sample of known weight is placed in a holder, and the amount of energy provided by an electrical heater, which is required to increase the temperature of the sample per degree is measured. The test is carried out at a constant heating rate during which the heater power is adjusted continuously to compensate for heat absorbed or evolved by the sample, to keep the sample holder temperature identical to that of the reference holder. The energy input by the electrical heater is measured and this is proportional to the specific heat.

Measurements were repeated on two samples of each composition and the results are plotted in Fig. 26. The room temperature values are the order of 20 % lower than the most often quoted value, although they do increase by approximately 15 % over the measured temperature range. The soft material has a slightly higher specific heat capacity than the hard, but the difference is not much greater than the sample-to-sample variation.

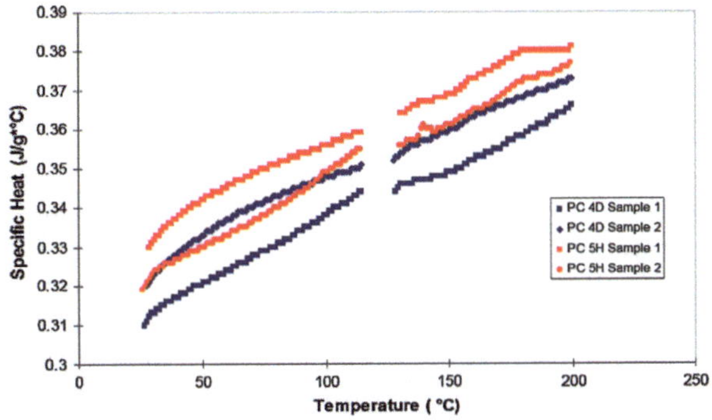

Fig. 26 Specific heat variation with temperature, for PZT 4D and PZT 5H

A.2 Thermal Conductivity, k (W/m K)

In order to determine the equilibrium state of a device, i.e. a steady state thermal solution, only the thermal conductivity of the material is required, which has units W/m K. Thermal conductivity is defined as the rate at which heat flows through a certain area of a body. The precise definition is given by Fourier's equation:

$$\frac{dq}{dt} = -kA\frac{dT}{dx} \tag{29}$$

where q is the heat energy flowing in the x direction through the area, A during time, t. dT/dx is the temperature gradient, and k is the thermal conductivity of the material.

Table 3 summarises the published values of thermal conductivity of PZT type piezoelectric materials, where values range from 0.8 to 2.3 W/m K. Again Yarlagadda et al. [37] have determined the thermal conductivity below 300 K and found that PZT-4S was almost twice as conductive as PZT-5H, and that there is a significant temperature dependence, however their room temperature thermal conductivity results fall well below the generally accepted value of between 1 and 2 W/mK. In general most thermal models assume the thermal conductivity is temperature independent, most probably because of a lack of published data on the temperature dependence.

A.2.1 Errors

As can be seen from Table 3, the published values of thermal conductivity varies by around 100 %, and the errors in the measured value will be reflected in the predicted thermal behaviour, depending on the exact details of the case. In general the errors in the thermal conductivity are propagated linearly to the results. For example, Sherrit et al. [24] predict an internal temperature rise of 36 °C in a 20 mm diameter, 2 mm thick disc, using a value of k = 1.25 W/m K, however this will be reduced to 23 °C when

Table 3 Thermal conductivity values for PZT type piezoelectric materials

Material	Temperature K	Thermal conductivity (W/mK)	References
PZT-4D	Room	1.8	[32]
PZT	Room	2.3*	[41]
PZT 5H	15	0.01	[37]
PZT 5H	300	0.14	[37]
PZT 4S	15	0.018	[37]
PZT 4S	300	0.34	[37]
PZT	Room	1.0–1.5	[25]
PMN-PT-BT	Room	3.59	[30]
PZT-4S	Room	2.1	[42]
PZT	Room	1.25	[36]
Modified Ba Titanate	Room	2.5	[36]
PZT[MT-18]	Room	1.5	[3]
PZT	Room	1.1	[39]
PZT	Room	0.8	[40]
PZT 4	Room	2.0	[43]
PMN-PT single crystal	20	0.2	[44]
PMN-PT single crystal	320	2.6	[44]

* published value 23, but presumed to be a typographical error

a value of $k = 2\,\text{W/m K}$ is used. This is evident from the first term in Eq. (15) where the temperature rise is proportional to the reciprocal of the thermal conductivity. In contrast, Hu [25] has shown that varying the thermal conductivity from 1 to 1.5 W/m K has little effect on the predicted temperature rise, and that power input was the dominant factor in modelling temperature rise in piezoelectric transformers.

A.3 Thermal Diffusivity, α (m²/s)

Thermal diffusivity is the material property governing heat flow when the temperature is varying with time, and has the dimension length^2/time, with units of m^2/s. The thermal diffusivity can be determined experimentally, and it is related to the thermal conductivity k, through the following relationship

$$\alpha = \frac{k}{\rho \cdot C_p} \tag{30}$$

where ρ is the density and C_p is the specific heat capacity.

Using the published values for k, ρ and C_p the thermal diffusivity of PZT is around $5 \times 10^{-7}\text{m}^2/\text{s}$. Lang [45] has measured the thermal diffusivity of bulk piezoelectric materials using the Laser Intensity Modulation Method (LIMM), and obtained values close to this estimate, Table 4.

Table 4 Thermal diffusivity values for PZT type piezoelectric materials

Material	Temperature (K)	Thermal diffusivity (m^2/s)	References
PZT (Plessey RM200)	Room	4.59E-7	[45]
PSZT	Room	4.53E-7	[45]

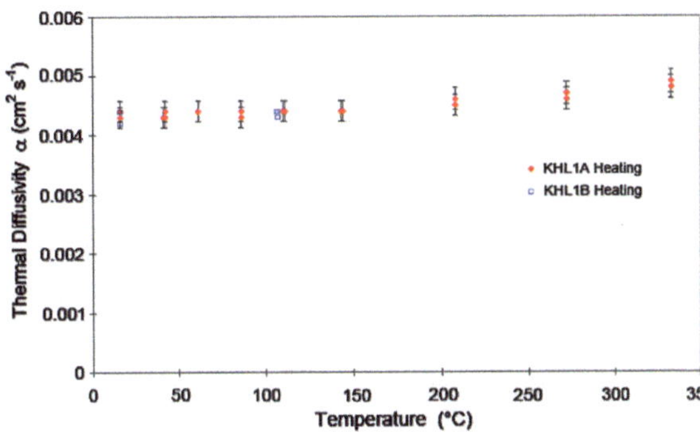

Fig. 27 Thermal diffusivity measurements of PZT-5H using laser flash

The thermal diffusivity of PZT was also measured at NPL, using a Netzsch Laser Flash 427. The measurement involves heating the front face of a disc-shaped sample of known dimensions, usually 12.5 mm diameter and 1.5–2.5 mm thickness using a high intensity Nd:YAG laser with a pulse width of between 0.2 and 1.2 ms. The temperature rise on the rear face is monitored using an InSb infrared detector. From the temperature rise with time, the thermal diffusivity can be calculated, applying appropriate corrections for radiation and the finite laser pulse length. Because of limited samples of the correct dimension only a soft PZT 5H composition was measured. The measured value for the thermal diffusivity of PZT-5H was around 4.5×10^{-7} m^2/s, with little measurable change over the range room temperature to 350 °C, this compares favourably with measured values reported in the literature, Fig. 27.

The thermal conductivity can also be derived from the measured thermal diffusivity and the measured values of specific heat capacity, and lie between 1.1 and 1.3 W/mK over the temperature range room temperature to 200 °C.

Acknowledgments This work was supported by the UK's National Measurement System programme—Measurement for Materials Processing and Performance, MPP. Thanks are also due to T Amato, (PURAC), F Rawson (FFR Ultrasonics Ltd.) and D Hazelwood (R&V Hazelwood Associates) for the loan of equipment, support and advice.

References

1. Berlincourt, D., Krueger H., Near C.: Technical publication TP-226 properties of Morgan Electro ceramic ceramics. Morgan Electro Ceramics, vol. 12. (2003)
2. Tokin, N.: Multileyer Piezoelectric Actuators. NEC/TOKIN vol. 07

3. Merker, U., Droste, E., Michaelis, A.: Thermal Runaway of Tantalum Capacitors, pp. 102–106. CARTS, Europe (2002)
4. Richard, C., Lee, H.S., Guyomar, D.: Thermo-mechanical stress effect on 1–3 piezocomposite power transducer performance. Ultrasonics **42**(1–9), 417–424 (2004)
5. Mitrovic, M., Carman, G., Straub, F.K.: Durability properties of piezoelectric stack actuators under combined electromechanical loading. Proc. SPIE **3992**, 217 (2000)
6. Takahashi, S., Sasaki Y., Hirose, S.: Driving electric field effects on piezoelectric transducers. Jpn. J. Appl. Phys. **36**(1), 3010–3015 (1997)
7. Technical Publication TP-221 Piezoelectric CeramicsVisit the Morgan Electro Ceramics. Web Site: www.morganelectroceramics.com. Power Capacities of Piezoelectric Ceramics in Sonar-type Acoustic Transducers. pp. 1–8, (1999)
8. Hu, J., Ho, S.-F., Ong, E.-L., Du, J.: An experimental investigation of the temperature field in small piezoelectric vibrators. Ultrasonics **41**(9), 731–735 (2004)
9. Uchino, K., Hirose, S.: Loss mechanisms in piezoelectrics: how to measure different losses separately. IEEE Trans. Ultrason. Ferroelectr. Freq. Control **48**(1), 307–321 (2001)
10. Richard, C., Goujon, L., Guyomar, D., Lee, H.S., Grange, G.: Selecting passive and active materials for 1.3 composite power transducers. Ultrasonics **40**(1), 895–901 (2002)
11. Lente, M., Eiras, J.: Interrelationship between self-heating and ferroelectric properties in PZT ceramics during polarization reorientation. J. Phys. Condens. Matter **12**, 5939 (2000)
12. Ronkanen, P., Kallio, P., Vilkko, M., Koivo, H.: Self heating of piezoelectric actuators: measurement and compensation. In: Proceedings of the 2004 International Symposium on Micro-Nanomechatronics and Human Science, 2004 and the Fourth Symposium Micro-Nanomechatronics for Information-Based Society, 2004, pp. 313–318 (2004)
13. Mignogna, R.B., Green, R.E., Duke, J.C., Henneke, E.G., Reifsnider, K.L.: Thermographic investigation of high-power ultrasonic heating in materials. Ultrasonics **19**(4), 159–163 (1981)
14. Incropera, F.P., DeWitt, D.P.: Fundamentals of Heat and Mass Transfer. Wiley, New York (1996)
15. Tashiro, S., Ikehiro, M., Igarashi, H.: Influence of temperature rise and vibration level on electromechanical properties of high-power piezoelectric ceramics. Jpn. J. Appl. Phys. **36**(1), 3004–3009 (1997)
16. Jordan, T., Qunaies, Z.: Piezoelectric ceramics characterization. Technical Report 2001–28 (2001)
17. Jordan, T.: Langley Research Center, U. S. N. A. Administration, and Space. Electrical properties and power consideration of a piezoelectric actuator (2000)
18. Zhang, Q.M., Wang, H., Zhao, J.: Effect of driving field and temperature on the response behaviour of ferroelectric actuator and sensor materials. J. Intell. Mater. Syst. Struct. **6**, 84–93 (1995)
19. Hooker, M., Langley Research Center: Properties of PZT-based piezoelectric ceramics between -150 and 250C. Technical Report (1998)
20. Gdula, R.A.: High field losses of adulterated lead zirconateG titanate piezoelectric ceramics. J. Am. Ceram. Soc. **51**(12), 683–687 (1968)
21. Berlincourt, D.A.: Clevite: Power Limitations of piezoelectric ceramics in radiating transducers (Technical paper TP-225). Piezoelectric Division/Clevite Corp. (1962)
22. Ando, E., Kagawa, Y.: Finite-element simulation of transient heat response in ultrasonic transducers. IEEE Trans. Ultrason. Ferroelectr. Freq. Control **39**(3), 432–440 (1992)
23. Lu, X., Hanagud, S.V.: Extended irreversible thermodynamics modeling for self-heating and dissipation in piezoelectric ceramics. IEEE Trans. Ultrason. Ferroelectr. Freq. Control **51**(12), 1582–1592 (2004)
24. Sherrit, S., Bao, X., Sigel, D.A., Gradziel, M.J., Askins, S.A., Dolgin, B.P., Bar-Cohen, Y.: Characterization of transducers and resonators under high drive levels. Ultrasonics Symposium, 2001 IEEE, vol. 2, pp. 1097–1100 (2001)
25. Hu, J.: Analyses of the temperature field in a bar-shaped piezoelectric transformer operating in longitudinal vibration mode. IEEE Trans. Ultrason. Ferroelectr. Freq. Control **50**(6), 594–600 (2003)

26. Kanayama, K.: Thermal analysis of a piezoelectric transformer. Ultrasonics Symposium, 1998, Proceedings., 1998 IEEE, vol. 1. pp. 901–904 (1998)
27. Piezoelectric properties of ceramic materials and components: Part 3: Methods of measurement: High power. BS En 50234–3 (2002)
28. Carslaw, H.S., Jaeger, J.C.: Conduction of Heat in Solids, Ser. Oxford science publications, Clarendon Press, New York (1959)
29. Stulen, F., Senapati, N., Gould, R.: Temperature distribution in an ultrasonic power transducer. Ultrason. Int. **83**, 301–306 (1983)
30. Shankar, N., Hom, C.: An acoustic/thermal model for self-heating in PMN sonar projectors. J. Acoust. Soc. Am. **108**, 2151 (2000)
31. Robinson, H.: Large signal dielectric losses in electrostrictive materials. Proc. SPIE **3992**, 91 (2000)
32. Abboud, N., Mould, J., Wojcik, G., Vaughan, D., Powell, D., Murray, V. MacLean, C.: Thermal generation, diffusion and dissipation in 1–3 piezocomposite sonar transducers: finite element analysis and experimental measurements. Ultrasonics Symposium, 1997. Proceedings., 1997 IEEE, vol. 2. pp. 895–900 (1997)
33. Zheng, J., Takahashi, S., Yoshikawa, S., Uchino, K., Vries, J.d.: Heat generation in multilayer piezoelectric actuators. J. Am. Ceram. Soc. **79**(12), 3193–3198 (1996)
34. Yao, K., Uchino, K., Xu, Y., Dong, S., Lim, L.C.: Compact piezoelectric stacked actuators for high power applications. IEEE Trans. Ultrason. Ferroelectr. Freq. Control **47**(4), 819–825 (2000)
35. Pritchard, J., Ramesh, R., Bowen, C.R.: Time-temperature profiles of multi-layer actuators. Sens. Actuat. A: Phys. **115**(1), 140–145 (2004)
36. Mason, W.P., Thurston, R.N.: Physical Acoustics: Principles and Methods, ser. Physical Acoustics. Academic Press, New York (1976)
37. Yarlagadda, S.: Low temperature thermal conductivity, heat capacity and heat generation of PZT. J. Intell. Mater. Syst. Struct. **6**, 757 (1995)
38. Weiland, L.M., Lynch, C.S.: Thermo-electro-mechanical behavior of ferroelectric materials part II: Introduction of rate and self-heating effects. J. Intell. Mater. Syst. Struct. **14**(10), 605–621 (2003)
39. P. I. Ceramic. Piezoelectric Ceramics Products, PI Ceramic GmbH. http://www.piceramic.com
40. Bauer, S., Ploss, B.: A method for the measurement of the thermal, dielectric, and pyroelectric properties of thin pyroelectric films and their applications for integrated heat sensors. J. Appl. Phys. **68**(12), 6361 (1990)
41. Zhou, S.W., Rogers, C.A.: Heat generation, temperature, and thermal stress of structurally integrated piezo-actuators. J. Intell. Mater. Syst. Struct. **6**(3), 372–379 (1995)
42. Lesieutre, G.A., Fang, L., Koopmann, G.H., Pai, S.P., Yoshikawa, S.: Heat generation of a piezoceramic induced-strain actuator embedded in a glass/epoxy composite panel. 1996 Symposium on Smart, Structures and Materials, pp. 267–275 (1996)
43. Dubus, B., Boucher, D.: An analytical evaluation of the heating of lowGfrequency sonar projectors. J. Acoust. Soc. Am. **95**, 1983 (1994)
44. Zhu, D., Han, P.: Thermal conductivity and electromechanical property of single-crystal lead magnesium niobate titanate. Appl. Phys. Lett. **75**, 3868 (1999)
45. Lang, S.: Technique for the measurement of thermal diffusivity based on the laser intensity modulation method (LIMM). Ferroelectrics **93**(1), 87–93 (1989)

Piezoresponse Force Microscopy

Serban Lepadatu and Markys G. Cain

1 Scanning Probe Microscopy

Scanning probe microscopy (SPM) is a very versatile technique allowing for a large range of sample properties to be measured and manipulated with nanometre spatial resolution. One important SPM mode is piezoresponse force microscopy (PFM). PFM is an invaluable tool for measuring the piezoresponse of functional materials at the nanoscale, allowing for high resolution measurements of the electromechanical coupling of thin films. In this chapter we will give a brief overview of SPM, following which PFM will be analysed in some detail.

The origins of current SPM setups can be traced back to the invention of the scanning tunnelling microscope (STM) by Binnig and Rohrer in 1982 [1, 2]. The importance of this measurement technique was quickly recognized and the Nobel prize was awarded for its discovery in 1986. With STM, using a X-Y-Z scanner stage, an atomically sharp tip is brought to a precisely controlled distance from the conductive sample surface and a bias voltage is applied between the sample and tip. Quantum tunnelling of electrons from the sample to the tip gives rise to an electrical current which is accurately measured. Several types of measurements are possible using the basic STM setup. When imaging in constant current mode, the sample-tip separation is adjusted using a feedback loop in order to maintain a constant tunnelling current as the tip is scanned over the sample surface. The variation in height arises due to the topography of the sample as well as the local density of electron states (LDOS) [3] When measuring in constant height mode, the sample-tip separation is kept constant and the variation in tunnelling current is recorded at a fixed tip-sample

S. Lepadatu (✉) · M. G. Cain
National Physical Laboratory, Hampton Road, Teddington, Middlesex TW11 0LW, UK
e-mail: serban.lepadatu@npl.co.uk

M. G. Cain (ed.), *Characterisation of Ferroelectric Bulk Materials and Thin Films*,
Springer Series in Measurement Science and Technology 2,
DOI: 10.1007/978-1-4020-9311-1_8, © Queen's Printer and Controller of HMSO 2014

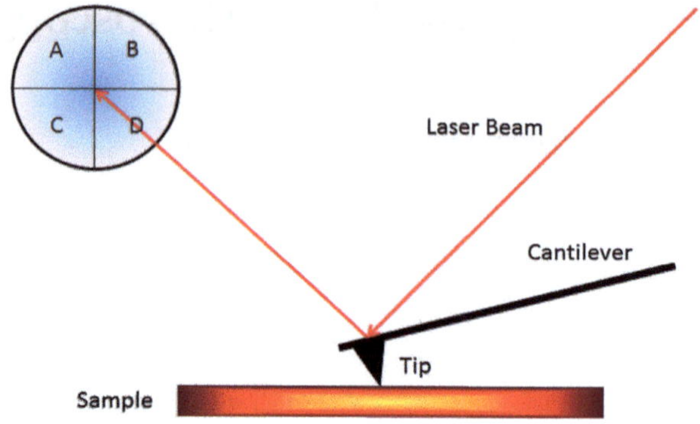

Fig. 1 AFM setup. A laser beam is reflected off the cantilever and into a position sensitive photodetector, allowing *lateral* and *vertical* tip displacements to be detected as the cantilever is scanned over the sample using a piezo scanner

bias, revealing the local charge density. Due to the nature of STM these measurements may be performed with Ångström lateral resolution.

The most important extension of STM was the atomic force microscope (AFM) in 1986 by Binnig et al. [1]. Typically, in contact mode AFM measurements a cantilever with a sharp tip is brought in contact with a sample and the deflection of the cantilever is monitored by reflecting a laser beam off into a position sensitive photodetector (PSD). The PSD is a quadrant type photodiode, thus it is split into four photodetectors, A, B, C and D, as indicated in Fig. 1.

Vertical cantilever deflection is proportional to (A+B)−(C+D) whilst lateral cantilever deflection is proportional to (B+D)−(A+C). Once the tip is in contact with the sample, further increasing the Z scanner position towards the sample results in a proportional vertical cantilever deflection, as described by Hookes law, Eq. 1, for a linear spring, where F is the contact force and k_Z is the normal cantilever stiffness.

$$F = -k_z \Delta Z \tag{1}$$

As we can see from Eq. 1, a constant vertical deflection setpoint results in a constant contact force. By rastering the tip over the surface of the sample and using a feedback loop to adjust the Z scanner position in order to maintain a constant contact force, the plot of ΔZ over the scanned area represents the topography of the sample surface. Modern AFMs have very good vertical resolution, able to measure accurately variations in sample topography with sub-nanometre precision, and lateral resolutions down to nanometre precision. In contact mode AFM the main forces acting on the cantilever are adhesion forces due to Van der Waals interaction and short-range repulsive forces due to atomic interactions. The combined attractive and repulsive forces cause the cantilever to deflect according to the sample topography. Since the introduction of AFM a large number of imaging modes have been developed, which can be largely categorized to reflect the different origins of the forces acting on the cantilever and different sample properties measured. Magnetic force microscopy (MFM)—forces due to magnetic fields, electric force microscopy (EFM)—forces due to electric fields, lateral force microscopy (LFM)—frictional forces, are just some of the myriad of existing SPM methods. For a comprehensive review of SPM modes, reference [4] can be consulted. For the remainder of this chapter we will concentrate on PFM, first giving an introduction to the setup and principles of PFM, artefacts and calibration of PFM measurements, advanced PFM modes and finally a tutorial on PFM measurements and procedures.

2 Piezoresponse Force Microscopy

PFM is based on the standard contact mode AFM setup with the cantilever and tip being electrically conductive, typically either through highly doped Si or metallic coating. The samples measured are piezoelectric and a voltage applied between the tip and a bottom electrode results in sample strains due to the inverse piezoelectric effect [5]. The sample strains cause vertical and lateral deflection of the cantilever which can be measured using the standard PSD. For example, two electric domains with polarization direction normal to the surface but opposite sign (c+ and c− domains) will strain in the vertical direction but with opposite sign, thus their orientation can be distinguished by monitoring the PSD vertical deflection signal. In order to separate the topography and piezoresponse signals and also to increase the signal to noise ratio a lock-in amplifier technique is used, with the voltage applied to the tip, $V_a \cos(\omega t)$, having a much larger frequency compared to the scanning frequency. The vertical and lateral PSD signals are measured using lock-in amplifiers at the excitation frequency, w, thus separation of the piezoresponse and topography signals is possible. A diagram of the PFM setup is shown in Fig. 2. Additionally, use of lock-in amplifiers for piezoresponse signal detection allows much larger signal to noise ratios by cutting broadband noise, allowing average displacements of just a few picometres to be detected.

From each lock-in amplifier we obtain two outputs, the magnitude and phase of the vertical or lateral response. The magnitude output is related to the size of the piezoelectric coefficients of the sample, whilst the phase is related to the orientation of the electric domains. Thus, taking again our example with the c+ and c− domains, the magnitude in both cases should be equal whilst the phase responses should differ by 180°, allowing their directions to be distinguished. When applying the sinusoidal excitation to the piezoelectric sample, we have two possibilities. The first method discussed above, whereby the voltage waveform is applied directly through the PFM tip, between tip and bottom electrode, is called the local excitation method. In this case the PFM tip is effectively a moving top electrode. Another possibility is to pattern or deposit a large top conducting electrode onto the sample and then apply the voltage through the tip in contact with the large top electrode. This case is called the global excitation method. The two methods are shown in Fig. 3. With the former method the electric fields generated are highly non-uniform, as seen in Fig. 3a, making any quantitative interpretation of PFM measurements very difficult, whilst with the latter the electric fields are uniform under the tip, as seen in Fig. 3b, but this comes at the cost of reduced lateral resolution.

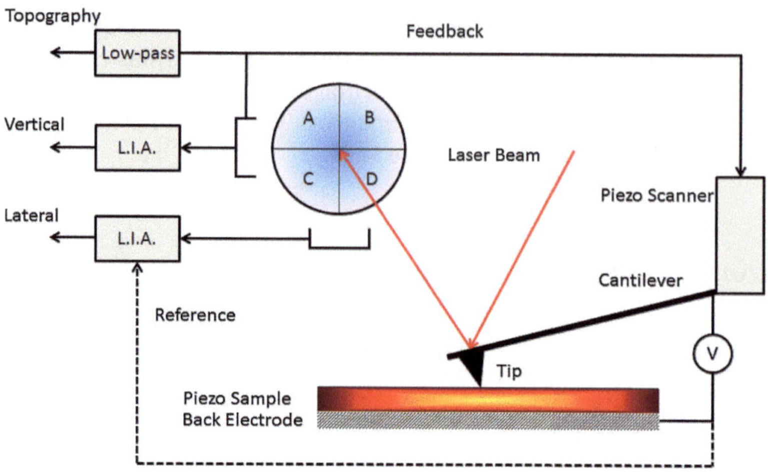

Fig. 2 PFM setup. A sinusoidal signal is applied across a piezoelectric sample between *tip* and *bottom* electrodes with the *vertical* and *lateral* displacement signals from the PSD detected using lock-in amplifiers. A feedback loop is used as in AFM to adjust the deflection setpoint and obtain topographic information. A piezo scanner is used to adjust the Z position and to raster in the X-Y plane

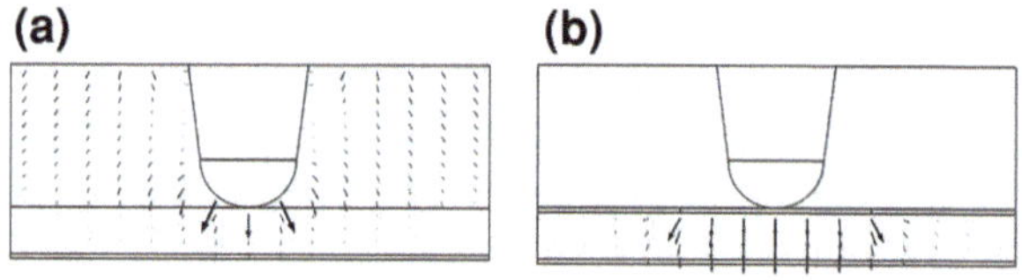

Fig. 3 Simulations showing electric field distribution for a model PFM tip on ferroelectric sample with *bottom* electrode and voltage applied between *tip* and *bottom* electrode in **a** local excitation method and **b** global excitation method with additional electrode sandwiched between tip and ferroelectric surface. In the latter case lateral resolution is sacrificed for uniformity of electric field under the *tip*

A typical PFM image is shown in Fig. 4, showing both amplitude and phase components. The sample imaged is a 100 nm thick Pb(Zr0.2Ti0.8)O3 (PZT) epitaxial layer on SrTiO3:Nb (1 at %) (STO), 400 μm thick. The STO substrate is electrically conductive due to the Nb doping, thus serving as the bottom electrode. The sample was imaged in the local excitation mode. By looking at both the amplitude and phase responses we can identify c+ and c− domains due to the phase contrast and equal amplitude response. Moreover, from the amplitude response we observe boundaries of zero response between the different domains. Between two electric domains with different orientations we have a transition region, called a domain wall. In ferroelectric materials the domain wall width is very small, typically only a few unit cells, making imaging domain walls directly very difficult. The tip diameter used to obtain the image in Fig. 4 is 40 nm, thus the transition region seen in Fig. 4a

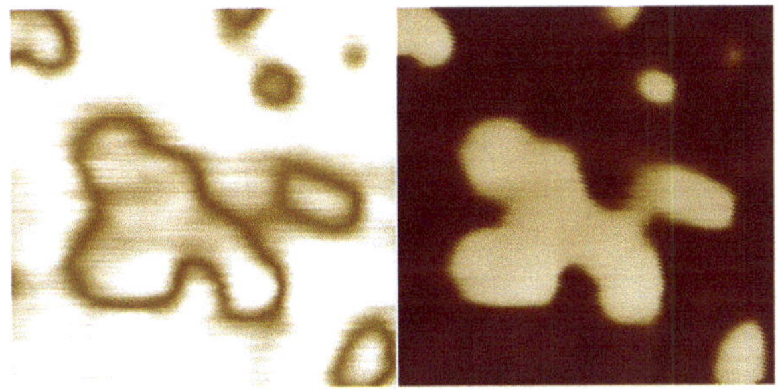

Fig. 4 PFM scan of 100 nm thick epitaxial PZT, 200 nm^2 scan size, showing **a** amplitude response and **b** phase response

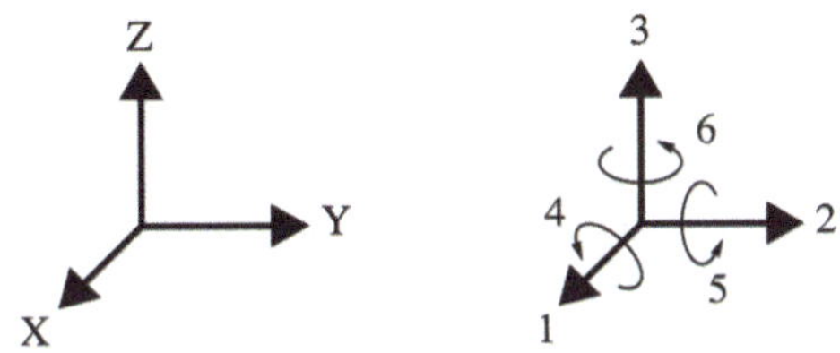

Fig. 5 Normal and shear components of *stress* and *strain* vectors

arises due to the opposite responses of the c+ and c− domains effectively cancelling each other as the tip is scanned across the boundary.

2.1 Strain-Charge Equations

For piezoelectric materials the effect of stresses and electric fields on the electric displacement field and sample strain is described by Eq. 2.

$$
\begin{aligned}
S &= s^E T + d^T E \\
D &= dT + \epsilon^T E
\end{aligned} \tag{2}
$$

Here, T and E are the stress and electric field vectors respectively, S and D are the strain and electric displacement field vectors respectively. s^E denotes the compliance for constant electric field, d is the piezoelectric coefficients tensor and ϵ^T is the permittivity under constant stress. The T superscript for the piezoelectric coefficients tensor denotes transposition. For a capacitor configuration with top and bottom electrodes, the electric displacement field at the surface gives the charge density on the top capacitor electrode. In general there are six components of stress and strains, as indicated in Fig. 5. D and E have three spatial components each.

For the special case of piezoelectric ceramics, the compliance, piezoelectric and permittivity tensors can be simplified due to crystal symmetries. Thus we obtain the simple forms in Eq. 3, valid for a piezoelectric ceramic with electric polarization oriented in the Z direction.

$$\mathbf{d} = \begin{pmatrix} 0 & 0 & 0 & 0 & d_{15} & 0 \\ 0 & 0 & 0 & d_{15} & 0 & 0 \\ d_{31} & d_{31} & d_{33} & 0 & 0 \end{pmatrix}$$

$$\mathbf{s} = \begin{pmatrix} s_{11} & s_{12} & s_{13} & 0 & 0 & 0 \\ s_{12} & s_{11} & s_{13} & 0 & 0 & 0 \\ s_{13} & s_{13} & s_{33} & 0 & 0 & 0 \\ 0 & 0 & 0 & s_{44} & 0 & 0 \\ 0 & 0 & 0 & 0 & s_{44} & 0 \\ 0 & 0 & 0 & 0 & 0 & 2(s_{11} - s_{12}) \end{pmatrix} \tag{3}$$

$$\epsilon = \begin{pmatrix} \epsilon_{11} & 0 & 0 \\ 0 & \epsilon_{11} & 0 \\ 0 & 0 & \epsilon_{33} \end{pmatrix}$$

The piezoelectric coefficients d_{31} and d_{33} are related through Poissons ratio of the material, ν, Eq. 4:

$$d_{31} = \nu d_{33} \tag{4}$$

where $0 < \nu < 1$. From Eq. 2 we distinguish two important effects, the direct and indirect piezoelectric effects. For the direct piezoelectric effect a stress T is applied and this results in a change in the dielectric displacement field in the sample and thus a change in the surface charge density, as described by the second part of Eq. 2. For the indirect piezoelectric effect an electric field is applied and this results in a sample strain, as described by the first part of Eq. 2. For PFM imaging this latter effect is exploited in order to characterize the surface domain structure and obtain quantitative information on the piezoelectric coefficients of the material.

2.2 PFM Theory and Quantification

The response of the cantilever on a piezoelectric sample is composed of not only the piezoresponse of the sample but is also influenced by capacitive forces arising from the tip, cantilever and bottom electrode configuration. In general the capacitive force is related to the stored energy, $E_{cap} = V^2C/2$, where V is the voltage applied and C the capacitance value, by Eq. 5:

$$F_{cap} = \frac{dE_{cap}}{dz} = \frac{V^2}{2}\frac{dC}{dz} \tag{5}$$

The applied voltage consists in general of d.c. and a.c. voltages, $V = V_{dc} + V_{ac}\cos(\omega t)$. Thus we obtain the total force as a combination of a constant force, F_{dc}, a force at the frequency ω, F_ω and a force at the second harmonic, F_{2w}, as shown in Eq. 6.

$$F_{cap} = \frac{1}{2}\frac{dC}{dz}\left(V^2 + dc + \frac{V_{ac}^2}{2}\right) + V_{dc}V_{ac}\cos(\omega t) + \frac{V_{ac}^2}{4}\frac{dC}{dz}\cos(2\omega t) \tag{6}$$

The constant force, F_{dc}, is not detected directly in the measurements, whilst the capacitive force at the measurement frequency ω can be eliminated by setting the d.c. voltage, V_{dc}, to zero. This latter force can be significant for non-zero d.c. voltages, thus it is important to obtain PFM measurements without a d.c. voltage applied also see Sect. 2.4 on PFM spectroscopy. We can distinguish two capacitive contributions, one due to the cantilever and one due to the tip. Regarding the constant force, F_{dc}, even though it is not detected at the measurement frequency ω, it can influence the piezoresponse as it contributes a contact force which can result in polarization changes due to the direct piezoelectric effect. However, in most cases these forces are typically quite small, usually under 1 nN, thus they can be neglected. Other contributions can arise due to Coulomb attractive forces between the tip and charges on the piezoelectric material surface, as shown in Eq. 7:

$$F_{Coulomb} = \frac{qCV}{2\epsilon_0} \tag{7}$$

where q is the surface change density. This force is estimated to be in the range of a few nN. The largest forces acting on the tip however, are the attractive forces due to Van der Waals interaction and short-range atomic repulsive forces. These forces are estimated to be in the range of 100 nN [6], thus as the sample surface moves due to the inverse piezoelectric effect, the cantilever is forced to deflect mostly due to the surface displacement.

Referring back to Eqs. 2 and 3, for the simplest case with full axial symmetry and PFM imaging of piezoelectric ceramics, we have the further simplifications $T_1 = T_2, E_1 = E_2, T_4 = T_5$ and $T_6 = 0$. Thus we obtain the following set of equations:

$$S_1 = S_2 = (s_{11} + s_{12})T_1 + s_{13}T_3 + d_{31}E_3 \tag{8}$$

$$S_3 = 2s_{13}T_1 + s_{33}T_3 + d_{33}E_3 \tag{9}$$

$$S_4 = S_5 = s_{44}T_4 + d_{15}E_1 \tag{10}$$

$$S_6 = 0 \tag{11}$$

$$D_1 = D_2 = \epsilon_{11}E_1 + d_{15}T_4 \tag{12}$$

$$D_3 = \epsilon_{33}E_3 + 2d_{31}T_1 + d_{33}T_3 \tag{13}$$

For PFM imaging Eqs. 8, 9 and 10 apply and we can distinguish two main contributions of the piezoelectric coefficients and sample polarization on the sample strain and displacement of the PFM tip: out-of-plane and in-plane displacements, as indicated in Fig. 6. If we consider just the out-of-plane electric field component we then

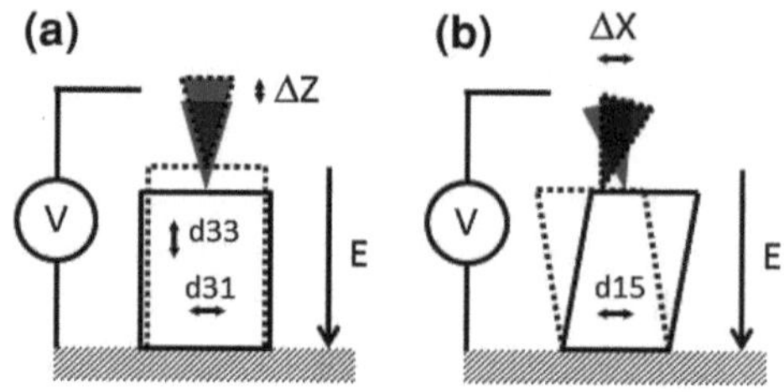

Fig. 6 Piezoelectric coefficients contributions to tip displacement for **a** out-of-plane and **b** in-plane

have two cases with orthogonal configurations of the electric polarization P: out-of-plane as in Fig. 6a and in-plane as in Fig. 6b. In the former case the out-of-plane sample strain, Eq. 9, and consequently tip displacement measured on the vertical PSD channel arises due to an effective d_{33} piezoelectric coefficient see below for a discussion of the effective d_{33} piezoelectric coefficient. In-plane strains also arise, Eq. 8, due to the d_{31} piezoelectric coefficient, however because of the axial symmetry of the problem these strains do not normally cause a lateral displacement of the tip. Moreover since the in-plane electric field, E_1, is negligible, as is certainly the case for the global excitation method, the shear strain predicted by Eq. 10 is also negligible. Thus, for out-of-plane polarization the main tip displacement is in the vertical channel and arises due to out-of-plane sample strain, S_3.

The other case is electric polarization in-plane, Fig. 6b. In this case, due to a rotation of the coordinate system in Fig. 5—90° rotation about the X axis—the matrices in Eq. 3 are also rotated. Therefore, the main component of electric field in Eqs. 8, 9, and 10 is now the E_1 component and E_3 is negligible. Thus, the sample strains S_1, S_2 and S_3 are negligible and the main tip displacement is an in-plane displacement due to the shear strain S_4 as described by Eq. 10. In the general case the electric polarization can have any orientation and in order to understand the relationship between tip displacement, sample strains and piezoelectric coefficients, a full numerical simulation based on Eq. 2 is necessary. This requires knowledge of the electric polarization orientation and methods for this are available. In the following sections we will discuss one such method which relies on combining crystallographic information obtained by electron backscattered diffraction (EBSD) with PFM.

2.3 *Effective d_{33} Coefficient*

From the previous section we can start to see some of the difficulties in quantifying PFM measurements, or even extracting the configuration of electric domain orientations. Quantifying PFM measurements becomes even more difficult in the local excitation method, since the electric field distribution tends to be non-uniform, as seen in Fig. 3a, and also dependent on the particular tip geometry. In the global excitation method and out-of-plane electric polarization we can considerably simplify our analysis and obtain some quantitative information about the sample properties.

Eqs. 8, 9, and 10 apply and as noted in the previous section the main tip displacement is a vertical displacement due to the S_3 strain. The in-plane electric field, E_1, is negligible and therefore we have $S_4 = 0$. The tip-surface contact force tends to be very small compared to the in-plane stresses in Eq. 8, and 9, and since the sample surface is free to move we can assume $T_3 = 0$. If we start our analysis for voltage applied to a large circular top electrode, with diameter larger than the film thickness as in the analysis given by Lefki and Dormans [7], the non-active part of the piezoelectric film, i.e. the film with zero electric field, tends to constrain the active part of the film, resulting in in-plane stresses. Away from the edges of the circular top electrode we then have $S_1 = S_2 = 0$ due to this constraining effect. Thus we obtain from Eq. 8:

$$T_1 = \frac{-d_{31}}{(s_{11} + s_{12})} E_3 \tag{14}$$

Substituting Eq. 14 into Eq. 9 we obtain:

$$T_1 = d_{33} - 2d_{31} \frac{s_{13}}{(s_{11} + s_{12})} = d_{33ef} \tag{15}$$

The expression in Eq. 15 is termed the effective d_{33} piezoelectric coefficient. To see the importance of this coefficient we will rewrite the expression S_3/E_3. The out-of-plane sample strain, S_3, is a ratio between out-of-plane sample deformation and sample thickness, i.e. $S_3 = \Delta t/t$. The out-of-plane electric field, E_3, can be written as $E_3 = V/t$. Thus $S_3/E_3 = \Delta t/V$. The vertical tip displacement is related to the sample deformation Δt through a calibration constant, thus for a potential of 1V, the measured sample displacement is a direct measurement of the effective piezoelectric coefficient. This relationship also holds to some extent for PFM measurements in the global excitation method due to the enhanced effective tip diameter and uniform electric field, although it becomes necessary to refer to a calibration standard or method in order to verify the validity of Eq. 15. For the local excitation method, however, it is certainly incorrect to use Eq. 15 in order to relate the measured sample deformation to the piezoelectric coefficients of the material.

2.4 PFM Calibration

The simplest type of calibration available is the cantilever sensitivity calibration. This is especially important for AFM measurements where it becomes necessary to relate the data obtained from the vertical PSD channel, measured in units of Volts, to the actual topography of the sample, measured in units of metres. As discussed in Sect. 1, for contact mode imaging at a constant deflection setpoint, or equivalently constant contact force, the cantilever deformation is directly proportional to the change in sample topography, thus we require a calibration constant to relate these two quantities. This method will be illustrated in Sect. 3. Briefly, this consists of

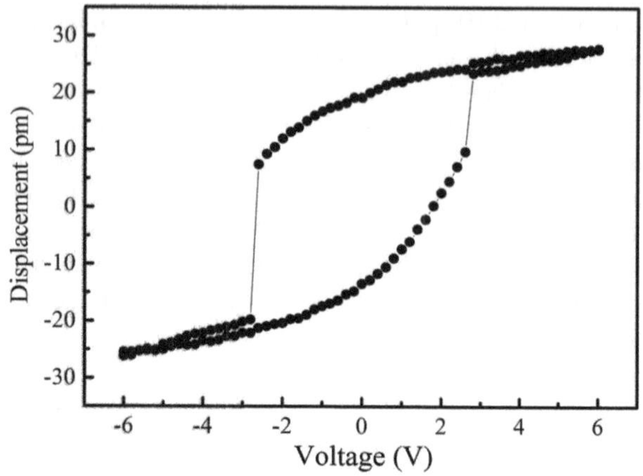

Fig. 7 Ferroelectric domain vertical displacement response loop

making contact with a very stiff surface which will resist deformation due to the tip-surface contact force, and plotting the Z piezo scanner displacement against vertical PSD deflection. After the tip has made contact with the surface this relationship is linear and its gradient is called the cantilever sensitivity.

Another closely related method involves causing cantilever deflections through vertical sample deformation rather than ramping the Z piezo stage. Typically a quartz crystal is used with electrodes deposited on opposite sides. A voltage applied between the electrodes will result in a vertical deformation dependent on the piezoelectric coefficients of the quartz crystal. The quartz sample is in turn calibrated using a traceable method such as a double-beam interferometer. Following this, the PFM tip displacement is calibrated by plotting the PSD vertical channel voltage against calibration sample voltage. This method has the advantage of allowing calibration at the frequency used in PFM measurements. As noted in Ref. [8] the frequency background due to the microscope used can contribute significantly to PFM measurements, thus it is important to take this into account by calibrating for the vertical tip displacement due to sample deformations alone.

2.5 PFM Spectroscopy

Apart from imaging the static domain configuration, the switching characteristics of electric domains can be studied by PFM. The switching of electric domains, similar to magnetic domains, is hysteretic and is characterized by a remanent polarization and coercive field amongst other features. For an example of a ferroelectric domain switching loop see Fig. 7.

In Fig. 7 the vertical displacement amplitude of a ferroelectric thin film subject to an applied d.c. voltage is plotted for both the positive and negative scanning directions of applied voltage. At zero applied voltage due to the remanent electric

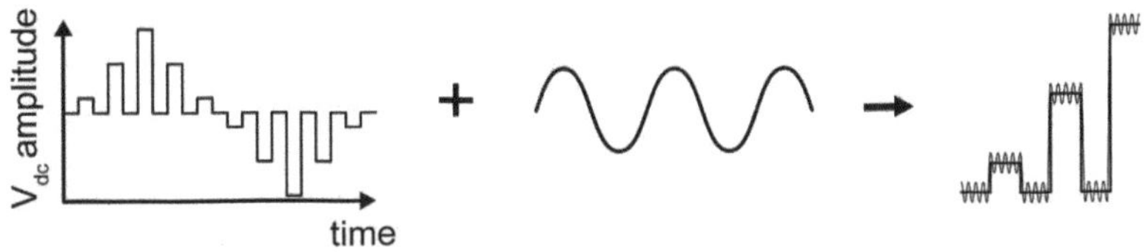

Fig. 8 Illustration of SS-PFM method. A series of voltage steps are applied and the PFM signal is measured at the detection frequency using a superimposed a.c. voltage at the point of zero d.c. voltage

polarization we obtain non-zero displacement amplitude with opposite phase (180° phase difference) on the two sides of the loop. This behaviour is due to the switching of electric polarization under the applied electric field acting in the opposite direction, hence the hysteretic behaviour observed in Fig. 7. The electric field at which the polarization, and thus the displacement amplitude, is reduced to zero is called the coercive field.

There are two main methods for studying the switching of electric domains by PFM. In one method the d.c. voltage is ramped back and forth and a superimposed a.c. voltage is used in the standard lock-in method discussed in the previous sections to obtain the tip displacement amplitude and phase. The main problem with this method is the presence of capacitive forces at the detection frequency due to the applied d.c. voltage, as seen in Eq. 6. This electrostatic interference can be significant and result in a distorted hysteresis loop, dependent on the cantilever used and even on the tip geometry. This problem can be mitigated somewhat by using stiffer cantilevers, which are less sensitive to the capacitive forces in Eq. 6, however this usually comes at the cost of decreased resolution. In order to eliminate this problem an alternative spectroscopic method was developed, termed switching spectroscopy PFM (SS-PFM) [9]. This method is illustrated in Fig. 8.

Here, rather than continuously ramping the d.c. voltage, a series of voltage steps are applied, as shown in Fig. 8, and the PFM signal is measured using a lock-in method at a given detection frequency, at the points of zero d.c. voltage. This has the advantage of eliminating any electrostatic effects, as the capacitive force at the detection frequency is reduced to zero, as seen in Eq. 6. Complications can occur however due to dynamic domain relaxation effects. In order to understand this we have to consider how the switching of polarization occurs under the PFM tip. Typically as the strength of the applied electric field becomes sufficient to switch the polarization, initially the surface of the film under the PFM tip is switched and this reversed region quickly grows vertically down through the thickness of the film [10]. Following the vertical switching the reversed domain starts to expand laterally and its final size is limited by both the strength of the electric field, the activation time and diameter of PFM tip. Domain relaxation processes may occur as the switching voltage is reduced back to zero, resulting in shrinking back of the reversed region. Thus, the measured hysteresis loop can depend on the time taken to acquire it. Indeed

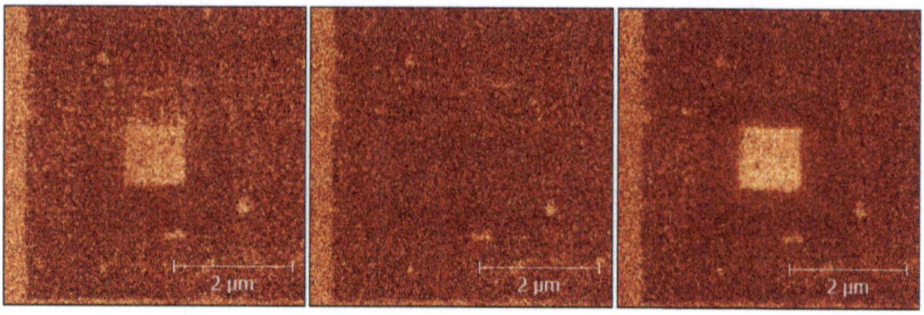

Fig. 9 Illustration of PFM lithography. From *left* to *right*, a reversed square domain is written, erased then written back

these dynamical effects can be studied by PFM in order to obtain a bigger picture of the physics governing to behaviour of materials under study [11, 12].

2.6 PFM Lithography

The ability to switch the polarization under the PFM tip this process is also referred to as poling can indeed by utilized in order to controllably pattern the domain structure in a ferroelectric thin film. Initially the sample is poled with a uniform polarization orientation, following which a defined reversed domain structure is imprinted on the material by scanning the PFM tip in a controlled pattern whilst applying a sufficiently large reversing voltage. A simple illustration of this method is shown in Fig. 9. A further development of this method allows for self-assembly of structures on the ferroelectric sample surface depending on the polarization pattern at the surface [13].

It should be noted that whilst the switching of polarization is a required feature in this case, during normal PFM imaging it is an unwanted effect. In order to increase the signal to noise ratio the simplest solution is to increase the amplitude of the excitation voltage. However, due to the coercive field of the sample, we are limited by the amount of voltage we may apply without significantly distorting the domain distribution we are trying to characterize. In the worst case damage to the sample due to excessive voltage is possible. Even below the coercive field of the material it is possible to distort the domain configuration, thus this becomes an important consideration. In order to obtain a reliable PFM image of the domain configuration it is important to perform the measurement with the minimum possible applied excitation voltage.

2.7 Imaging Using Resonance Methods

So far we have considered PFM imaging at a single excitation frequency. Whilst the mechanism for single frequency PFM is clear and due to its simple nature we can obtain quantitative information in certain cases, as discussed in Sect. 2.3, the main drawback is a relatively poor signal to noise ratio. Indeed, some samples have such a small piezoresponse that PFM imaging becomes almost impossible. Whilst increasing the amplitude of the excitation voltage is a simple remedy, there is a limit to this approach as discussed in the previous section.

Another method relies on the use of cantilever resonances. Due to the deflection and restoring forces acting on the tip and cantilever we have a harmonic oscillator formed which is characterized by a number of resonance frequencies. At these resonance frequencies the cantilever motion is greatly amplified, increasing its sensitivity to dynamic sample deformations and therefore increasing the signal to noise ratio. The amplitude and phase of cantilever motion around the resonance frequency, ω_0, is shown in Eqs. 16, and 17 where A_{max} is the deflection amplitude at resonance and Q is the quality factor of the harmonic oscillator configuration [14].

$$A(\omega) = \frac{A_{max}\frac{\omega_0^2}{Q}}{\sqrt{(\omega_0^2 - \omega^2)^2 + \left(\frac{\omega_0\omega}{Q}\right)^2}} \tag{16}$$

$$\tan\phi(\omega) = \frac{\omega_0\omega}{Q(\omega_0^2 - \omega^2)} \tag{17}$$

At the resonance frequency the motion of the cantilever is amplified by the quality factor Q, which can increase the response by one or even two orders of magnitude. In the simplest case we can obtain PFM images at a fixed frequency, which was determined to be the resonance frequency prior to acquiring the PFM image. The problem with this approach is the variation in resonance frequency depending on the local tip-surface contact conditions and frictional forces. Thus, with the single frequency resonance imaging approach the image quality can vary greatly within the same scan. Two methods have been developed to solve this problem, the dual a.c. resonance tracking (DART-PFM) [15] and band-excitation PFM (BE-PFM) [16] methods.

With DART-PFM two excitation frequencies are used, ω_1 and ω_2, one slightly below the resonance frequency and the other slightly above. The sum of these two voltage signals is applied through the cantilever to the sample. The resulting cantilever response due to the inverse piezoelectric effect is analysed using two lock-in amplifiers, one at the reference ω_1 and the other at the reference ω_2. We obtain two sets of amplitude and phase outputs from the two lock-in amplifiers, A_1, ϕ_1 and A_2, ϕ_2 respectively. The difference term $A_2 - A_1$ is a measure of the change in resonance frequency and can be used in order to adjust the excitation frequencies ω_1 and ω_2, thus maintaining a consistent image quality.

The other possibility is to use the BE-PFM method. Here, a frequency domain boxcar function is used such that the resonance frequency and its range of variation are encompassed. Boxcar functions in the frequency domain are generated by sinc or chirp time domain excitation functions. These are applied as voltages to the sample through the cantilever and the response of the cantilever is recorded in the time domain. Using a suitable physical model of the cantilever and sample configuration, such as the simple harmonic oscillator model, the response of the cantilever is fitted in order to extract the amplitude and phase at the resonance frequency, amongst other parameters. Thus by encompassing the range of variation of resonance frequency, the resulting BE-PFM images are not affected by changes in resonance as for single frequency PFM at the contact resonance. Indeed, the change in resonance frequency and determination of Q-factor of the cantilever-sample system can be used to obtain additional information about sample properties, such as energy dissipation [16].

2.8 Vector PFM

The vertical and lateral PFM response may be combined to obtain a full mapping of the polarization vectors in the sample, termed vector PFM (V-PFM) [17]. Recalling our discussion in Sect. 2.2, for out-of-plane polarization we have vertical tip displacement due to the effective d_{33} piezoelectric coefficient whilst for in-plane polarization we have horizontal tip displacement due to the shear strain of the sample as a result of the d_{15} piezoelectric coefficient. The horizontal tip displacement results in a torque on the cantilever which is detected as a lateral signal on the PSD. This may also be calibrated in much the same way as the vertical sensitivity is calibrated as discussed in Sect. 2.3. In the general case the electric polarization direction and in order to fully characterize it three components are required: the vertical component, VPFM, and two lateral components, x-LPMF and y-LPFM. The two lateral components are obtained in two separate measurements of the same area by rotating the sample with respect to the cantilever by 90 degrees. It is not sufficient to simply change the scanning angle in order to obtain two orthogonal components of the lateral displacement. This approach is called 3D-PFM as it allows in principle the mapping of the polarization vector in any given direction. By combining just two components, VPFM and LPFM which are obtained simultaneously in a single scan, the projection of the polarization vector onto a plane may be obtained, termed 2D-PFM.

2.9 EBSD-PFM

In order to obtain quantitative information from the measured sample strains, it is necessary to not only know the generated electric fields but also know the crystal structure and orientation of the various grains in the sample under measurement. This information can then be used in Eq. 2 in order to reproduce the PFM images using finite element simulations. For epitaxially grown single crystal samples this task is fairly straightforward as the crystal structure and orientation can be easily checked using x-ray diffraction measurements after sample growth. For polycrystalline samples this task becomes considerably more difficult and a method capable of characterizing the individual grains is necessary. This can be achieved using electron back-scatter diffraction (EBSD) [18]. EBSD is an SEM-based method where

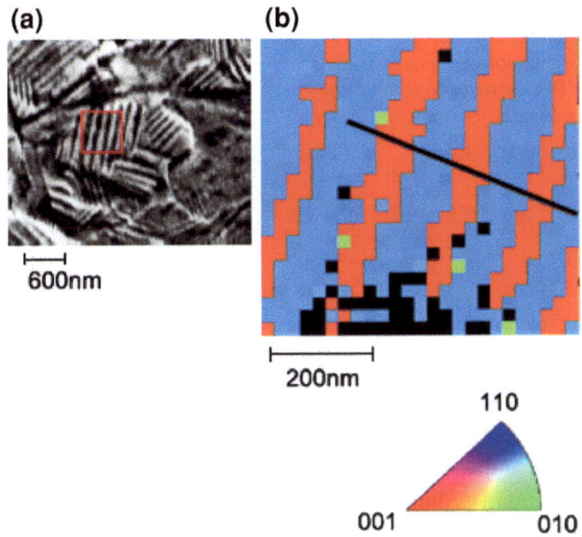

Fig. 10 The region selected for subsequent analysis: **a** SEM image with the EBSD scan area marked in *red* and **b** EBSD orientation map. The legend line shows the location of the orientation profile shown in Fig. 11

the diffraction patterns of electrons incident on the crystallographic planes of the sample are recorded. These are called Kikuchi patterns and they can be indexed in order to obtain the local crystallographic structure and orientation, [18]. Thus, combining EBSD and PFM polycrystalline samples may be analysed. This technique was recently demonstrated at the National Physical Laboratory [19] where by combining textural analysis, through electron backscattered diffraction, with piezoresponse force microscopy, quantitative measurements of the piezoelectric properties can be made at a scale of 25 nm, smaller than the domain size, see Fig. 10. A domain structure closely matching the topography in the secondary electron image is clearly seen in the EBSD image. The line profile across the domains shown in Fig. 11 confirms that these domains are related by a domain boundary at an angle very close to 90° measured value = 89.3° as would be expected for a tetragonal ferroelectric. The combined technique was used to obtain data on the domain-resolved effective single crystal piezoelectric response of individual crystallites in $Pb(Zr_{0.4}Ti_{0.6})O_3$ ceramics. These and similar results offer insight into the science of domain engineering and provide a tool for the future development of new nanostructured ferroelectric materials for memory, nanoactuators, and sensors based on magnetoelectric multiferroics.

3 PFM Imaging Tutorial

One of the roles of this book is to provide the experimentalist routes into performing the measurement methods described in each chapter, practically in their own laboratory. Hence, in this section, a tutorial on PFM imaging is given, illustrating some of

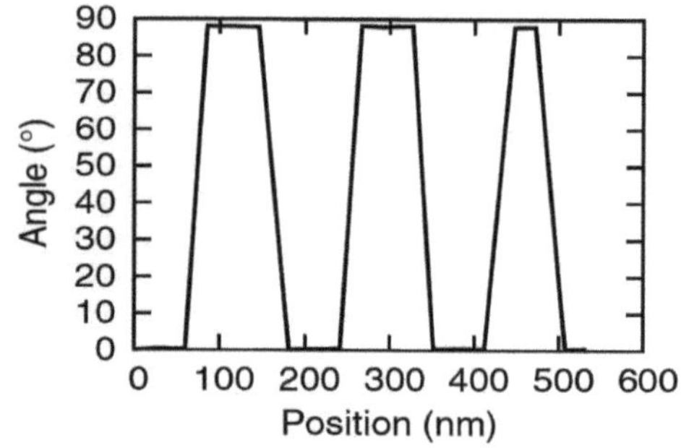

Fig. 11 Line-scan of the relative crystallographic orientations measured using EBSD across the *legend line* in Fig. 10

the issues discussed in the previous sections. The instrument used was the Dimension ICON Scanning Probe Microscope. Two types of PFM probes were used, the "SCM-PIT" and "SCM-PIC". Both have doped Si cantilevers with Pt/Ir tip coatings. The nominal tip diameter is 40 nm. The SCM-PIC has a very low stiffness, around 0.1 N/m, thus it is suitable for contact mode measurements even for soft samples, whilst the SCM-PIT has a higher stiffness, around 3 N/m, thus it is designated as a tapping mode probe. In the following sections we will illustrate the setting up a PFM session including calibrations, imaging using both resonance and off-resonance, discussion of artefacts and poling.

3.1 PFM Setup

Setting up a PFM measurement session is similar to contact mode AFM imaging setup. The SPM probe is placed on the scanner and the laser beam is aligned to reflect off the cantilever as shown in Fig. 12. Here the cantilever is shown in the middle of the image, with the laser spot reflecting from the top part. The tip is underneath the cantilever, its position indicated by the cursor. Initially the position of the laser spot is adjusted to obtain maximum total signal, indicated by the scale bar in the right in Fig. 12. The next step is to adjust the position of the laser spot on the PSD in order to reduce both the vertical and lateral signal as close as possible to zero, as indicate in the right of Fig. 12. The alignment procedure is typically done above a reflecting surface in order to allow an image of both the cantilever and laser spot to be obtained.

After alignment, the next step is to calibrate the cantilever sensitivity. This is most easily done by using the force curve approach. A typical force curve is shown in Fig. 13. This is composed of two branches, the retract and approach curves. Here we plot the deflection error, which is linearly related to the actual vertical cantilever displacement, versus the Z scanner position.

Initially, with the tip far away from the sample surface, there is no cantilever displacement, thus the deflection error remains constant. Following the red curve, as we

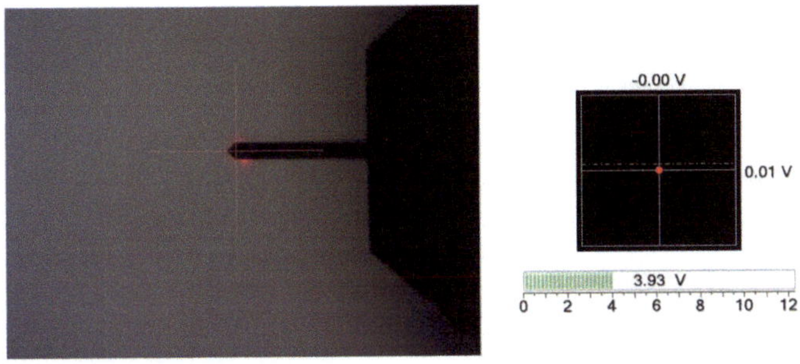

Fig. 12 Laser alignment showing laser spot on PFM cantilever and PSD signal, including total intensity, *horizontal* and *vertical* signals

Fig. 13 *Force curve* for SCM-PIC probe on Al_2O_3 substrate showing approach and retract

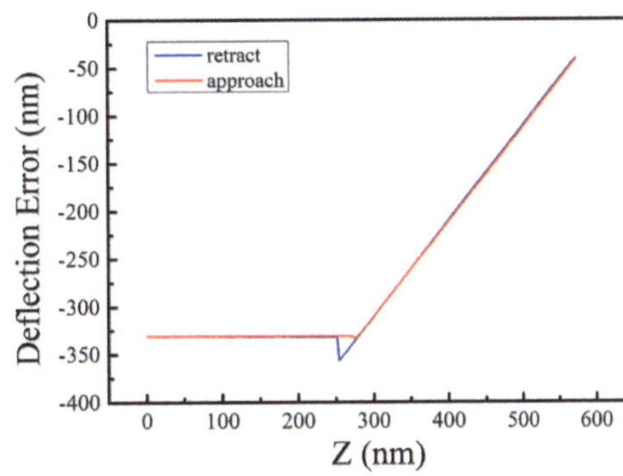

approach the substrate, the tip is suddenly pulled in due to attractive forces between the tip and the substrate surface, causing a slight cantilever deflection towards the substrate. After the tip has made contact, further increasing the Z scanner position results in a linear deflection of the cantilever and the inverse of the slope of this curve gives the vertical sensitivity of the cantilever. On the retract curve attractive forces between tip and sample surface tend to retain the tip causing a negative cantilever deflection and eventually the tip snaps off the surface as the Z scanner position is further reduced. An alternative method to vertical cantilever sensitivity calibration is to use either a calibrated piezo stack or quartz crystal, similar to the method discussed in Sect. 2.4, with a known displacement amplitude, in order to cause a known vertical cantilever displacement. As discussed in Sect. 2.4 this method can also be used for lateral sensitivity calibration. For the force curve approach a hard substrate surface should be used in order to minimize any indentation of the tip into the sample surface. A suitable substrate for this is Al_2O_3.

In order to check the cantilever properties it is also useful to perform a thermal tune of the cantilever. With this method the cantilever is held above the surface and oscillated with fixed amplitude in a range of frequencies. At certain frequencies the oscillation amplitude is enhanced due to resonance, as shown in Fig. 14, and by using a geometrical model of the cantilever the stiffness constant may be obtained

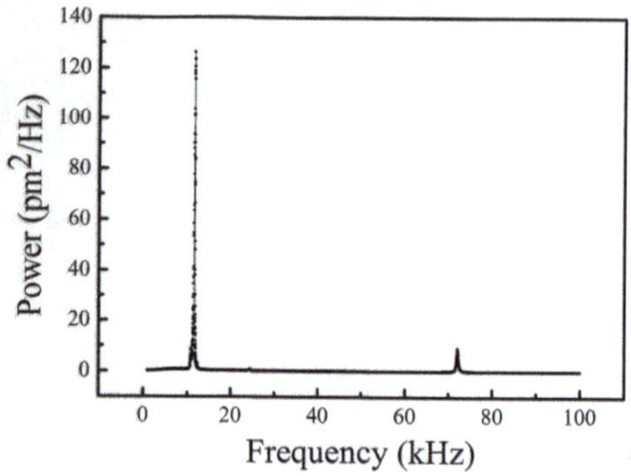

Fig. 14 Thermal tune for SCM-PIC. The main free resonance peak is observed at around 12 kHz

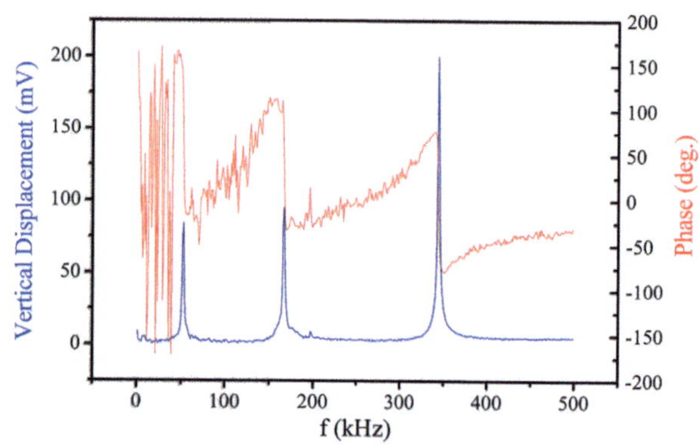

Fig. 15 Contact resonances for SCM-PIC on PZT thin film, showing both the vertical displacement *amplitude* and *phase* as a function of frequency

as detailed in Ref. [20]. This may then be compared with the quoted manufacturer cantilever stiffness.

3.2 PFM Imaging

For this tutorial we are going to obtain images of a high quality epitaxial thin film sample. The sample is 100 nm thick tetrahedral 20–80 PZT on single crystal $SrTiO_3$ (STO) substrate. The STO substrate is doped using 1 % Nb which makes it electrically conductive. This allows the use of the STO substrate directly as a back electrode. In order to increase the signal to noise ratio we are going to obtain PFM images at resonance. The cantilever resonances on the PZT thin film up to 500 kHz are plotted in Fig. 15. Here the tip is placed in contact with the PZT surface and a voltage with fixed amplitude is applied between the tip and back electrode, varying the frequency. The resulting vertical displacement amplitude and phase are plotted in Fig. 15.

The most pronounced resonance peak is the third peak observed at around 350 kHz. After choosing the operating frequency in the centre of the third reso-

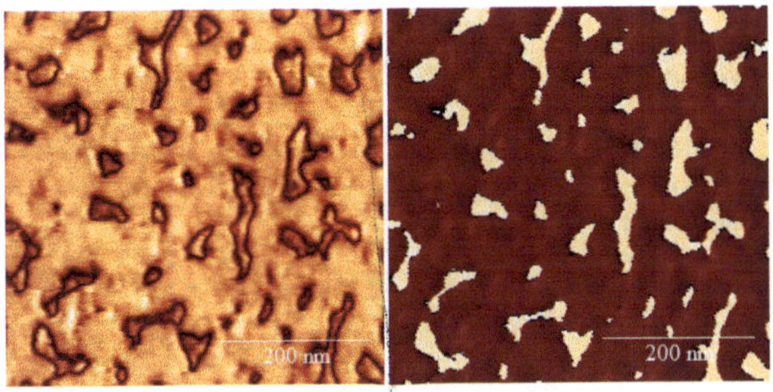

Fig. 16 PFM image at the first resonance peak of as-grown epitaxial PZT thin film showing, from *left* to *right*, amplitude and phase components

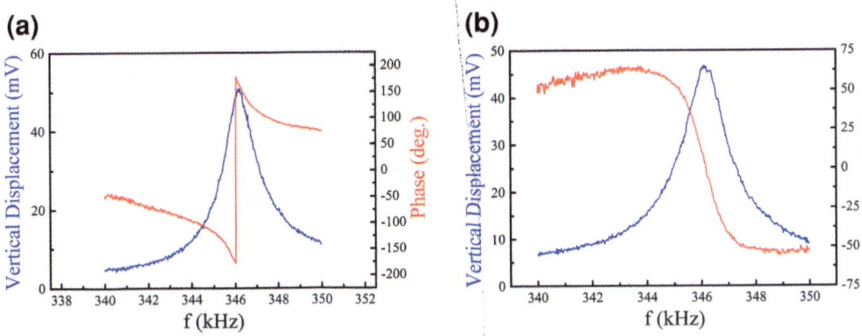

Fig. 17 Resonance peaks for two out-of-plane domains, **a** c+ and **b** c− domains

nance peak, PFM images are obtained as shown in Fig. 16. Here the amplitude, left and, phase, right are plotted. The resulting domain pattern shows a series of c+ and c− domains, characterized by a similar amplitude response and a change in phase between them. The domain boundaries show a decrease in the response amplitude, as discussed in Sect. 2.

It is interesting to investigate the change in resonance peaks between these two types of domains. As shown in Fig. 17, the amplitude peak is the same for both c+ and c− domains, however the phase response changes. This is due to the different response of the c+ and c− domains, in particular the shift in phase between the piezoresponse and excitation voltage.

To illustrate the difference between PFM imaging at resonance and off-resonance several PFM images are shown in Fig. 18, taken from the same area but with different excitation frequencies. Figure 18a, b are both taken at resonance, the first and third resonance peaks respectively. The quality of these images is comparable, although a better contrast is observed for imaging at the third resonance peak. On the other hand, when imaging off-resonance, for example at 25 kHz, as shown in Fig. 18c, the quality is seen to drop significantly. For this particular sample, imaging off-resonance is not sufficient to obtain a surface domain pattern. On the other hand, off-resonance

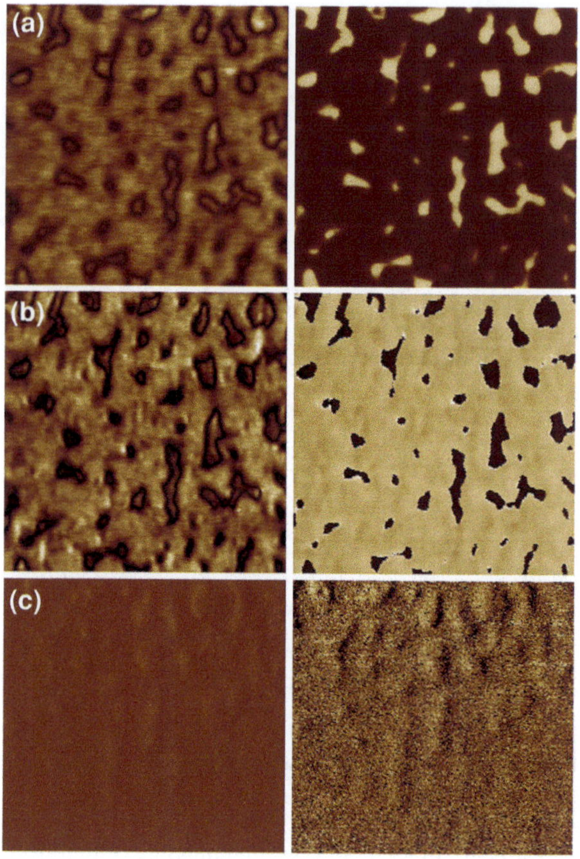

Fig. 18 PFM images at different frequencies showing amplitude on the *left* and phase on the *right* for **a** 3rd resonance peak at 345 kHz, **b** 1st resonance peak at 53 kHz and **c** off-resonance at 25 kHz

PFM measurements have the advantage of allowing quantitative measurements under certain conditions, as discussed in the previous section.

3.3 Poling

As an example of sample poling, we perform the following simple experiment. Initially the PZT thin film is scanned using a voltage offset well above the coercive voltage. The poled sample is shown in the first image in Fig. 19. Here the PFM phasee scan is shown after poling using +5 V. Next the tip is held in the centre of the scanning area and a 1 ms voltage pulse of −5 V amplitude is applied. The resulting PFM amplitude image is shown in the second scan in Fig. 19. The area underneath the tip has been reversed. Following this further pulses are applied. The reversed domain is seen to increase as more reversal voltage pulses are applied. Eventually

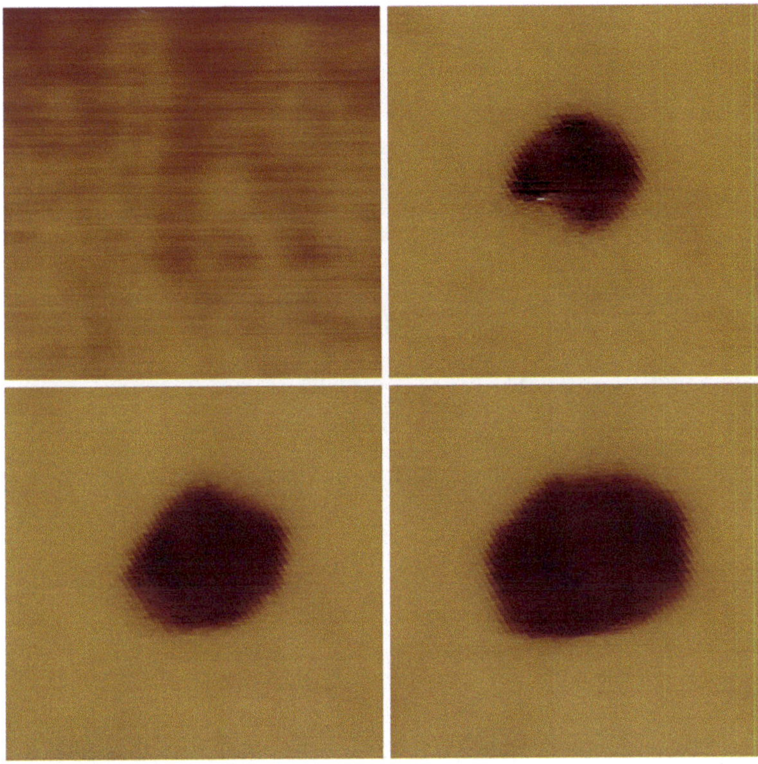

Fig. 19 PFM phase images showing reversal of domain under tip. Initially sample is poled uniformly then voltage pulses of fixed amplitude are applied. After each pulse a PFM image is taken. The reversed domain grows as more pulses are applied and eventually reaches a maximum dimension dependent on the tip diameter

the reversed domain reaches a maximum dimension dependent on the tip diameter, as would be expected.

3.4 Setpoint Variation and Imaging Artefacts

Choice of deflection setpoint, or equivalently tip-surface contact force, can have a significant effect on imaging quality. Also using a large voltage amplitude to obtain PFM images can result in domain distortions. The effect of different deflection setpoints and large excitation amplitudes are illustrated in Fig. 20.

The series of images in Fig. 20, from left to right and top to bottom show PFM phase scans taken of nominally the same area, using a 3 V excitation amplitude and increasing tip-surface contact force. Two effects are observed: the imaged domains change in shape and size on subsequent scans and their relative position is seen to

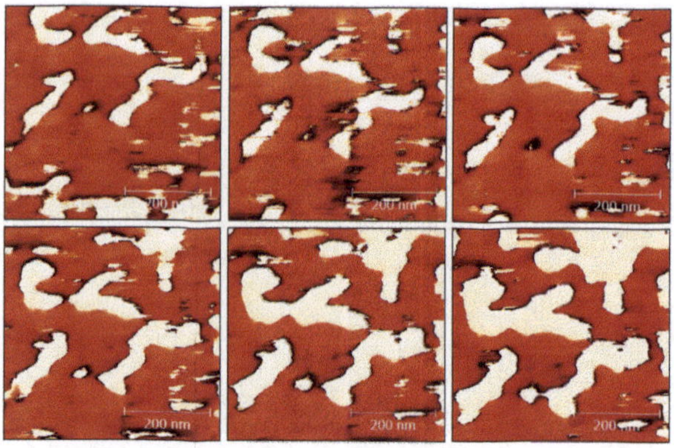

Fig. 20 PFM phase images showing effect of different deflection setpoints and large excitation amplitude. From *left* to *right*, *top* to *bottom* consecutive images are taken with decreasing deflection setpoint. A shift in scan position is observed due to the different deflection setpoint and a growth of domains due to the large excitation amplitude

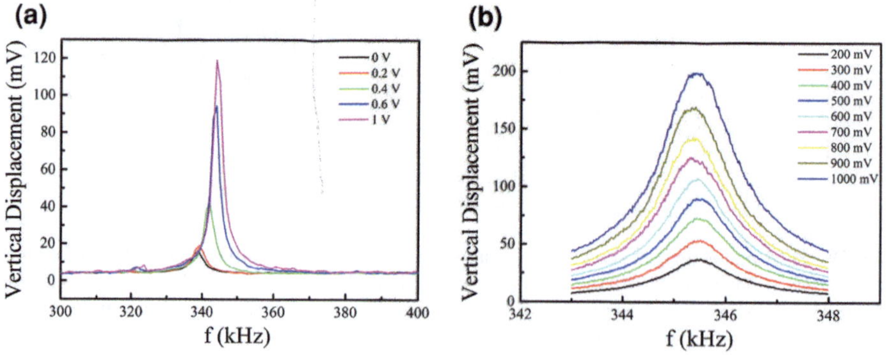

Fig. 21 Changes in resonance frequency due to variation in **a** contact force and **b** excitation amplitude

shift. The change in domain structure is due to the large imaging voltage amplitude used, resulting in movement of domain walls as the tip is dragged across the sample surface. Also the relative position of all the domains is seen to shift. As the tip-surface contact force is increased the cantilever is deflected and the tip slides across the surface. This results in a shift in the imaging area, as observed in Fig. 20.

Changing the deflection setpoint and imaging voltage amplitude also changes the position of the resonance peak, as shown in Fig. 21, thus for single-frequency resonance imaging the excitation frequency must be adjusted depending on the tip-surface contact force and excitation amplitude. For the sample imaged here, single-frequency resonance imaging can be used, as the sample surface is of very high quality, with a small topography variation. For rough samples, variations in tip-surface contact conditions can result in significant variations of resonance frequency,

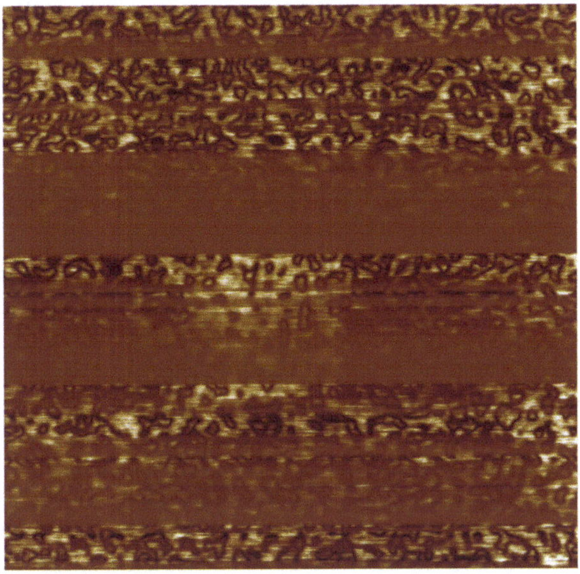

Fig. 22 PFM amplitude image showing artefacts due to change in tip-surface contact conditions resulting in changes in resonance frequency

making single-frequency imaging inadequate. In this case either DART-PFM or BE-PFM should be used. To illustrate the effect of varying resonance frequency, a long duration scan is taken for Fig. 22. Changes in resonance frequency result in some areas being blurred as seen in Fig. 22.

3.4.1 Surface Contamination

Surface contamination comes from a variety of sources including initial film formation, absorbed water from the surrounding environment, and even conductive layers deposited by contact with other objects and oxidation. Often it is stated that there is no surface preparation required before PFM is carried out [21] with most experiments being undertaken at atmospheric pressures and humidity.

Desheng [22] proposed that one of the reasons why the measured piezo coefficients, using PFM, were so low was because an ultra-thin air gap could exist between the tip and the sample. At nanometre scales this could have a noticeable affect on the E-field. An alternative explanation is that, as most of these experiments are operated in atmospheric conditions, absorbed water fills any space between tip and sample creating a meniscus on the tip, introducing a thin dielectric layer between tip and sample. PFM is ideally suited to explore these issues because; by changing the tip force interaction (via changing the bias voltage on the AFM cantilever and scanning the tip) one is able to 'wipe' successive layers of contaminant material from the surface of the ferroelectric thin film.

Several papers deal with surface contamination issues (such as [23], and in this section we explore the effects of surface environmental chemical stability on a range of sol-gel derived ferroelectric thin films deposited onto two substrate types:

- ITO/Glass, coated with PZT (30, 70) at 210 nm thickness, which formed a rosette like structure surrounded by an amorphous matrix on the surface [24]
- Pt/Ti/SiO2/Si, coated with PZT (30, 70) at 200 nm thickness and formed a very fine grain structure [25].

On each sample, one corner of the PZT was carefully scraped off using a scalpel and fine wire anchored in place with conductive epoxy resin. This enabled the bottom electrode to be grounded during experimentation. In the experiments described in this section, no separate top electrodes were deposited onto the ferroelectric thin film.

The AFM was configured for PFM operation, utilising a digital lock-in amplifier and signal generator, as described in this chapter. The grounding wire of the sample was connected to the ground of the signal generator output. The output from the signal generator was set to a frequency of 18 kHz and amplitude of 4 Vpk-pk. For all the samples used, an initial scan of $25 \times 25\ \mu$m at 0V deflection set-point was undertaken. On completion of the initial scan an area of interest was selected, and a series of scans then followed, using an initial deflection set point of 0V and ending with 6V, in increments of 1 V. This had the effect of increasing the force between the Si-AFM tip and the ferroelectric thin film, see above sections. Two sets of materials were investigated. One set was several years old that had been stored in a normal laboratory environment. The second set was the same material type but had been cleaned using a standardized si-wafer cleaning process. The effect of removal of surface contamination on PFM image contrast was then established.

Un-cleaned (aged) samples The first scans at 0V deflection set point (low tip force) resulted in poor PFM image contrast. Increasing the tip force resulted in an improvement in image contrast up to a certain level beyond which the contrast did not improve noticeably. The difference between the contained area and the scrubbed area can be seen clearly in Fig. 23. The image on the left of the figure is a topographic AFM image and the image on the right is the PFM image, where bright regions indicate a high degree of piezoelectric induced strain.

The magnified rosette in Fig. 23ii shows how the demarcation between the scrubbed and un-scrubbed areas affects the contrast. The top half of the rosette is in the scrubbed and the bottom in the un-scrubbed areas. The initial scans for the contaminated samples gave an improved contrast as the tip force increased. However, the improvement with tip force reached a point of saturation beyond which no further improvements were observed. In addition, on reducing the tip force the contrast did not diminish but stayed much the same. The initial improvement in image quality is therefore presumed to be due to the thick contamination layer being scrubbed from the sample surface by the scanning tip. This either allowed the tips electrical field to make better contact to the ferroelectric material and/or allowed an enhanced mechanical coupling between tip and surface (resulting in a better measurement of

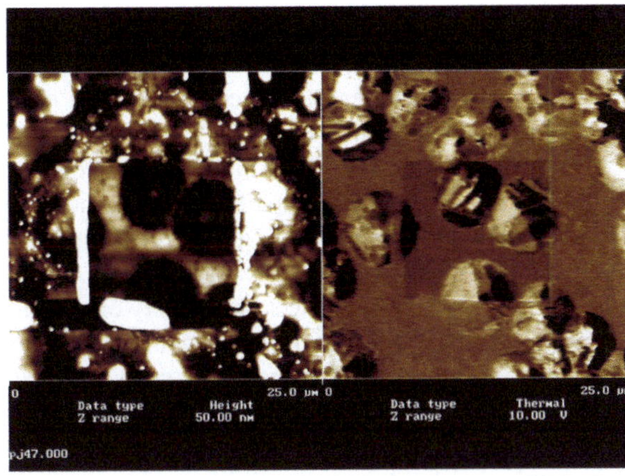

Fig. 23 **i** PFM Image after surface has been scrubbed clean by tip and inset **ii** zoomed in area across the scrubbed un-scrubbed interface

resultant piezo-strain). Both improvements would result in a clearer higher contrast PFM image.

The processing of the ferroelectric thin films is known to leave a residue of a surface contamination layer of lead oxide and lead hydroxy-carbonate on all the samples. When manufacturing the thin films by the sol-gel method it is normal to add excess lead to the mixture in order to guarantee that there are enough lead atoms to fill the perovskite structure (loss of volatile lead is a known challenge affecting processing of lead based perovskites). The top surface of the lead oxide film will eventually react with carbon dioxide to form a thin layer of lead carbonate, which, when exposed to water, will also form lead hydroxy-carbonate. Thermodynamic analysis carried out using NPLs MTDATA (http://www.npl.co.uk) software shows that only trace amounts (10^{-8}atm) of CO_2 are required for the formation of lead carbonate. The lead oxide/carbonate contaminant layers are masking the piezoresponse of the thin films. The masking effect could be either electrical and/or mechanical. Mechanically, the layer of lead carbonate can be thought of as a hard crust on top of a softer layer of lead oxide. This layer would act as a buffer allowing only a certain amount of coupling between the PFM tip and the piezoresponse of the film, reducing the effective contrast of the images obtained. The initial improvement in scan quality would be due to the PFM tip scrubbing the layers of organic contamination from the top of the lead carbonate layer. The tip would stop scrubbing at the lead carbonate layer and this would mean that the contrast obtained from scans would be the same no matter what tip pressure was applied. Electrically, both the inhomogeneous E-field due to the PFM scanning tip and the dielectric properties of the lead oxide/lead carbonate layers would influence the observed piezoresponse during PFM. It has been observed that due to the shape of the PFM tip the E-field would be inhomogeneous and that the greatest flux density would be within the first few nanometres of the film [26, 27]. The thickness of the contamination layer would therefore dictate what strength of E-field the ferroelectric film would experience. If the contamination is thick then the ferroelectric will experience a greatly reduced

E-field compared to that of one with a very thin contamination layer. Finite element models can be set-up to examine the effect of the PFM tips inhomogeneous E-field within the film. The different dielectric properties of the contamination layers would also add to this variation in E-field.

In this section, we have shown that the condition of the ferroelectric sample surface has a great impact on the quality of image obtained using Piezo Force Microscopy. It was also found that some samples gave a PFM contrast that was tip to surface pressure dependent and others tip pressure independent. By testing various samples of thin film PZT on different substrates of a variety of ages it was shown that age, handling and fabrication methods has an effect on the PFM image quality. During investigation it was found that samples that had been extensively handled since fabrication showed a distinct layer of contamination on their surfaces. This contamination had the effect of reducing the effective piezoresponse of the thin film possibly due to a masking of the applied E-field from the PFM tip. The contamination was probably organic and inorganic material from sources handled prior to the sample being touched. This layer was easily scrubbed away with the AFM tip operated at a high tip to surface pressure, the resulting PFM image becoming clearer and showing a contrast that became tip pressure independent.

3.5 Tip Deterioration

Because PFM imaging is a contact mode SPM method, after repeated scans, especially if large tip-surface contact forces are used, tip deterioration can become a significant problem. The tip geometry used for the work in this chapter is shown in the SEM image of Fig. 24. This image was taken with the cantilever tilted by about 20° in order to see the tip in profile. The tip has a tetragonal structure with a small sphere on top, with diameter of 40 nm. An SEM image of a new tip, taken head-on is shown in Fig. 25a. In Fig. 25b an SEM image of a used tip is shown after 1 day of PFM scanning work. Initially the tip has a 40 nm diameter, however after repeated scans this is seen to increase to around 120 nm due to deformation of the sphere on the imaging tip.

The deterioration of the imaging tip will result in a loss of resolution compared to a new tip. In some cases, especially after many days of scanning using the same tip, the deterioration can become so pronounced, the electric field distribution produced at the tip is completely changed. As an example, Fig. 26 shows a PFM image after poling, obtained using a heavily used tip under large tip-surface contact forces. The tip imprint in Fig. 26 may be compared to that shown in Fig. 19 for a good tip. In the case of Fig. 26, the tip has deteriorated to the point that the central region is no longer reversed, indicating a highly non-uniform electric field distribution. Consideration of the tip quality is important for quantification of PFM measurements.

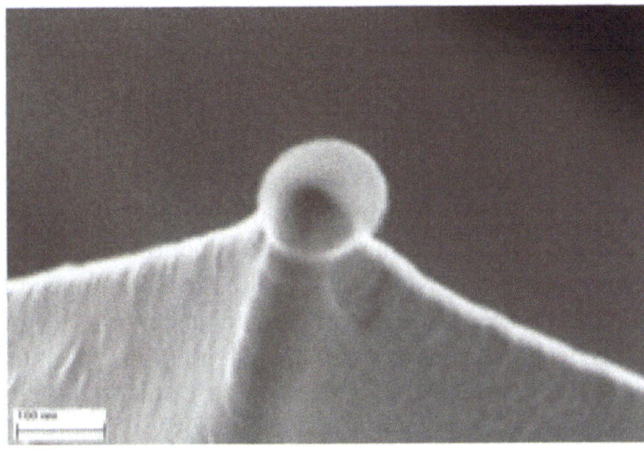

Fig. 24 SEM image of a tilted SCM-PIT tip used for PFM imaging

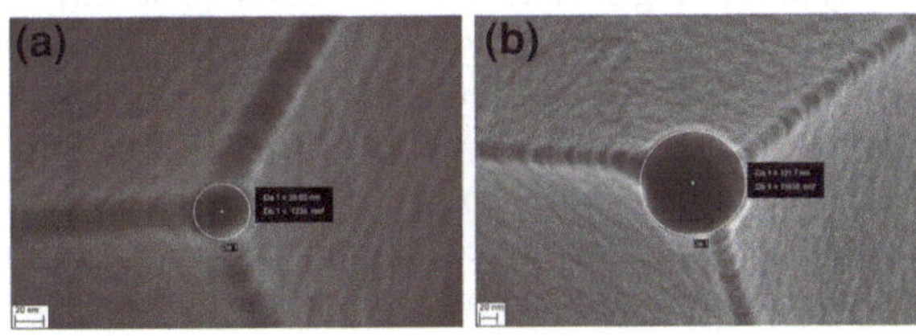

Fig. 25 SEM images of SCM-PIC tips, **a** new and **b** used

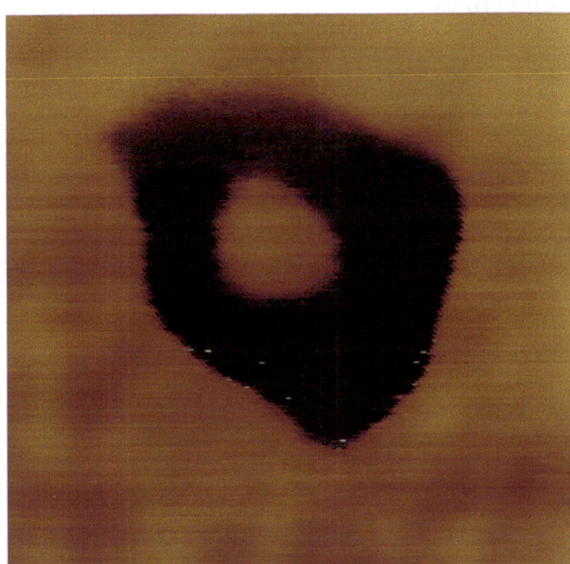

Fig. 26 PFM phase image after domain reversal under a heavily used SCM-PIT tip

4 Conclusions

In this chapter the piezoresponse force microscopy method of characterizing piezoelectric thin films was discussed. PFM is currently the only available tool for characterizing the surface electric domain structure of piezoelectric thin films with nanometric resolution. During the past two decades there have been many advances in PFM imaging, particularly the introduction of resonance imaging techniques, scanning spectroscopy studies as well as more advanced PFM modes, including vector PFM and EBSD-PFM. In order to advance the usefulness of this method even further it has become necessary to establish methods for obtaining quantitative information using PFM imaging. For piezoelectric materials their response is dictated by the piezoelectric coefficients, and the ability to precisely measure these coefficients at the nanoscale using PFM measurements will have important implications for a large number of applications and devices, including ferroelectric random access memory, energy harvesting devices, micro actuators, microwave phase shifters as well as opening up the possibility of designing completely new devices. To achieve this aim it is important for PFM measurements to be standardized and the interactions between PFM tips and piezoelectric sample surfaces to be fully understood. The introduction of PFM-based methods capable of obtaining not only information about the sample surface but also volumetric information, as well as the design of new standardized cantilevers and tips could be an important step in this direction.

References

1. Binnig, G., Rohrer, H.: Scanning tunnelling microscopy. Helv. Phys. Acta **55**, 726–735 (1982)
2. Binnig, G., Rohrer, H., Gerber, C., Weibel, E.: 7 × 7 reconstruction on si/111/resolved in real space. Phys. Rev. Lett. **50**, 120–123 (1983)
3. Binnig, G., Rohrer, H.: Scanning tunneling microscopy. Surf. Sci. **126**(1), 236–244 (1983)
4. Colton, R.J.: Procedures in Scanning Probe Microscopies. Wiley, New York (1998)
5. Holterman, J., Groen, P.: An Introduction to Piezoelectric Materials and Components. Stichting Applied Piezo, Apeldoorn (2012)
6. Rabe, U., Janser, K., Arnold, W.: Vibrations of free and surface-coupled atomic force microscope cantilevers: theory and experiment. Rev. Sci. Instrum. **67**(9), 3281 (1996)
7. Lefki, K., Dormans, G.: Measurement of piezoelectric coefficients of ferroelectric thin films. J. Appl. Phys. **76**(3), 1764–1767 (1994)
8. Jungk, T., Hoffmann, Á., Soergel, E.: Quantitative analysis of ferroelectric domain imaging with piezoresponse force microscopy. Appl. Phys. Lett. **89**(16), 163507 (2006)
9. Jesse, S., Baddorf, A.P., Kalinin, S.V.: Switching spectroscopy piezoresponse force microscopy of ferroelectric materials. Appl. Phys. Lett. **88**(6), 062908 (2006)
10. Kalinin, S.V., Gruverman, A., Bonnell, D.A.: Quantitative analysis of nanoscale switching in srbi[sub 2]ta[sub 2]o[sub 9] thin films by piezoresponse force microscopy. Appl. Phys. Lett. **85**(5), 795 (2004)
11. Tybell, T., Paruch, P., Giamarchi, T., Triscone, J.M.: Domain wall creep in epitaxial ferroelectric Pb(Zr0.2Ti0.8)O3 thin films. Phys. Rev. Lett. **89**(9), 097601 (2002)
12. Gruverman, A., Rodriguez, B.J., Dehoff, C., Waldrep, J.D., Kingon, A.I., Nemanich, R.J., Cross, J.S.: Direct studies of domain switching dynamics in thin film ferroelectric capacitors. Appl. Phys. Lett. **87**(8), 082902 (2005)

13. Kalinin, S., Gruverman, A.: Scanning Probe Microscopy: Electrical and Electromechanical Phenomena at the Nanoscale. Springer, New York (2007)
14. Sader, J.E.: Frequency response of cantilever beams immersed in viscous fluids with applications to the atomic force microscope. J. Appl. Phys. **84**(1), 64 (1998)
15. Rodriguez, B.J., Callahan, C., Kalinin, S.V., Proksch, R.: Dual-frequency resonance-tracking atomic force microscopy. Nanotechnology **18**(47), 475504 (2007)
16. Jesse, S., Kalinin, S.V., Proksch, R., Baddorf, A.P., Rodriguez, B.J.: The band excitation method in scanning probe microscopy for rapid mapping of energy dissipation on the nanoscale. Nanotechnology **18**(43), 435503 (2007)
17. Kalinin, S.V., Rodriguez, B.J., Jesse, S., Shin, J., Baddorf, A.P., Gupta, P., Jain, H., Williams, D.B., Gruverman, A.: Vector piezoresponse force microscopy. Microsc. Microanal. **12**(03), 206 (2006)
18. Engler, O., Randle, V.: Introduction to Texture Analysis: Macrotexture, Microtexture, and Orientation Mapping, 2nd edn. Taylor & Francis, New York (2010)
19. Burnett, T.L., Weaver, P.M., Blackburn, J.F., Stewart, M., Cain, M.G.: Correlation of electron backscatter diffraction and piezoresponse force microscopy for the nanoscale characterization of ferroelectric domains in polycrystalline lead zirconate titanate. J. Appl. Phys. **108**(4), 042001 (2010)
20. Green, C.P., Lioe, H., Cleveland, J.P., Proksch, R., Mulvaney, P., Sader, J.E.: Normal and torsional spring constants of atomic force microscope cantilevers. Rev. Sci. Instrum. **75**(6), 1988 (2004)
21. Lehnen, P., Dec, J., Kleemann, W.: Ferroelectric domain structures of PbTiO3 studied by scanning force microscopy. J. Phys. D: Appl. Phys. **33**, 1932 (2000)
22. Fu, D., Suzuki, K., Kato, K.: Local piezoelectric response in bismuth-based ferroelectric thin films investigated by scanning force microscopy. Jpn. J. Appl. Phys. **41**(Part 2-10A), L1103–L1105 (2002)
23. Cain, M.G., Dunn, S., Jones, P.: The measurement of ferroelectric thin films using piezo force microscopy. In: Laudon, M., Romanowicz, B. (eds.) Technical Proceedings of the 2004 NSTI Nanotechnology (2004)
24. Roy, S.S., Gleeson, H., Shaw, C., Whatmore, R.W., Huang, Z., Zhang, Q., Dunn, S.: Growth and characterisation of lead zirconate titanate (30/70) on indium tin oxide coated glass for oxide ferroelectric-liquid crystal display application. Integr. Ferroelectr. **29**(3–4), 189–213 (2000)
25. Zhang, Q., Whatmore, R.: Sol-gel pzt and mn-doped pzt thin films for pyroelectric applications. J. Phys. D: Appl. Phys. **34**, 2296 (2001)
26. Rodriguez, B.J., Gruverman, A., Kingon, A.I., Nemanich, R.J.: Piezoresponse force microscopy for piezoelectric measurements of III-nitride materials. J. Cryst. Growth **246**(3), 252–258 (2002)
27. Abplanalp, T., Günter, P.: Imaging of ferroelectric domains with sub micrometer resolution by scanning force microscopy. In: Proceedings of the Eleventh IEEE International Symposium on Applications of Ferroelectrics, ISAF 98, pp. 423–426 (1998)

Indentation Stiffness Analysis of Ferroelectric Thin Films

C. Chima-Okereke, M. J. Reece and Markys G. Cain

1 Introduction

The physics of size effects in ferroelectric materials influences finite, measurable, changes in the macroscopic functional behaviour of small-scale systems. In these confined geometries the ceramic microstructure, and in particular grain size and domain switching properties, often enhances the importance of extrinsic effects in the ferroelectric response by comparison with the properties of the bulk material. Whilst ferroelectric scaling effects have been discussed for many years [1], the recent technologically-based drive for sub-micron scale ferroelectric memory applications has resulted in many investigations into thin film systems and 3D micro- and nano-structures. Static scale-size effects are a fundamental, as yet unsolved, problem in nano-ferroelectrics in which destabilization of the spontaneous polarization in ultra-thin films (down to the size of a few unit cells) is predicted theoretically [2] but has not yet been conclusively observed [3]. Whilst this phenomenon occurs on the smallest length scales, size effects relating to the dynamic behaviour of ferroelectrics, such as variation in the Curie temperature [4], collapse of the dielectric constant [5] and phonon hardening [6], can be seen at the micron level and are caused by changes in both sample thickness and lateral size.

This chapter demonstrates the applicability of a depth sensing indentation technique for the measurement and interpretation of the elastic properties of ferroelectric thin filmmaterial. Ferroelectric thin films are of interest as the active materials in actuators and sensors in MicroElectroMechanical Systems (MEMS). For their appli-

C. Chima-Okereke · M. J. Reece (✉)
Materials Department, Queen Mary, University of London, Mile End Road, London E1 4NS, UK
e-mail: m.j.reece@qmul.ac.uk

M. G. Cain
National Physical Laboratory, Hampton Road, Teddington, Middlesex TW11 0LW, UK
e-mail: markys.cain@npl.co.uk

M. G. Cain (ed.), *Characterisation of Ferroelectric Bulk Materials and Thin Films*, Springer Series in Measurement Science and Technology 2,
DOI: 10.1007/978-1-4020-9311-1_9,

cations as actuators it is necessary to know their elastic properties in order to design the devices. For example, the design of many actuators is based on bending cantilevers, such as the active arm of an Atomic Force Microscope (AFM). Because of the lack of availability of data, designers are often relying on the elastic coefficients of bulk materials with the same nominal coefficients. This may lead to large errors not only because of the differences in composition and microstructure, but also because the mechanical and functional behaviour of films can be very different to that of bulk materials.

The film-substrate interaction limits the usefulness of certain measurement techniques—such as surface acoustic wave detection for example. The film thickness typically range between 50 nm and 1 μm on Pt–Ti–SiO_2–Si substrates. The determination of the thin film's transverse piezoelectric coefficient, d_{31}, for example, is currently made by fabricating special cantilever structures or diaphragms and measuring the piezo induced charge upon bending or the strain upon application of electric fields [7–9]. This method relies on the knowledge of the materials elastic coefficients (from the substrate and the film) and although useful techniques exist (based on pressure induced deflection of diaphragms [10] etc.), a method that does not rely on the fabrication of complicated samples remains desirable. This study explores one alternative method, based on depth sensing indentation. The method has high spatial resolution for obtaining elastic properties [11], is non-destructive and can be readily automated to measure film properties at a number of points relatively quickly. The technique provides the so-called indentation stiffness coefficient, and for a thin film on a substrate, both the film and substrate respond to the applied load. The main intricacy of the method is that the geometry results in a relatively complex stress field. In this work issues of indentation analysis (plastic and elastic response), materials data scatter, variation of film thickness and indenter radius, and elastic anisotropy have been investigated.

2 Nanoindentation: A Mechanical Measurement Tool

Nanoindentation is a powerful technique for determining the mechanical properties of thin films because of its high spatial resolution over their surface and also through their thickness [11]. Among the different geometries available, the spherical geometry provides a smooth stress field for which analytical solutions for the elastic contact problem exists [12]. When a thin film on a substrate is indented both the film and the substrate deform elastically, even at small penetrations of the indenter. Therefore, the indentation stiffness coefficient that is obtained is not that of the film but an effective one for the film/substrate heterostructure. It is possible to determine the stiffness coefficient of the film alone by measuring the coefficient of the heterostructure at different penetration depths and extrapolating to zero penetration [13]. A procedure has been developed that involves the stepwise indentation with partial unloading of a film that allows the separation of elastic and plastic contributions to deformation, and yields the required stiffness coefficient as a function of the penetration depth

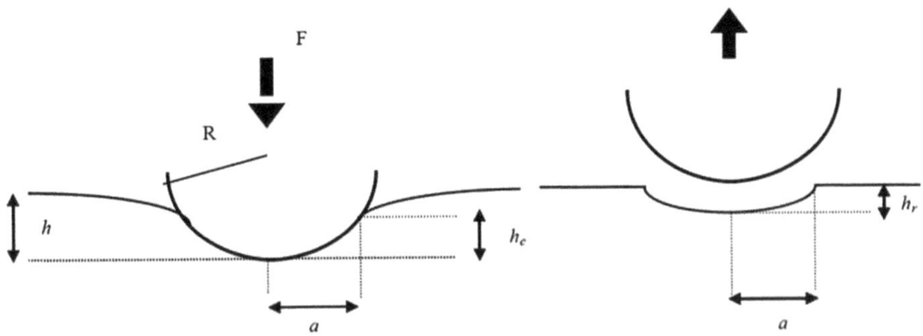

Fig. 1 Spherical indentation geometry and parameters: F indentation force, h penetration, R radius of the sphere, h_c depth of contact, a radius of contact, h_r residual depth

[14]. In this technique, the unloading is assumed to be completely elastic. However, ferroelastic domain walls in ferroelectric thin films are moved by stresses and may move back (or partially back) to their original position (recover) when the stresses are removed [15]. This phenomenon produces an anelastic contribution to the deformation of a ferroelectric when indented that cannot be treated with the procedures of Ref. [14]. In this chapter we describe an indentation method that allows for the anelastic contribution to the deformation of ferroelectric thin films and yields an effective value of the materials elastic coefficient [16].

3 Experimental Set Up and Data Collection

A scheme of the spherical indentation geometry with the most important parameters is shown in Fig. 1. Measurements were performed with a UMIS 2000 nanoindentation system (Australian Scientific Instruments Ltd. (ASI), Canberra, Australia) and diamond indenters with nominal spherical tip radius of 3 and 7 μm. The Young's modulus and Poisson's ratio of the indenters were taken as 1,150 GPa and 0.07 respectively. Diamond indenters are never ideal spheres, and a calibration is needed to obtain the effective indenter radius as a function of the depth of contact (see Fig. 1). The calibration procedure is detailed elsewhere [17]. The effective indenter radius versus depth of contact obtained from the calibration for the diamond indenter is shown in Fig. 2.

PZT ferroelectric thin films of 50:50 and 30:70 (Zr/Ti) composition were supplied by Cranfield University, UK with preferred <111> orientation. The films have a range of thickness: 70, 140, 350 and 700 nm. PLT films ($Pb_{0.88}La_{0.08}TiO_3$) films were also studied [16]. We have investigated unpoled films to avoid the complication of having to consider the electrical boundary conditions. The example data shown in the Figs. 2, 3, 4, 5 are for the PLT films.

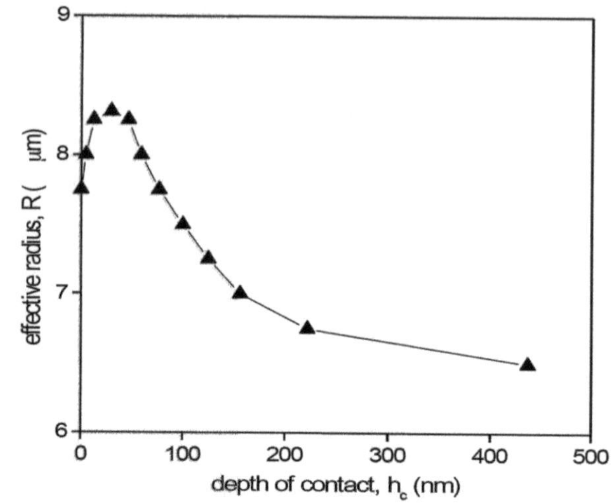

Fig. 2 Effective indenter radius as a function of the depth of contact for the diamond sphere of nominally 7-mm radius (manufacturers data)

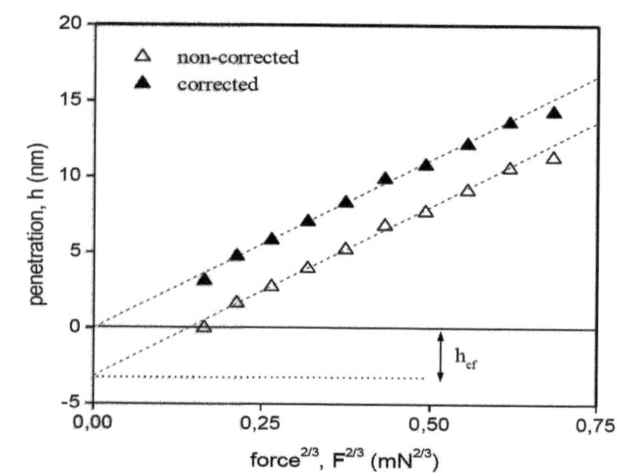

Fig. 3 Evaluation of the indenter penetration at the contact force h_{cf}, and correction of the (h, F) data from its initial neglect

The procedure presented here consisted of a series of load-unload cycles or "indentation loops" with each loop in the series proceeding to a higher maximum force [18]. The individual loops consisted of 40 penetration-force (h, F) data pairs on each of the load and unload half cycles. At the beginning of each loop in the series the depth datum is set to zero as the indenter makes contact with the surface and applies a small "contact force" (typically 0.1 mN). The raw data must be corrected for the displacement associated with the contact force post facto as described below. Each experiment consisted of a series of at least 20 loops up to a final maximum force of 16 mN. The procedure was repeated at least 5 times at different locations 50 μm apart.

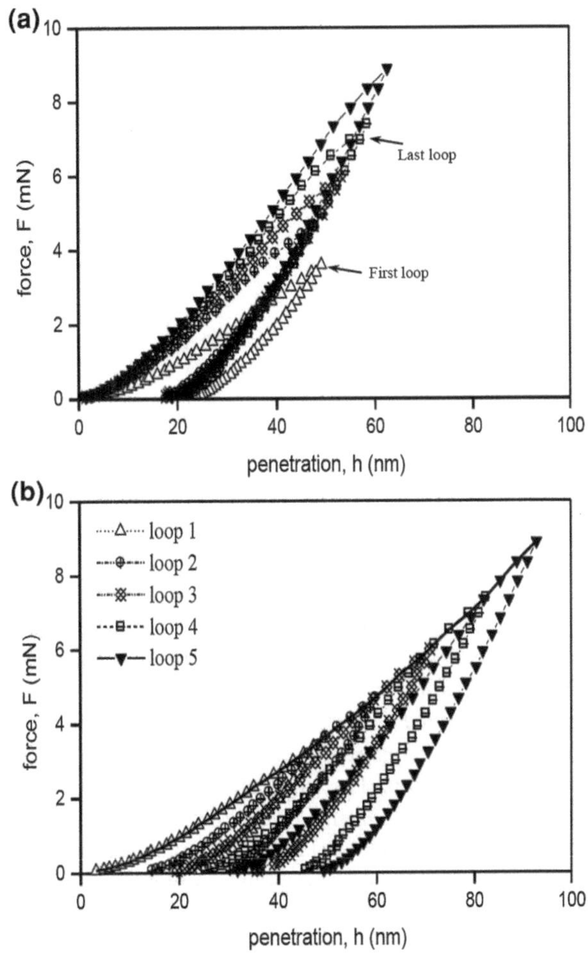

Fig. 4 Reconstruction of the penetration history for the first five loops of a test: **a** initial data all starting at origin; **b** reconstructed history

4 Penetration Data Correction

4.1 Load Frame Compliance

The measured raw data (h, F) must be corrected for the frame compliance, i.e. the measured displacement due to the deflection of the instrument frame. This is accomplished by subtracting from the penetration data a value, D_{hfs} given by Eq. 1.

$$\Delta h_{fs} = C_f F \tag{1}$$

where C_f is the frame compliance that takes a constant value of 0.24 nm/mN for our system.

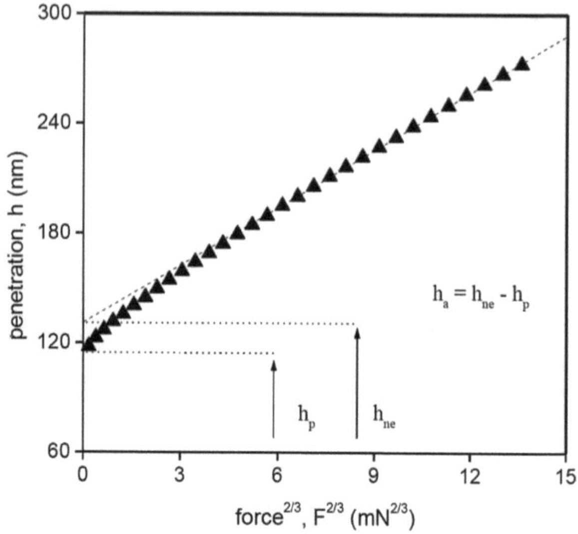

Fig. 5 Isolation of the upper part of the unloading half cycle that is elastic and determination of the anelastic penetration at maximum force, h_a

The penetration data must also be corrected for the penetration at the contact force. This was achieved by assuming that the initial contact was purely elastic. This is a reasonable assumption for blunt indenters. For spherical indentation the relationship h a $F^{2/3}$ Hertzian dependence is then expected [12]. The first point of the loop was disregarded, and the seven following ($F^{2/3}$, h) data were fitted with a linear regression. The intercept took a negative value (typically a few nm) as a consequence of assuming zero penetration at the contact force. Then, the absolute value of the intercept was taken as the penetration at the contact force, h_{cf}, and added to the penetration data along the complete loop. The result of this procedure is that each successive loop starts at the origin. An example is shown in Fig. 3.

4.2 Indentation Experiment History

Finally, the penetration history must be reconstructed. Since each loop started from the origin, the permanent displacement associated with the previous loop, i.e. the depth of the residual impression after the loop, h_r, was not automatically added to the penetration of the next loop. This reconstruction was accomplished such that subsequent loops were data-shifted to the right (along the x-axis) so that the value of penetration, associated with the previous loops maximum force (F_m), became coincident with the original penetration at F_m. This is illustrated in Fig. 4 for the five initial loops of a test. By this procedure it is assumed that (i) no further plastic deformation takes place without increasing the force above that of the previous loop, and (ii) any anelastic contribution recovered on unloading is reversed on loading to the same level as the previous loop. The procedure appears to be reasonable from the evidence that the part of the curves associated with first loading are coincident, to within a few percent, after reconstructing the load history (see bold line in Fig. 4b).

5 Extraction of the Materials Parameters

The following analysis is based on the fully corrected data, as described in Sect. 3. Two parameters were directly extracted from a loop: the maximum penetration, h_m, and the maximum force, F_m. The remaining parameters were extracted from the analysis of the unloading half cycle of the loop as follows. This half cycle was expressed as ($F^{2/3}$,h) and fitted to $y = \alpha + \beta x$. This is the expected behaviour for an elastic-plastic contact where the unloading is assumed to be entirely elastic. For this case, α corresponds to the residual depth, h_r, and βx to the Hertzian elastic behaviour. However, the data deviated from this behaviour for the PTL films. This implies an anelastic contribution to deformation, which recovered during unloading. To analyse the data, it was assumed that this recovery did not start until the indentation force had decreased below a certain value. In other words, it was assumed that there was an initial part of the unloading curve at high forces that could be considered as purely elastic, i.e. fitted to $y = \alpha + \beta x$. To isolate this part, the three data points from the low force end of the curve were disregarded, and the linear regression fit to $y = \alpha + \beta x$ repeated. If the analysis did not improve the first solution was kept. If it did, another three points were disregarded and the procedure repeated. This was done iteratively until the linear regression fit to $y = \alpha + \beta x$ stopped improving, which happened for the PTL films after removing between 1/3 and 1/2 of the data points at low forces. An example of the procedure is shown in Fig. 5. The parameter α of the final fit is in this case the non-elastic penetration at F_m, h_{ne}. This non-elastic penetration comprises the plastic penetration at F_m, h_p, and the anelastic penetration at F_m, h_a. Any plastic deformation will remain after unloading, and therefore, the plastic penetration generated at maximum load, F_m, will correspond to the residual depth, h_r, at the end of the loop. The anelastic penetration recovers during unloading, but not in the upper part of the curve that gives the good fit to the elastic-plastic behaviour. It can be evaluated as the difference between h_{ne} and h_r, or in other words, as the recovery of penetration during unloading. The residual depth was evaluated by fitting the five last ($F^{2/3}$,h) points of the unloading, disregarding the last one, to $y = \alpha + \beta x + \delta x^2$, and taking h_r as α.

Once h_m and h_{ne} were known, both the depth of contact, h_c, and the radius of contact, a, were readily obtained by Eqs. 2 and 3 [19]:

$$h_c = \frac{h_m + h_{ne}}{2} \tag{2}$$

$$a^2 = 2Rh_c - h_c^2 \tag{3}$$

where R is the effective radius of the indenter corresponding to the depth of contact h_c taken from Fig. 2. Once a was known, the mean pressure acting over the circle of contact, P_m, also called indentation stress, was calculated by means of Eq. 4.

$$P_m = \frac{F_m}{\pi a^2} \tag{4}$$

The isolated upper part of the unloading is generally considered as the elastic reloading of the residual impression plus the anelastic penetration at Fm, which has a curvature R′ given by Eq. 5 [17].

$$R' = \frac{a^2 + h_{ne}^2}{2h_{ne}} \tag{5}$$

As a result, the slope β of the linear regression was related to the indentation stiffness coefficient of the indenter/sample composite, E*, by Eq. 6.

$$\beta = \left[\frac{9}{16E^{*2}}\left[\frac{1}{R} - \frac{1}{R'}\right]\right]^{1/3} \tag{6}$$

R is the effective radius of the indenter corresponding to the depth of contact at maximum load. R is not constant during the unloading curve but changes with the depth of contact, following the curve shown in Fig. 2. However, it can be assumed that it does not change significantly in the isolated upper part of the curve that gives the good fit to a linear behaviour (if it changed a small error would be introduced that is of order $R^{-1/3}$). Once E* has been calculated, Eq. 7 is used to evaluate the stiffness coefficient of the sample, E:

$$\frac{1}{E^*} = \frac{1 - \nu^2}{E} + \frac{1 - \nu_i^2}{E_i} \tag{7}$$

where ν is the Poisson's ratio, and the subindex i refers to the indenter. In the case of a thin film, E is the effective coefficient of the film/substrate heterostructure. A value of 0.3 was taken for the Poisson's ratio of the samples [20].

6 Experimental Data for Partial Unloading Experiments

The complete procedure for analysing the Force-Penetration data involves a considerable amount of processing. If only the elastic data is required the effort can be considerably reduced by doing partial unloading experiments. Figure 6 shows the results for unpoled PZT films for which a partial unloading of 20 % of the maximum load was used. It has been shown that the anelastic response only becomes apparent at much higher unloading (>50 %) [16]. The data extrapolated to zero penetration gives values of indentation stiffness for the films of 100–110 GPa, which are comparable

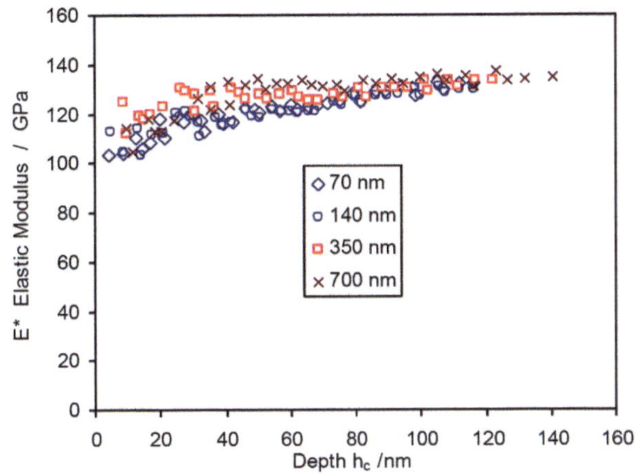

Fig. 6 Example data for <111> textured PZT films of different thickness on Pt/–Ti–SiO_2–Si substrates. The data represent the average of ten measurements in different parts of the films. (Modulus-Depth unpoled films for 3 μm spherical indenter)

to values for bulk materials [16]. This assumes that the elastic properties are constant through the thickness of the films. This was not confirmed, but Scanning Electron Microscopy (SEM) cross sectional images of the film's microstructure revealed a homogeneity consistent with this interpretation.

7 Discussion and Recommendations

The indentation stiffness data, E*, for the films (Fig. 6) varied as a function of indentation load/indenter penetration because of the different relative contributions from the substrate and film. At small values it approached the indentation stiffness of the film. At higher values it approached that of Si. There is a dispersion of the data with the thickest films (700 nm) having a stiffness of about 15 % greater than the thinnest film (70 nm) when plotted against depth of indenter contact. Attempts to normalize the data in terms of a/R, a/t and a/Rt did not improve the coincidence of the data with respect to the indenter/substrate interactions. The rationale for attempting these plots is that for a monolithic, elastically isotropic material the stresses scale to a/R [19]. The inability to normalise the plots may be a consequence of material pile-up at the surface, but this is still to be proved.

The indentation stiffness of the films is not the same as its elastic modulus in the direction of indentation but some average of its anisotropic elastic properties in all directions. This is a consequence of the hydrostatic stress field generated by the indenter. While it is not possible to obtain the five independent elastic coefficients to completely define the elastic properties of a transverse isotropic material, it may be possible to make estimates with associated errors. This can be done by modelling the behaviour of a range of PZT materials using Finite Element Analysis (FEA). Modelled and measured indentation stiffness values for single crystal silicon are within 3 % of each other. The experimental values for <100> and <111> crystal orientation (156 and 178 GPa respectively) lie between the elastic moduli in these directions (130 and 188 GPa respectively).

The problem of characterizing the elastic properties of ferroelectric thin films is extremely demanding because of the complexity and scale of their structures. In this chapter we have applied the Hertzian analysis for an isotropically elastic solid. The work has shown that we can obtain indentation stiffness data that is some average of the elastic coefficients of the film. The method is also sensitive to local variation in elastic properties.

References

1. Kittel, C.: Theory of the Structure of Ferromagnetic Domains in Films and Small Particles. Phys. Rev. **70**(11/12), 965–971 (1946)
2. Junquera, J., Ghosez, P.: Critical thickness for ferroelectricity in perovskite ultrathin films. Nature **422**(6931), 506–509 (2003)
3. Chang, L.W., McMillen, M., Morrison, F.D., Scott, J.F., Gregg, J.M.: Size effects on thin film ferroelectrics: Experiments on isolated single crystal sheets. Appl. Phys. Lett. **93**(13), 132904 (2008)
4. Jiang, Q., Cui, X.F., Zhao, M.: Size effects on Curie temperature of ferroelectric particles. Appl. Phys. A **78**(5), 703–704 (2004)
5. Saad, M., Baxter, P., Mcaneney, J., Lookman, A., Sinnamon, L., Evans, P., Schilling, A., Adams, T., Zhu, X., Pollard, R., Bowman, R., Gregg, J., Jung, D., Morrison, F., Scott, J.: Investigating the effects of reduced size on the properties of ferroelectrics. IEEE Trans. Ultrason. Ferroelectr. Freq. Control **53**(12), 2208–2225 (2006)
6. Sirenko, A., Bernhard, C., Golnik, A., Clark, A., Hao, J., Si, W., Xi, X.: Soft-mode hardening in $SrTiO_3$ thin films. Nat. Lond. **404**(6776), 373–376 (2000)
7. Luginbuhl, P., Racine, G.-A., Lerch, P., Romanowicz, B., Brooks, K.G., De Rooij, N.F., Renaud, P., Setter, N.: Piezoelectric cantilever beams actuated by PZT sol-gel thin film. Sens. Actuators A: Phys. **54**(1), 530–535 (1996)
8. Shepard, J.F., Moses, P.J., Trolier-McKinstry, S.: The wafer flexure technique for the determination of the transverse piezoelectric coefficient (d31) of PZT thin films. Sens. Actuators A: Phys. **71**(1), 133–138 (1998)
9. Muralt, P., Kholkin, A., Kohli, M., Maeder, T.: Piezoelectric actuation of PZT thin-film diaphragms at static and resonant conditions. Sens. Actuators A: Phys. **53**(1), 398–404 (1996)
10. Tuchiya, T., Itoh, T., Sasaki, G., Suga, T.: Preparation and properties of piezoelectric lead zirconate titanate thin films for microsensors and microactuators by sol-gel processing. J. Ceram. Soc. Jpn. **104**(3), 159–163 (1996)
11. Swain, M.V., Menčik, J.: Mechanical property characterization of thin films using spherical tipped indenters. Thin Solid Films **253**(1), 204–211 (1994)
12. Lawn, B.R.: Indentation of ceramics with spheres: a century after Hertz. J. Am. Ceram. Soc. **81**(8), 1977–1994 (1998)
13. Menc-k, J., Munz, D., Quandt, E., Weppelmann, E.R., Swain, M.V.: Determination of elastic modulus of thin layers using nanoindentation. J. Mater. Res. **12**(9), 2475 (1997)
14. Field, J.S., Swain, M.V.: Determining the mechanical properties of small volumes of material from submicrometer spherical indentations. J. Mater. Res. **10**(1), 101–112 (1995)
15. Schäufele, A.B., Heinz, K.: Härdtl, "Ferroelastic properties of lead zirconate titanate ceramics". J. Am. Ceram. Soc. **79**(10), 2637–2640 (1996)

16. Algueró, M., Bushby, A.J., Reece, M.J.: Direct measurement of mechanical properties of (Pb, La) TiO_3 ferroelectric thin films using nanoindentation techniques. J. Mater. Res. **16**(04), 993–1002 (2001)
17. Bushby, A.J.: Nano-indentation using spherical indenters. Nondestruct. Test. Eval. **17**(4–5), 213–234 (2001)
18. Algueró, M., Bushby, A.J., Reece, M.J., Calzada, M.L., Pardo, L.: Mechanical characterisation of ferroelectric thin films for MEMS. Integr. Ferroelectr. **32**(1–4), 83–92 (2001)
19. Sneddon, I.N.: The relation between load and penetration in the axisymmetric Boussinesq problem for a punch of arbitrary profile. Int. J. Eng. Sci. **3**(1), 47–57 (1965)
20. Jaffe, B., Cook, W.R., Jaffe, H.L.: Piezoelectric ceramics. In: Non-metallic Solids. Academic Press, London (1971)

Losses in Piezoelectrics via Complex Resonance Analysis

Markys G. Cain and Mark Stewart

1 Methods of Resonance Analysis and Losses in Piezoelectrics

Historically, the method defined by IEEE has been used extensively to analyse the resonance spectra of piezoelectric materials. Here, the resonance frequencies from a set of samples of different geometries are measured and along with their dimensions and sample density the piezoelectric coefficients determined. This method is simple to carry out in principle, and the calculations are relatively straightforward (although the wave equations for some of the geometries must be solved empirically through the use of Bessel functions).

In our original analysis of resonance in Sect. 1 losses were only briefly described. It is now relevant to discuss losses prevalent in piezoelectric ceramic compositions since these values are often as important as the functional, dielectric and elastic constants that resonance analysis yields. In reality, a piezoelectric material comprises losses originating from its dielectric response to an electrical field, mechanical response to applied stress or following piezoelectric motion and its piezoelectric (strain) response to an electric field. The impact of these losses on a resonance sweep is a reactive and resistive part to the measured impedance. A material with zero losses would exhibit zero impedance at resonance. The significance of loss results in sample heating or noise production and this is why for many applications an understanding of loss mechanisms and absolute values becomes important. Normally, the mechanical loss at resonance is calculated from the width of the resonant peak and is labelled the mechanical Q or Quality factor. The narrower the resonant peak, the higher its Q. Dielectric losses are normally calculated from the phase angle between observed capacitance and applied field, labelled tan δ. Piezoelectric loss may not normally

M. G. Cain (✉) · M. Stewart
National Physical Laboratory, Hampton Road, Teddington, Middlesex TW11 0LW, UK
e-mail: markys.cain@npl.co.uk

M. G. Cain (ed.), *Characterisation of Ferroelectric Bulk Materials and Thin Films*, Springer Series in Measurement Science and Technology 2,
DOI: 10.1007/978-1-4020-9311-1_10,

be calculated from resonance data but may be assessed through strain—electric field response whereby any hysteresis present may be tentatively ascribed to this loss alone—of course, if strain is produced then mechanical loss may also have an additive effect. This issue is contentious.

A method adapted by Sherrit and co-workers from earlier work by Holland [1, 2] and ultimately commercialised by TASI Technical Software (© PRAP), was to treat the entire piezoelectric matrix as complex. Again, assuming ideal linear piezoelectric behaviour and small field level perturbations, then the same formulation used for ideal loss-less resonators could be utilised, yielding similar equations but of complex form. Here, the impedance, resonant frequencies, piezoelectric, stiffness and dielectric coefficients are all complex quantities now and the forms of the equations may be iteratively fitted to the various shapes of resonance curves for each of the different geometries. PRAP makes extensive use of complex elastic, dielectric, and piezoelectric properties in fitting the resonance curves. Since the method fits to the data around resonance peaks an improved estimate of the resonant and anti-resonant frequencies is obtained through interpolation. Additionally, this method automatically takes into account the losses by virtue of its complex notation. Alternative iterative methods have also been developed including [3], and [4], yielding accurate determination of the piezoelectric and dielectric properties of ceramic samples.

The use of complex coefficients to represent the three types of loss in piezoelectric materials has its merits, and it is important to be able to verify the values given by this analysis by independent experimental means. This section of the chapter has attempted to explore these complicated issues. The losses may be identified and measured as follows:

1. Dielectric loss—or $\tan\delta$. This may be calculated by measuring the phase angle between applied electric field and current, for example with a capacitance bridge.
2. Mechanical loss—or Q. This may be determined by drawing graphs of stress against strain and measuring the hysteresis between increasing and decreasing stress. This area is related to the energy absorbed during the cycle wherein mechanical loss may be calculated according to the following equation [5]:

$$W_m = \pi S^E X_0^2 \tan\phi \tag{1}$$

   with W_m = work done per half cycle, S^E = compliance at constant E, X_0 = strain, ϕ = loss angle. Alternatively, measuring the phase angle between stress and strain using gain phase analysers would also yield this loss. Another method would be ultrasonic attenuation techniques or DMTA methods.
3. Piezoelectric loss. This is the most difficult to measure independently because when the material deforms under the action of the piezoelectric effect it also must endure mechanical loss. However, a method to investigate the piezoelectric loss would be to measure the piezoelectric strain as a function of applied field. The hysteresis is indicative of the loss. Alternatively, the use of a gain phase analyser in the measurement of the phase angle between strain and field can also be used perhaps to an advantage since over many cycles an average may be calculated and

signal to noise thus increased. This is contentious though, because most of the literature points out that *piezoelectric* loss is not an independent quantity at all, but is rather simply a compounded assembly of dielectric and mechanical loss.

The material coefficients can be described in either complex format, with real and imaginary parts or in the form of an amplitude and a phase lag or loss. The concept of a phase lag is easier to understand and so the term loss is more prevalent. However, for the description of impedance behaviour, that includes lossy behaviour, the complex notation is simpler, so that any coefficient can be represented as:

$$\text{real part} - j.\text{imaginary part} \tag{2}$$

These can be simply converted to a loss or a Q by the ratio of the real and imaginary component:

$$\text{Mechanical } Q = \frac{\text{real part (elastic coefficient)}}{\text{Imaginary part (elastic coefficient)}} \tag{3}$$

$$\tan\delta = \frac{\text{imaginary part (permittivity)}}{\text{Real part (permittivity)}} \tag{4}$$

The symmetry of the ∞mm structure also leads to several restrictions on the possible values of the imaginary parts of the coefficients. The diagonal elements of all the imaginary coefficients are positive, [6]:

$$\text{coefficient}_{ij} \geq 0 \tag{5}$$

Also:

$$(s_{11}^E)'' \geq |(s_{12}^E)''| \tag{6}$$

$$(s_{11}^E)''(\epsilon_{33}^T)'' \geq (d_{31}'')^2 \tag{7}$$

1.1 Comparison of Methods

The losses associated with dielectric constant, piezoelectric constant and elastic constant have been measured independently of one another (but see notes earlier) and most importantly using methods that do not include resonance. This is to try and explore the validity for the use of complex notation—vis. Real and Imaginary components of the parameters with loss given by the ratio real part/imaginary part.

Three losses associated with piezoelectric material have been classified:

1. Dielectric loss
2. Piezoelectric loss
3. Mechanical loss

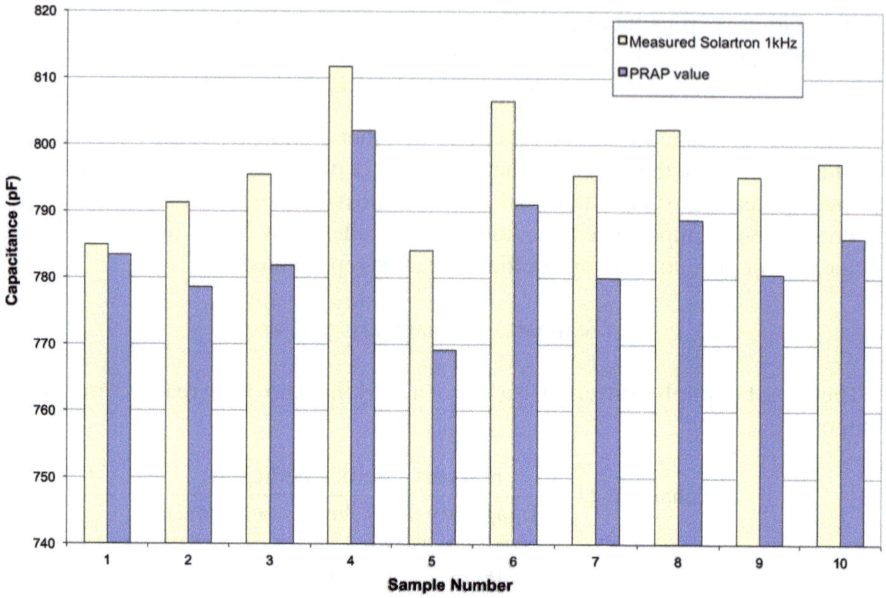

Fig. 1 Capacitance values are approximately 3 % lower when measured at resonance for PZT 4D hard materials

Each has been determined using the resonance analysis performed by PRAP. In this software, the piezoelectric equations have been uncoupled through the measurement of well-defined sample geometries and the resultant spectra have been fitted to the equations describing the electro-mechanical piezoelectric equations. The result is a value for the real and imaginary parts of dielectric permittivity, piezoelectric strain coefficient, and mechanical stiffness coefficient. In PRAP the real and imaginary parts of the fitted coefficients are refined independently and these results used to calculate the real and imaginary parts of some of the other coefficients. For example, in length extensional analysis, the fitted parameters are K_{33}, f_p, ϵ_{33}^S. The derived real and imaginary coefficients are s_{33}^E, d_{33}, ϵ_{33}^T and s_{33}^D.

Dielectric Loss

The first set of alternative experiments to determine the materials dielectric loss was performed on a hard (PZT 4D) and a soft (PZT 5A) composition PZT ceramic rod. Here, the data from a dielectric analyser (Solartron 1296/1260) were compared to the PRAP data, in Figs. 1, 2, 3, and 4.

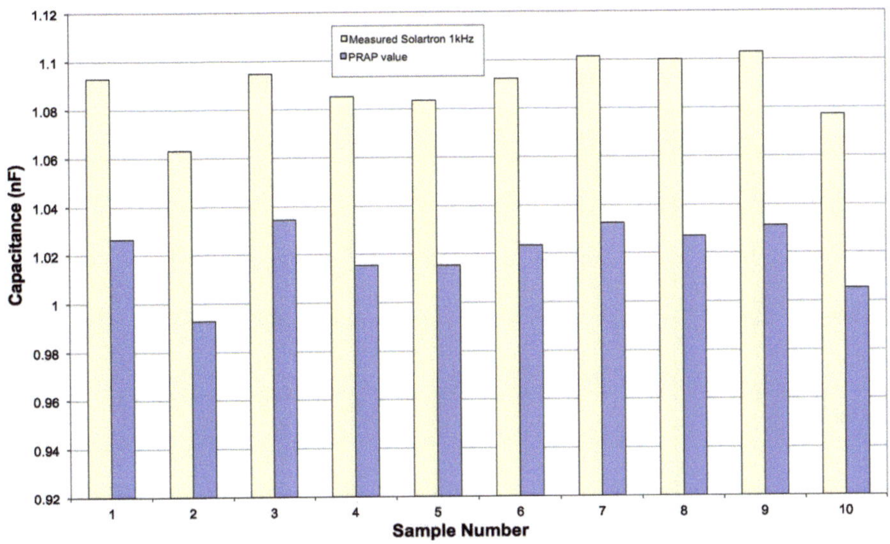

Fig. 2 Capacitance values are approximately 5 % lower when measured at resonance for soft PZT 5A materials

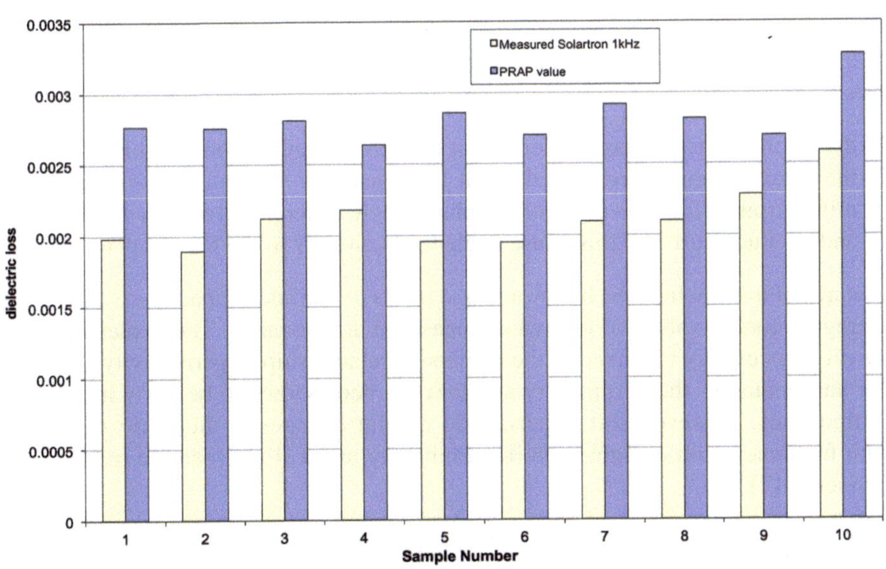

Fig. 3 PZT 4D: Loss values are approximately 25 % higher when calculated using PRAP compared to the Solartron values

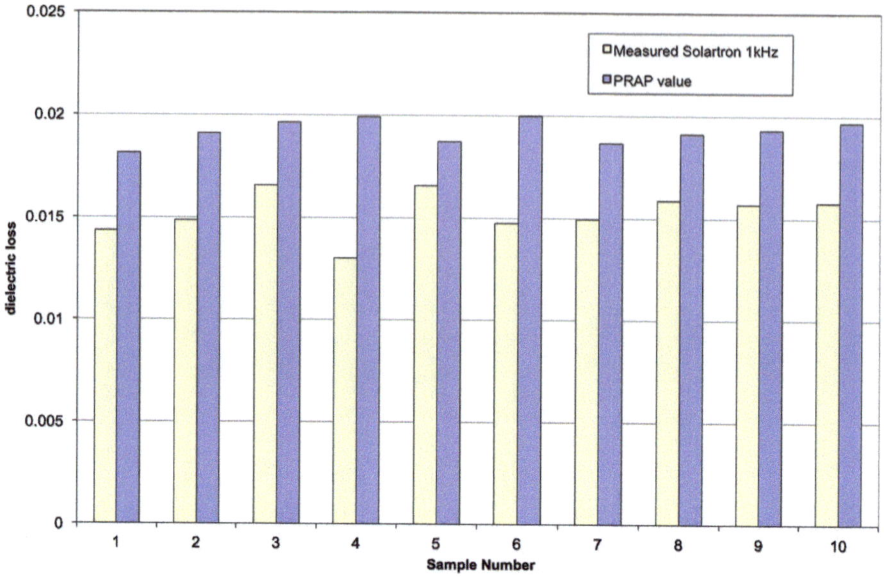

Fig. 4 PZT 5A: Loss values are approximately 25 % higher when calculated at resonance using PRAP than measured with the Solartron

It is important to note a few differences in the measurement parameters for these tests:

1. the resonance tests were carried out at high frequency—about 180 kHz
2. the Solartron tests were carried out at 1 kHz and low field but the loss was determined from a direct measure of the phase angle difference between applied field and resultant current. This phase angle leads directly to the loss tangent.

Features of note include the below average levels of recorded capacitance and above average recorded values of loss when measured at resonance. The explanation for the differences may be linked to the dispersive relationship of permittivity and loss. For hard materials the permittivity and loss has been shown to be fairly insensitive to frequency. However, soft materials do exhibit changes in dielectric properties with frequency with a slight drop in permittivity and a slight increase in loss with frequency [7].

Piezoelectric Loss

The piezoelectric coefficient was measured using two additional complimentary techniques. The first was to measure, in one complete driven cycle, the strain response to an applied electrical field. The strain was measured using a fibre optic probe and field with an oscilloscope. The strain and field were acquired and analysed using software specially written for hysteresis loop measurements [8]. The area enclosed

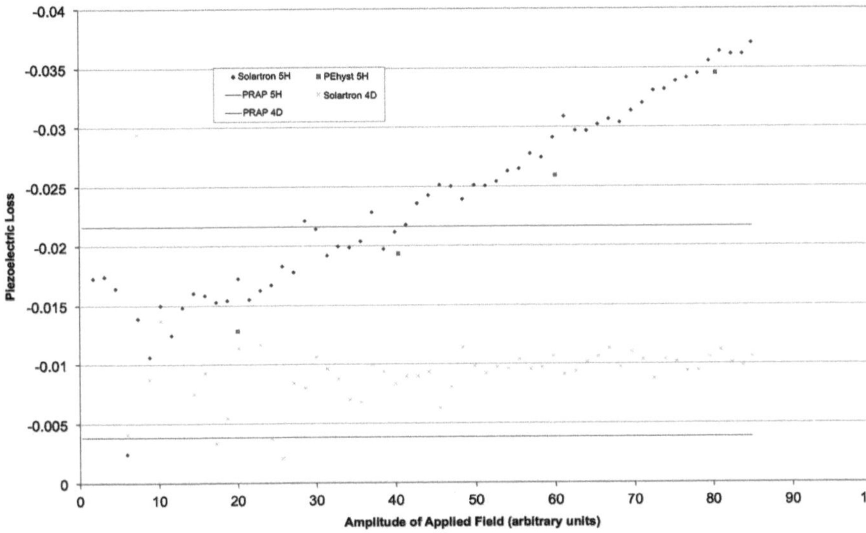

Fig. 5 Piezoelectric loss with electric field measured using PRAP (*horizontal line*), P-E hysteresis loop area, and gain phase analyser

by the loop is representative of the piezoelectric loss; these are represented by the "PEHyst" data points on the graph. Additionally the strain field data was measured using a Solartron Gain Phase analyser, providing a direct measure of the phase angle between strain and electric field. This data was also compared to the resonance PRAP data, in Fig. 5.

Direct piezoelectric coefficient and loss was also measured using the Berlincourt method (see p. 39). In this technique, the sample was placed between two ball bearings under a small static pre-load and a small alternating mechanical force was applied. The sample electrodes were connected to the input of a charge amplifier (virtual earth and hence short circuit conditions prevailed). A Solartron 1260 gain phase analyser captured the amplitude and phase difference between applied mechanical load (ac) and resultant charge developed. The data is reproduced in Fig. 6, which clearly indicates the increase in loss for soft materials with alternating mechanical load and an insensitivity with load for hard materials. This experiment provides some confidence in the indirect displacement measurements in Fig. 19.

The piezoelectric loss value for hard materials determined from measurement of displacement is seen to be approximately two times that calculated by PRAP. A potential explanation might involve phenomena that are only active at lower frequencies such as domain wall motion that becomes diminished at resonant frequencies (but this is improbable for hard materials). For soft materials the values of loss are in agreement with the resonance value only at a certain applied field (and hence strain) with loss exhibiting a linear increase with amplitude, as expected. Again, the non-resonant measurements are performed at much lower frequencies, kHz rather than

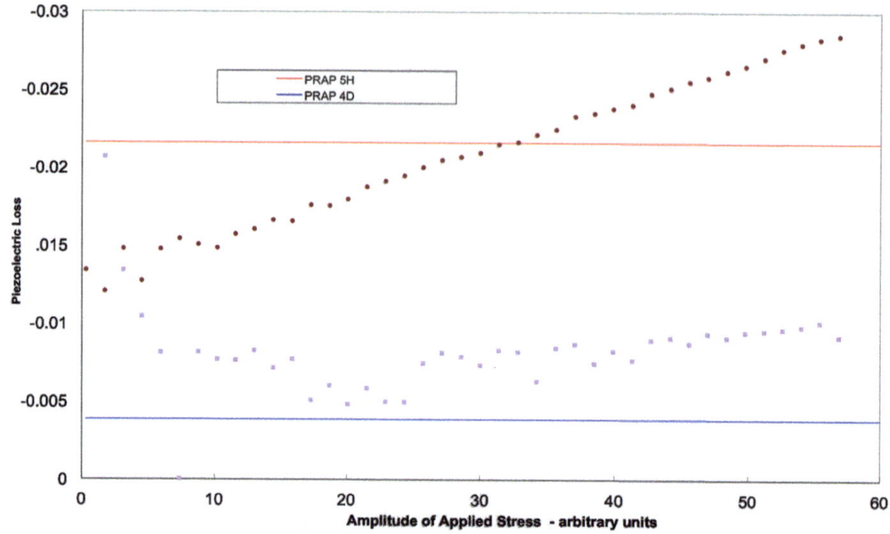

Fig. 6 Piezoelectric loss with electric field measured using PRAP (*horizontal line*) and impedance gain phase analyser for direct (charge) effect

hundreds of kHz. Also the non-resonant methods involve measurements conditions well into the non-linear behaviour regime.

Mechanical Loss

The mechanical loss of a series of length mode resonators was measured using conventional strain gauge techniques to measure the stress-strain behaviour up to around 10 MPa, and these were then compared with data derived from resonance measurements with subsequent complex analysis using PRAP software. The real part of the compliance was determined from the slope of the stress strain curve and measuring the enclosed area in the stress-strain hysteresis yielded the imaginary part.

The stress strain measurement technique is difficult to carry out as the displacements involved are very small and can easily be "lost" in small misalignments. Strain gauge experiments do not suffer from this "lost" strain, however care must be taken to apply a uniform stress field across the sample. Three strain gauges were placed around the circumference of the samples, length 15 mm, and diameter 6.35 mm. Care was taken to achieve good alignment, but often there was over 100 % difference in the outputs from the three gauges. However considering that the total displacement in any experiment is only 1 μm, it is not surprising that there is such a large scatter.

A summary of the results determined by the independent methods on the same samples is given in Table 1. In order to get some idea of the validity of the results since neither method is definitive, the results can be tested against some empirical generalisations. The mechanical behaviour of soft and hard PZT piezoelectric materials can be generalised by the following statements.

Table 1 Summary of mechanical loss measured using stress strain experiments and PRAP comparisons

Material	Method	Electrical conditions	Compliance real part (m^2/N)	Compliance imaginary part (m^2/N)	Mechanical Loss
PZT 5A	Stress-strain	open	1.30E-11	−5.87E-13	−4.51E-02
	Stress-strain	short	2.82E-11	−1.46E-12	−5.19E-02
	Resonance	open	1.13E-11	−5.17E-14	−4.57E-03
	Resonance	short	2.09E-11	−2.04E-13	−9.75E-03
PZT4D	Stress-strain	open	1.22E-11	−1.20E-12	−9.82E-02
	Stress-strain	short	1.78E-11	−4.30E-13	−2.42E-02
	Resonance	open	8.77E-12	−7.95E-15	−9.07E-04
	Resonance	short	1.52E-11	−1.84E-14	−1.21E-03

- Soft materials are more compliant than hard materials.
- Mechanical loss of soft materials is greater than in hard materials (often by an order of magnitude).
- Compliance and mechanical loss are greater for short circuit conditions than in open circuit.

For the resonance method all the results comply with these simple tests, as do the stress strain results except for the anomalously high loss value for the hard material in open circuit conditions. Intercomparison of the two methods show the compliance values for the stress strain technique are between 15 and 40 % lower than those determined by resonance. However, the mechanical loss values determined by stress-strain are between one and two orders of magnitude greater than calculated from the resonance technique.

There are many possible explanations for these anomalies; the frequency for the stress-strain measurements is much less than 1 Hz; the stress strain loss measurements are close to the noise level limits for the technique; the stress levels are very different for the two measurement methods; the stress-strain measurement is compression only, whilst the resonance is compression and tension. However, considering the limitations of the stress strain measurement technique the general agreement is remarkable.

The usual way of measuring mechanical losses in transducers is to measure the mechanical Q_m, i.e. the full width at 3 dB of the resonance peak. If this is done and compared with the values obtained from the PRAP fit there is excellent agreement. However, this is not surprising since PRAP produces an almost perfect fit to the experimental data, although it does verify that the mechanical loss terms in the PRAP fit produce the expected behaviour.

2 Conclusions

The imaginary coefficients derived from complex fitting of the dielectric, mechanical and piezoelectric coefficients to the resonance impedance spectrum is undertaken in the PRAP software. Alternative methods of directly measuring these losses were suggested and results compared.

The results indicates:

1. PRAP (and other iterative methods) yields loss values that are broadly consistent with our alternative measurements
2. PRAP (and other iterative complex methods) assumes that the materials behave in a linear fashion with zero frequency dispersion. Our non-resonant experiments were conducted in a non-linear regime and the material parameters will almost certainly exhibit frequency dependence
3. More work is required to understand the origins of loss, to adequately measure these losses independently and accurately and to understand the interdependence of these losses in piezoelectric material.

Acknowledgments The authors also wish to express their thanks to Professor Mike Reece of Queen Mary University of London for the mechanical stress strain measurements. The authors would also like to acknowledge the work of Will Battrick who worked on this project whilst completing his one-year industrial experience for his undergraduate degree, at Manchester University.

References

1. Holland, R.: Representation of dielectric, elastic, and piezoelectric losses by complex coefficients. IEEE Trans. Son. Ultrason. **14**(1), 18–20 (1967)
2. Smits, J.: Iterative method for accurate determination of the real and imaginary parts of the materials coefficients of piezoelectric ceramics. IEEE Trans. Son. Ultrason. **23**(6), 393–401 (1976)
3. Alemany, C., Pardo, L., Jiménez, B., Carmona, F., Mendiola, J., González, A.M.: Automatic iterative evaluation of complex material constants in piezoelectric ceramics. J. Phys. D: Appl. Phys. **27**(1), 148 (1994)
4. Alemany, C., González, A.M., Pardo, L., Jiménez, B., Carmona, F., Mendiola, J.: Automatic determination of complex constants of piezoelectric lossy materials in the radial mode. J. Phys. D: Appl. Phys. **28**(5), 945 (1995)
5. Uchino, K., Hirose, S.: Loss mechanisms in piezoelectrics: how to measure different losses separately. IEEE Trans. Ultrason., Ferroelectr. Freq. Control **48**(1), 307–321 (2001)
6. Holland, R., EerNisse, E.P.: Accurate measurement of coefficients in a ferroelectric ceramic. IEEE Trans. Son. Ultrason. **16**(4), 173–181 (1969)
7. González, A.M., Alemany, C.: Determination of the frequency dependence of characteristic constants in lossy piezoelectric materials. J. Phys. D: Appl. Phys. **29**(9), 2476 (1996)
8. Stewart, M., Cain, M. G. Hall, D.: "Ferroelectric hysteresis measurement and analysis". Technical Report, NPL Rep. CMMT **152**(A), 3–5 (1999)

Dielectric Breakdown in Dielectrics and Ferroelectric Ceramics

Markys G. Cain

1 Dielectric Breakdown

Under the application of a large enough electrical field a dielectric material can respond through motion of free charge carriers, injection of mobile charge carriers from electrodes, space charge generation, and dissipation of energy in the material, all of which may contribute to the electrical failure of the material. Various theories have been published going back several decades and include two fundamentally separate materials responses to an electric field; 1. An intrinsic dielectric breakdown phenomena and 2. A defect mediated breakdown phenomena. The former is likely only ascribable to very thin films or some polymeric materials. The latter is the more commonly regarded theory that details the weakest link statistics argument for electrical breakdown in porous materials, for example. The type and source of the free carriers can provide for a profound influence on the nature of the breakdown events and so any discussion or analysis of breakdown in solid materials especially must have an appreciation of the conduction processes occurring in the material [1]. If breakdown of the sample is not determined by some defect or imperfection in the material then one can consider what is the materials' intrinsic electrical breakdown strength. If breakdown is determined by defects or imperfections then this implies a weak link mechanism. In practice, even high density, carefully prepared ceramic alumina exhibits a Weibull weak link statistical response signifying defect induced (pores, voids, inclusions etc.) breakdown [2].

In this chapter we introduce some of the concepts that appear to govern dielectric breakdown in dielectric and ferroelectric ceramics and cite various excellent reviews and books which explore these issues. Then, we will present the various standards

M. G. Cain (✉)
National Physical Laboratory, Hampton Road, Teddington, Middlesex TW11 0LW, UK
e-mail: markys.cain@npl.co.uk

M. G. Cain (ed.), *Characterisation of Ferroelectric Bulk Materials and Thin Films*, Springer Series in Measurement Science and Technology 2,
DOI: 10.1007/978-1-4020-9311-1_11,

that already exist and ways in which measurement good practice has been developed at the UK's National Physical Laboratory (NPL) in the past few years. The focus is on the measurement of dielectric breakdown of bulk (or thin section bulk) ceramic materials. The subject of dielectric breakdown in thin films is the subject of current (2013) research at NPL.

1.1 Breakdown: Standards and Methodology

Existing standards and test methods for dielectric properties of advanced technical ceramics that are contained within the CENELEC standards have been shown to be not always appropriate to the forms in which the materials are currently used industrially, and certainly do not take into account the above considerations. Under a separate programme of study at NPL, CERAMELEC, a European consortium led by NPL, assisted the CENELEC BT-TF 63-2 standards committee in the task of updating existing methods to make them more relevant to modern materials and components. The principal objectives of this task were to evaluate traditional (e.g. IEC 672 [3]) and improved methods for determining the dielectric breakdown of advanced technical ceramics, and to examine the potential for using statistical methods for describing breakdown phenomena and potential material reliability. The breakdown test experiments described in this chapter have been conducted under guidelines developed from that CENELEC study. Note that, International standard ASTM D149 [4] also refers to the electrical breakdown of insulating materials.

These measurement notes are aimed at those who wish to familiarise themselves with the concept of dielectric breakdown in ceramic materials, the standard tests that have been developed and the adaptation of these measurement methods to piezoelectric ceramic materials. The electric strength is defined as the maximum electrical stress (MV/m) that can be applied to an electrically insulating material before breakdown occurs. A knowledge of the electrical breakdown strength is required for the efficient poling of ferroelectric material. Theoretical considerations and calculations of electrical breakdown exist but are only validated by work on single crystal materials with defined geometries, electrode arrangement and under ambient conditions. Electrical breakdown on real industrial ceramic materials, however, rarely follows a simple or theoretical prediction. This is due to the dependence of breakdown stress on materials properties such as homogeneity, porosity, sample dimensions and geometries, electrode sizes and arrangements, environmental conditions, stressing mode (DC, AC) and in practice control of these parameters is rarely achieved.

It is clear that dielectric breakdown data is only meaningful if the precise test conditions are adequately described; such as operating mode (DC, AC, impulse), sample and electrode geometry, stress application characteristics (rise time etc.), breakdown detection sensitivity and method, oil immersion and contact techniques and oil type. The dependence of dielectric breakdown on sample geometry can be described with reference to the two methods of sample preparation defined in IEC 672-2 [3], Fig. 1. The two methods do not give identical results and, according to the

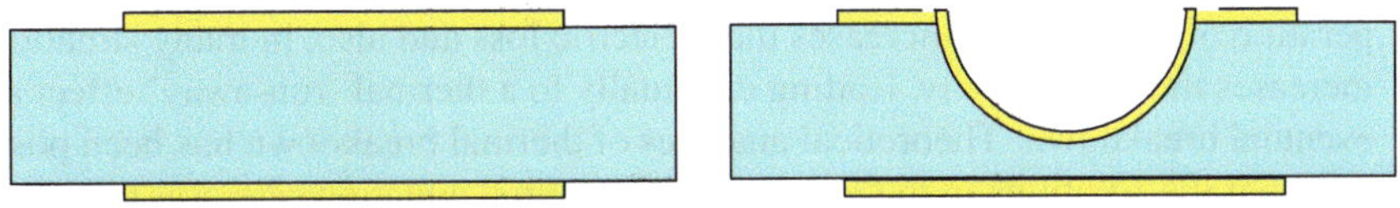

Fig. 1 Test piece geometries used in dielectric breakdown tests. *Flat disk* Hemispherical recess

standard, depending on the manner in which the recess is made, the results from the recessed material are likely to be more variable than those obtained by the flat disc. However, although a larger volume of the material is stressed in the first configuration the breakdown is likely to be initiated at the extreme electrode edges due to stress amplification at electrode discontinuities. Modelling this phenomenon based on a finite difference approach, has been conducted at NPL and results identify a rapid rise in electric field as the electrode edge is approached. The electric field enhancement is approximately 4–6 times that of the uniform applied field. However, results on flat alumina substrate materials have shown that breakdown always occurred within the electroded areas signifying perhaps that breakdown was dominated by the presence of pores and defects within the bulk of the material, that the oil immersion method significantly reduced the stress enhancement and that the high stress field does not penetrate the bulk to any appreciable extent.

1.2 Dielectric Breakdown: Mechanisms

There are three basic mechanisms by which an insulating material breaks down under the application of an electric field; intrinsic breakdown, thermal breakdown and discharge breakdown. During real breakdown it is expected that all of these may occur in combination and, perhaps in the case of sintered ceramics, will be dominated by residual porosity or surface asperities present within the material, rather than properties characteristic of the bulk material.

1. Intrinsic breakdown, whatever the bulk conductivity of a material may be, when a steadily increasing voltage is applied to it, a well defined current will flow until a saturation level is reached after which the current will increase rapidly (in around 10^{-8} s) initiating breakdown. This intrinsic property of the material is clearly electronic in character. The increase in current or electron generation may arise from events including collision ionisation of conduction electrons, emission from bulk impurity centres or field emission from the electrodes. Whatever mechanism prevails the result may be an electron avalanche and breakdown failure.
2. Thermal breakdown thermal failure of an insulator occurs when the internal heat generated via either Joule heating (current flowing through heats the surrounding volume) or other dielectric losses increases with a rate exceeding that which heat

may escape through radiation, conduction or convection. The increase in temperature of the material increases the dielectric loss and also, in many situations, increases the conductivity, leading eventually to a thermal 'run-away' effect and eventual breakdown. Theoretical analyses of thermal breakdown has been postulated but the solutions to the governing differential equations are non-trivial and validating data are hard to find. The thermal conductivity of the material is clearly an important factor when assessing the likelihood of this method of breakdown since it controls the balance between heat generation and heat dissipation.

3. Discharge breakdown materials, which exhibit porosity and inhomogenous microstructures, typically found in ceramics, have been found to exhibit breakdown characteristics which are dominated by the presence of their internal voids and inclusions. Porosity has been shown to initiate breakdown and is the key factor in most common ceramics with 1–3 % porasity. A close analogy with mechanical strength is observed where the breakdown strength and mechanical strength may be described using failure type statistical methods, such as that offered by Weibull analysis. The dependence of the breakdown strength on specimen size, and thickness in particular, can be explained by virtue of the fact that as thinner material is sampled there exists a smaller chance of the electric field 'finding' a critical defect at which discharge breakdown may be initiated. The mechanism by which the material then breaks down is open to debate but may include the idea of the propagation of a charge 'streamer' (in a similar manner to that described for lightening) progressing from pore to pore via conventional conduction routes [5]. This route to breakdown would then depend on the rate at which charges may flow within the material which, for materials like alumina, which have a charge leakage time of around 10^2 s, would place limits on the time dependent nature of breakdown. It is also highly probable that the localised heating caused by the discharge events will contribute to the breakdown event in a manner described previously. Certainly, in high permittivity materials such as ferroelectric ceramics, temperature rises of around 10^3 °C may occur.

Effect of Permittiivty

The permittivity of the sample has been shown to play a role in the breakdown phenomenon [6] but models to describe the effects have not been satisfactorily developed. Longer term effects such as a gradual reduction in resistivity at a continuous applied high stress may lead to breakdown, as well as environmental effects such as moisture, temperature, electrochemical reactions between electrodes and the material, and structural effects such as sharp and point discontinuities which might have the effect of amplifying the electric stress in certain regions.

Space Charge and Breakdown

Space charge may be defined by Eq. (1), [7]. Here χ is the space charge, J is the current density, x is the cartesian coordinate within the material, ρ is the electrical resistivity (which is a function of x) and ϵ is the absolute permittivity. The origins of space charge include local materials heterogeneity (leading to local variations in resistivity), ionization of materials within the dielectric to form heterocharge, 'Charge injection' from a stress enhancement (electrode, edge effects, pinholes, composite with radically different permittiivty for example), and polarization in structures such as water trees.

$$\chi(x) = J\epsilon\frac{d\rho(x)}{dx}. \quad (1)$$

Various materials and geometrical issues can affect and often dominate the generation or presence of space charge and they may be summarised as:

1. Interface and surface effects (oxide layers and electrodes) can have a large effect on space charge;
2. Current density in a material increases linearly at low fields and exponentially at higher fields. The cross over point is thought to be associated with thermal excitation of carriers (electrons) hopping from one sink to another;
3. Surface conductivity changes (brought about from oxidation of the surface for example) can affect the space charge penetration and it has been shown that this penetration is dependent on the conductivity enhancement and can extend many hundreds of micrometers into dielectric materials;
4. Available charges from point like charge sources;
5. Water tress. At very high electrical fields, space charge 'waves' propagate throughout a dielectric. Very little experimental or theoretical evidence is found for this phenomena however.

Microscopic Space Charge and Dielectric Breakdown

Most if not all dielectric polymers and possibly also ceramic dielectrics are not homogenous on the micro scale or nanoscale. This means that current density variations would exist within the material under applied DC field leading to possibly substantial space charge effects. (Conductivity is generally greater along polymer backbones than between them, so that the conductivity through the laminar crystallites of biaxially oriented polypropylene should be much less than that of the amorphous regions between.) Since the breakdown strength tends to increase as the inverse square root of thickness, probably for fundamental energy-related considerations, then it can be shown that the energy available following breakdown is proportional to the cube of the channel length and the square of applied electric field, Eq. (2) [7].

$$W_a \sim K_1 d^3 E^2. \tag{2}$$

The energy required to ionise the material as it breaks down is proportional to the breakdown volume and is given by approximately Eq. (3)

$$W_a \sim K_2 d^2 \tag{3}$$

with K_1, K_2 constants, d is the breakdown channel length and E the electrical field. This then yields the field needed for electrical breakdown:

$$W_{\text{breakdown}} \sim \left[\frac{K_1}{K_2 d}\right]^{1/2}. \tag{4}$$

This dependency of breakdown strength on the inverse square of the dielectric's thickness is observed in many experimental systems, including the examples that follow. For this study, the important observation is that very thin layers of dielectric exhibit much greater dielectric strength than the same materials but of thicker cross section.

Barrier Layer Dielectrics

It has been shown that the inclusion of a third dielectric layer placed between two layers of dielectric allows the breakdown voltage to be increased [6, 8]. The maximum value of the breakdown voltage enhancement is usually located at an optimum barrier position, $\xi \approx 0.25$. This anomalous increase in E_B in three-layer dielectrics in a quasi uniform electrical field may be caused by the local non-uniformity of electric field at the boundaries, edges and electrodes.

1.3 Dielectric Breakdown in Ferroelectric Thin Films

Generally, as described in some detail above, breakdown in ferroelectrics follows a multi-path mechanism whereby initially we observe electrical breakdown followed by thermal runaway. For thin films, the earliest observations were that of dendritic paths arising from electrodes penetrating into the material. Field emission injection of electrons from the electrodes can cause avalanche collision ionisation and subsequent breakdown, with effects of electrode work function being taken into account in Schottkby barrier analysis:

$$E_B = Ad^{-w} \tag{5}$$

with $1/4 \lesssim w \lesssim 1/2$, and A is the electrode area, and

$$eE_B\lambda = h\left(\Phi_M - \Phi_{FE}\right) \tag{6}$$

with Φ being the work functions (Metal and Ferroelectric), λ is the electron mean free path and h is a constant order unity. For ferroelectric thin films, the Schottky barriers at interfaces dominates all electrical breakdown behaviour [9]. If we follow the work of Nelson (in book [10]), then we can write the following equation which relates the impulse thermal breakdown (where the thermal conduction losses are ignored because breakdown happens quickly) to thermal and transient properties, and material conductivity:

$$E_B(T) = \left[\frac{3C_v K T^2}{\sigma_0 b t_c}\right]^{1/2} \exp\left(\frac{b}{2k_B T}\right) \tag{7}$$

with, σ_0 is the DC conductivity, C_V the specific heat capacity, K the thermal conductivity, t_c is the time to breakdown, T is temperature (K), k_B Boltzmann constant, b is a conduction activation constant for the material. In this solution it is assumed that $b \gg kT$, and that the time to breakdown $t_c < \frac{C_v d^2}{K}$ with d being the inter electrode distance. This equation estimates breakdown in a variety of ferroelectric perovskite thin films up to around 800 MV/m, predicted and measured [11].

More recent evidence for a deterministic basis for dielectric breakdown in $BaTiO_3$-based multilayer capacitors [12] has shown that breakdown fits a model based on thermionic emission from the cathode causing collision ionisation in connected defects in the ceramics. In Eq. (5) their value w was calculated to be 0.5–compatible with macroscopic DC thermal breakdown mechanisms. Interestingly, no significant effect of electrode area, A was found on E_B. Breakdown in ferroelectric thin films has been the subject of much academic and industrial research—see the work of Scott for example [11]—and is beyond the scope of this chapter. Future publications in this Springer Metrology Series are envisaged that will explore metrology of thin film breakdown.

1.4 Measurement Techniques for Dielectric Breakdown

From the earliest experiments it was noted that the breakdown stress was influenced by the experimental arrangement and electrode geometries (apart from any material effects). The observations of discharges occurring at the edges of electrodes were explained by the presence of electric stress concentrations at the edges arising from the bending of the equipotential field lines. Indeed it has been shown (also see previous section) that the field enhancement magnitude is of the order of twofold for equal electrode areas and up to sevenfold enhanced for unequal sized electrodes [10]. Attempts to reduce this stress enhancement included the use of oil immersion measurements where the oil, possessing a higher permittivity than air and also a much higher breakdown strength, significantly reduced this edge effect. The primary role of the oil, however, is to suppress surface discharge since ceramic breakdown in air would not be possible due to the inherently low (2 kV/mm) breakdown strength of air.

Variations in electrode geometries have also been investigated with the objectives of reducing the electrode effects, and the recessed electrode (Fig. 1) was shown to produce results with a much higher dielectric breakdown strength than those produced using planar electrode geometries. The recessed arrangement should ultimately reveal the intrinsic breakdown strength of the material, but only when sufficient tests and statistical analysis can reveal those breakdown events which occurred from pores or inclusions for example. It is clear that, whatever method is chosen, the breakdown stress must be placed entirely within the experimental context. The electrode arrangements that are used conventionally to measure the breakdown of real materials essentially do not permit a measure of a true intrinsic breakdown voltage but rather enable a measure of thermal breakdown or discharge breakdown phenomena. The thermal run-away may give rise to enhanced electronic or ionic conduction losses which generate heat more rapidly than can be dissipated, with ultimate failure through conductive breakdown. Similarly the discharge at the electrode interface or within trapped voids and pores within the material may result in serious enlargement of the discharge volumes leading to breakdown of the insulative properties of the material (especially within an oil medium where discharge sites are characterised by carbourised, and thus conducting, surfaces).

1.5 Weibull Analysis and the Mechanical and Electrical Strength of Piezoelectric Ceramics

Weibull Analysis

The Weibull distribution is an example of extreme value statistical distribution, and has been frequently used to describe the statistical variation in the mechanical strength of ceramics. The Weibull distribution is a special case of the exponential distribution and is empirical [13]—see also [14] and [15] for more recent advances in the statistical theory of breakdown. The concept that underlies the Weibull distribution as applied to the strength of ceramics is that the strength of ceramics is governed by the presence of failure initiating defects. Failure occurs when components are mechanically stressed as the stress at the failure initiating defect rises to the level defined by the Griffiths criterion.

$$\sigma = AK_{Ic}\sqrt{a} \tag{8}$$

where K_{ic} is the materials fracture toughness, a the defect size and A is a constant.

The strength of a component is then governed by the statistical *probability* that the failure initiating defect exists at a given position within the component. It can be shown [3] that if the size distribution of defects is given by $Q(x) = Px^{-n}$, where x is the size of the defect, and P and n are constants, then the Weibull distribution is found to describe the strength of components when combined with the Griffith's criterion. The Weibull modulus m is then given by $m = 2(n-1)$.

Thus the statistical variation in mechanical strength for ceramics is based on the likelihood of encountering failure initiating defects in the body. The underlying basis for the electrical strength of ceramics is not so clear, but there is little doubt that it will be dependent on the statistical distribution of defects in the material in a similar way to mechanical strength. For this reason, the use of the Weibull (or related Gumbel [14]) distribution to describe electrical strength is reasonable. In recent comparison of mechanical and electrical strength measurements, for the piezoelectric ceramics, it is likely that related features is the microstructure cause failure under both electrical and mechanical loading. Thus in both types of loading, the hard piezoelectric ceramic is stronger than the soft piezoelectric ceramic, supporting the microstructural link between the two types of test. It is noticeable that the Weibull moduli determined for the electrical breakdown experiments are lower than those obtained for the mechanical tests. This difference is not unexpected, since whilst the functional dependence of the Weibull modulus on the defect population through the Griffith criterion is well defined for mechanical testing, the equivalent failure criterion for electrical strength is not known.

2 Measurement Good Practice for Dielectric Breakdown of Ceramic Substrate Materials, Including Piezo Ceramics

This measurement good practice section covers the results from Work Package (WP) 10 of an EC Standards, Measurement and Test (SMT) project CERAMELEC, completed in 1999 [2]. This had the purpose of evaluating approaches to electric strength measurement that are more relevant to modern uses of ceramic insulators, specifically substrate materials, than accounted for in IEC standard 60672, part 2. This standard was originally devised in the 1970s to determine material properties for the purposes of applications as large insulators, strictly as a means of product classification and minimum specification. The dielectric breakdown test incorporated into this standard involved the use of a 5–7 mm thick plate of ceramic material into which a spherical cap-shaped depression was ground using diamond tooling, reducing the thickness at the pole of the depression to 1.5–2 mm. The test-piece is coated with an electrically conducting layer inside the depression and also on the back face, and is then sandwiched between spherical electrodes. Two types of testing regime can be applied. Either the applied voltage is ramped until the test-piece conducts ('breakdown test'), or a given voltage can be applied for a given period of time ('withstand test'). The former is used to determine the ultimate performance of the material, the latter maybe used as a pass/fail test for a specification.

A number of problems arise with this test:

1. Machining a depression can introduce flaws which may influence the result of the test compared with an annealed or green-state formed test-piece;
2. Green-state direct forming the test-piece by pressing may result in different properties in the material beneath the depression than surrounding it as a result of

pressure gradients in pressing, so testing specially shaped test-pieces may not be realistic;
3. Many modern materials are not produced in a sufficient thickness to machine such a test-piece; for example, one of the most widespread products is thin ceramic substrates, which may be as thin as 0.2 mm or even less for some applications. It is well known that dielectric breakdown voltage gradients are a function of test-piece thickness, but no studies have been performed to compare the performance of such materials with conventionally sized test-pieces to guide the user;
4. Dielectric breakdown is a statistical event; breakdown is initiated at strong electric field concentrations, such as those occurring within pores or cracks, or even grain boundaries. The tested volume of the material will play some role in determining the risk of breakdown, but existing tests use only 3–5 test-pieces for general quality purposes. Statistical assessment is therefore impossible.

The purpose of this work was thus to examine the behaviour of typical electronic substrate materials (not discussed in this chapter) and commercial piezoelectric ceramic materials in different thicknesses and to evaluate the role of various test-piece preparation, coating and testing parameters on a statistical basis in order to provide background information for the preparation of a standard more appropriate to substrate-like products than the existing method in IEC 60672. Standard piezoelectric ceramics (PZT) were used as materials for test.

2.1 Experimental Test Equipment

High Voltage Facility

In order to perform dielectric breakdown tests safely at the high voltages required in an open laboratory environment, a testing device is required which is designed to operate within an interlocked cabinet, such that no connections with the internal wiring of the cabinet can be made. NPL had previously purchased such a facility from SEFELEC, France, through the UK agents, Wessex Electronics Ltd, Bristol, UK. This equipment comprised a rampable high-tension (HT) DC or AC voltage generator beneath an insulated interlocked cabinet into which live and earth terminals protruded. The ramp function was controlled by a low voltage control signal from an external programming unit. The maximum rated voltage was 50 kV. The instrument was designed automatically to switch off the applied voltage when the breakdown current exceeded 3 mA. The peak breakdown voltage level was retained on a display.

For making the breakdown test, a test-jig was designed in polymethylmethacrylate (PMMA) which enabled a test-piece to be inserted between two spring-loaded opposing spherical metal electrodes, which were connected to the cabinet electrodes using metal braiding. This arrangement was immersed in transformer oil (Shell Diala BG, supplied by Senator Lubricants Ltd of Crawley, W Sussex, UK).

Measurement and Calibration

The design of many commercial instruments impacts on the way in which the breakdown test can be carried out. In the experimental systems developed in this work the following issues were noted and solutions provided:

1. The voltage level on such higher voltage sources is often indicated on a conventional front panel meter, which is not always output as a digital signal. Thus accurate readings are sometimes difficult to record. Various solutions exist if a digital system is not available, including the use of a digital voltmeter integrated with the meter in order to provide a digital read-out.
2. Many instruments have no direct means of traceable calibration of the high-tension voltage. There is implicit reliance on the relationship between HT voltage and the external low-voltage meter reading provided by the manufacturer. It would have been unsafe (as well as invalidating health and safety legislation) to have brought HT leads out from the sealed environment in order to make such measurements with a high-voltage calibrated meter, or a low-voltage meter connected via a high calibrated resistance chain. The possibilities of using a battery operated meter viewed through the glass window of the cabinet were investigated, but do not provide for a useful digital output. Optically isolated digital outputs are another possible solution to taking a calibrated HT voltage reading out of an enclosed safety shielded box.
3. The ramping mechanism employed in some HT supplies is mechanical in origin and sometimes will not provide for consistent voltage ramping rate to breakdown. For example, depending on the maximum voltage setting, at a given fraction of this setting, the ramp rate may change automatically. It was understood from the manufacturers that this was a safety precaution to avoid overshoot. Occasionally this interfered with a test series, some test-pieces breaking down while ramping fast, and others after the change of rate. Since the breakdown voltage under ramping conditions may depend on the ramp rate, and hence the time under electrical stress, this could interfere with the statistical description of the results. The problem was not directly soluble, but its effect was minimised by setting the upper limit breakdown voltage at a level such that the change in ramp rate was mostly clear of actual breakdown values.

Breakdown jig Design

In order to apply a high voltage across the test-piece thickness, a test jig as shown in Fig. 2 was designed and manufactured. The PMMA frame supported the two opposed spherical electrodes, which were 10 mm diameter. One of these electrodes was spring-loaded against the other so that a force was applied to the test-piece between them. This arrangement was immersed in a beaker containing transformer oil. All tests were performed under oil in order to prevent arcing around the test-piece edges.

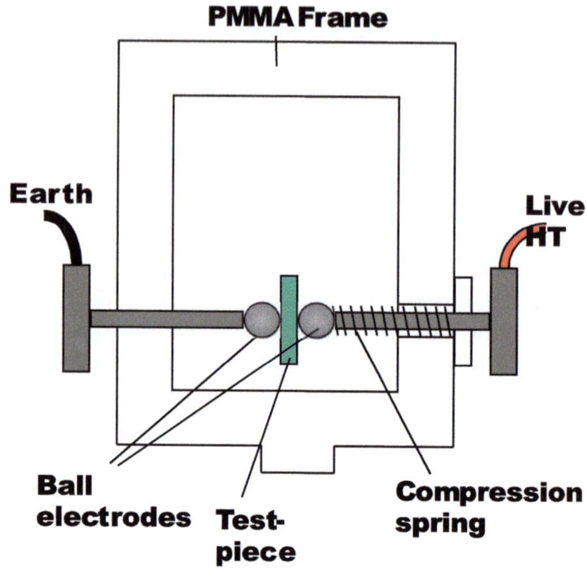

Fig. 2 Test jig for applying high tension voltage to test-pieces

2.2 *Test Materials and Test Piece Preparation*

Test Material

A variety of piezoelectric ceramic materials have been employed for this work, summarised in Table 1. In order to apply the test method to a high permittivity material, some hard lead zirconium titanate (PZT) discs were purchased from Morgan Matroc Ltd, Transducer Products Division, UK, in both electroded and unelectroded conditions.

Test-Piece Preparation

Individual pieces were either used in the as-received condition, or after grinding a spherical cap depression, the latter with and without subsequent annealing at high temperature in order to anneal out grinding damage.

In order to machine the depression, 10 mm diameter ball-ended electroplated diamond grinding tools were purchased from Diagrit Ltd of Stapleford, Kent. The mounting of the tool was specially designed to be accommodated in the high-speed horizontal spindle attachment of a Jones and Shipman toolroom grinder, operating at typically 10,000 rpm. In order to rotate the test-piece under the grinding ball, a milling machine rotary dividing head placed on the surface grinder bed was used as the test-piece support. The axis of the dividing head was visually aligned with the centre of the grinding ball using a lathe centre placed in the centre of the dividing

Table 1 Ceramic materials used in this study

No	Material	Code (NPL)	Supplier	Dimensions		No
				Lateral (Φ)	Thickness (mm)	
5	PZT-5A*	KXB	Morgan Matroc	10	1	100
6	PZT-4D*	KXC	Transducer	10	1	100
5U	PZT-5A**	KZAU	Products	10	1	20
6U	PZT-4D**	KZBU		10	1	20

*Supplied electroded and unpoled **Supplied unelectroded and unpoled

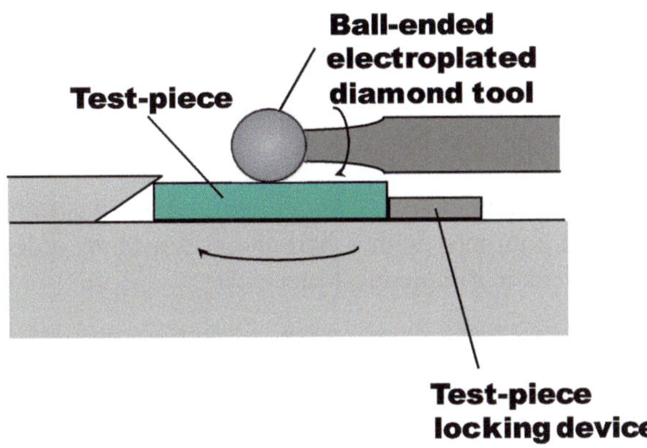

Fig. 3 Arrangement for grinding depressions into test-pieces

head bed as a marker. The grinding machine bed was then clamped into position. A special jig was made to permit substrate test-pieces to be clamped onto the dividing head (Fig. 3). The required depth of depression was achieved by slowly lowering the wheel-head of the grinding machine down into the test-piece while this was rotated using the hand-wheel. Using this arrangement, specimens could be quickly machined to reasonably repeatable depths. Initially, this procedure led to the possibility that a pip was left in the centre of the ground depression because of difficulties in accurately positioning the axis of rotation of the dividing head. Consequently, for later batches the dividing head was not rotated, at the risk that the dimple was not perfectly round.

All test-pieces, whether machined or not, were cleaned in solvents (5 min ultrasonic cleaning immersed first in acetone and then IMS) and dried in a vacuum oven before applying the conducting coating. In order to apply the coating to a limited area of each test-piece, yet process a large number of test-pieces in batches, a method of masking was required. A polymethylmethacrylate jig was designed to accommodate a number of specimens. Each specimen position incorporated a mask on each face with holes in them of the required electrode size, typically 5–20 mm in diameter (Fig. 4). The holes were well countersunk in order to prevent shading near the hole edges, which would otherwise mean variable thickness coatings. Coatings were of aluminium applied by evaporation. The jig was double sided, permitting test-pieces to be coated on both faces with the electrodes coaxial simply by turning the jig over and repeating the evaporation process.

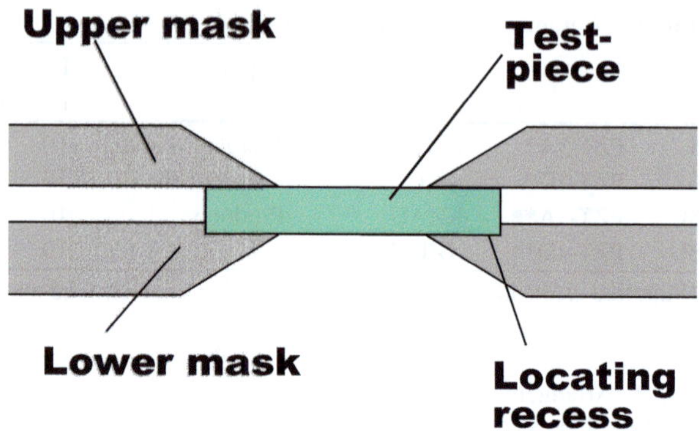

Fig. 4 Schematic of mask for defining the electrode areas on the test-piece faces

Test-piece thicknesses were measured using a digital dial gauge reading to 0.001 mm and equipped with a ball-ended sensor in order to locate the bottom of the depressions in dimpled test-pieces.

Analysis of Data

Weibull Analysis

As discussed, in Sect. 1.5, it is well recognised that dielectric breakdown is a stochastic process that has some parallels with mechanical strength testing in brittle materials, i.e. the weakest link is the one that ultimately fails. In the case of mechanical strength, the largest flaw in the highest stressed position is the most likely site of failure. A key element of this research was to investigate the applicability of Weibull statistics to electric strength measurements.

The Weibull distribution, in its simplest two-parameter form is given by, Eq. (9) [13]:

$$P_f = 1 - \exp\left(\frac{\sigma}{\sigma_0}\right)^m \tag{9}$$

where P_f is the cumulative probability of ranked results, σ is the strength, σ_0 is the characteristic strength, and m is the so-called Weibull modulus, which is a characteristic of the width of the distribution. To visualise the distribution, the series of test results are ranked in ascending order, assigned a probability of failure, and are plotted as $\ln\ln\left(1/(1 - P_f)\right)$ versus $\ln\sigma$. The distribution should then approximate to a straight line. The fitting parameters m and σ_0 can be obtained either by a least-squares fitting routine of a regression line to the plot, or by the preferred maximum likelihood method (as used in ENV 843-5 for mechanical strength testing [16]). For simplicity, this chapter uses primarily the least squares method using the probability ranking parameter, 10:

$$P_f = \left(\frac{i - 0.5}{n}\right) \tag{10}$$

where i is the rank number in ascending order and n is the total number of test data. Other ranking parameters are available with varying degrees of complexity, but the one given above has only a small bias, and is recommended in ENV 843-5.

In order to compute Weibull parameters for the batches of test results, a routine was developed using an Excel spreadsheet designed for the purpose. The parameters were fitted as a straight line relationship (i.e. two-parameter). In some cases parallel calculations were made using the maximum likelihood method. Often the results were rather different, because this second method places less emphasis on the shape of the extremes of the distribution.

Reliability and Significance of Weibull Parameters

The consistency of determining Weibull parameters depends on the sample size. The larger the number of test data, the more reliable the fit. For small numbers of test data, the computed parameters become much less reliable. The 90 % confidence interval on m becomes quite large, e.g. about ±20 % for n = 20, and ±40 % for n = 10. In contrast the computed values of mean strength and characteristic strength σ_0 have much smaller widths of confidence interval, typically less than 5 % for small numbers of test-pieces.

Breakdown Position

After breakdown, breakdown positions were clearly visible, and were measured relative to the centre of the electrode. It was of particular interest to evaluate whether the breakdown occurred predominantly around the edge where a field concentration might be expected, or randomly over the area of the electrode. If the breakdown positions are random, it would be expected that there would be a single value of probability of breakdown per unit area, p. The cumulative probability of failure P in an annulus of radius r and width dr is thus $2\pi p r dr$. Integrating this over the area of the electrode, the total probability of failure to radius r is thus, Eq. (11):

$$p = \int_0^r 2\pi pr\, dr = \pi pr^2 \tag{11}$$

Since $p = 1$ for $r = a$, the radius of the electrode, this reduces to, Eq. (12):

$$P = (r/a)^2 \tag{12}$$

thus, plotting the cumulative probability of failure against (normalised radius)^2 should produce a straight line through the origin with a slope of unity. Failure to approximate to this would imply that the breakdown positions are not randomly distributed, but are concentrated in some way indicated by the plot.

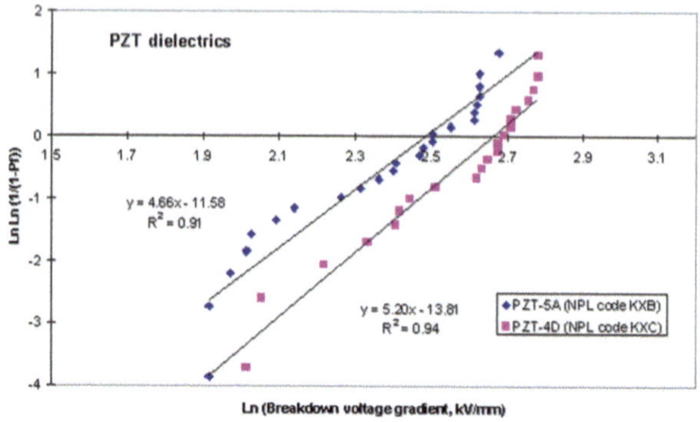

Fig. 5 Weibull distributions for PZT-5A and PZT-4D after removal of supplied electrodes

2.3 Test Results on PZT Ceramics

In the full report [2] both insulating dielectric substrate materials and piezoelectric ceramics were investigated. In this chapter we have focused on the piezo ceramic tests based on the nature of this book. The full series of tests undertaken in this project are summarised in Annex A.

Test 5 was made on the 10 mm diameter, 1 mm thick disc samples after removal of the supplied electrodes which covered the entire disc area, and application of evaporated aluminium electrodes of 5 mm diameter. Tests 5A (PZT-5A), and 5B and 5C (PZT-4D) were made on small numbers of test-pieces. It was found that the electric strength was typically 10 kV/mm, and no edge tracking breakdowns occurred. Test 5C used 1200 grit grinding on one surface before evaporating electrodes, but the results cannot be considered to be any different from those in the as-fired condition, from which it was deduced that internal defects controlled the electric strength. Subsequently, Test series 8A and 8B were made with larger numbers of test-pieces, and the Weibull distributions are plotted in Fig 5. For both materials, the Weibull modulus is rather low, about 5, compared with a high quality alumina substrate for example (also testing in this programme but not reproduced in this chapter), which is presumably a consequence of the different manufacturing processes.

A further batch of the same materials were received from the supplier, this time without electrodes. These were electroded and tested in the same way, and the Weibull distributions are shown in Fig. 6. The Weibull moduli are slightly greater than obtained in the earlier tests, but the mean electric strengths are significantly lower. PZT-4D is still electrically stronger than PZT-5A, as expected.

2.4 Discussion

General Conclusions

Overall, the series of tests have shown that electric strength measurements can have some meaning when applied to thin ceramic test-pieces (<1 mm) with permittivities between about 8 and several 1000. When undertaken in an oil environment, the tests

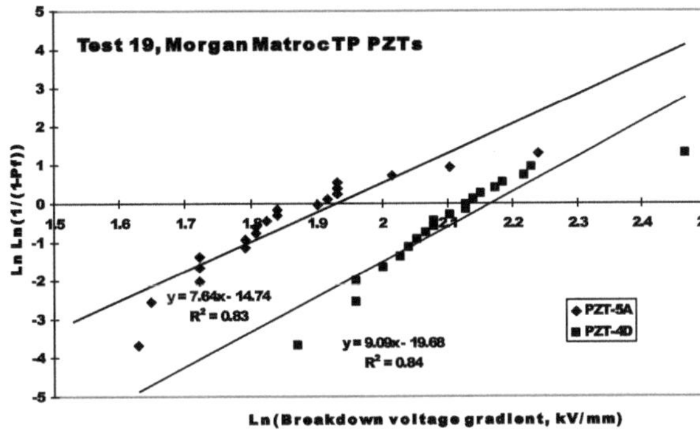

Fig. 6 Weibull distributions for PZT-5A and PZT-4D supplied without electrodes

can be performed without risk of edge discharge even when the electrode extends as close as 2.5 mm from the edge of the test-piece. This work has shown that meaningful differences can be obtained between material types, and even between similar materials treated in different ways, e.g. with different electrode sizes, or with and without annealing treatments.

The use of Weibull statistical analysis has shown that in most cases, the distribution of electric strength data can be represented by a two-parameter Weibull distribution. However, the sizes of the test batches used, typically 16 test-pieces per condition, were too small for consistency of determination of Weibull modulus. As the analysis within ENV 843-5 for mechanical strength demonstrates, the limits of, for example, the 90 % confidence interval on Weibull modulus are large for small sample populations, and the experimental results were scattered even outside these limits, despite good consistency on mean electric strength.

An analysis of alumina ceramic disks (not reproduced in detail in this chapter—but refer to original report [2] identified that the effect of electrode area on breakdown measured is marked, and in both of the cases examined, the electric strength increased with decreasing electrode area. The cumulative probability plot concept developed in this work has value in demonstrating whether a set of results is the consequence of random puncture position, or whether the position is biased in some way. The effect of poor electrode coating produced a very marked example of the value of the analysis, and underlined the need to ensure that the applied coating is continuous with low resistance, i.e. is thickly enough applied. Further, the analysis has also shown that the edge of the electrode is not a preferred site of breakdown. Electric stress concentrations derived in analyses of fields associated with electrode edges do not therefore appear to dominate the site of failure.

The traditional method of undertaking electric strength and withstand tests described in IEC 60672-2 [3] involved the use of dimpled test-pieces. Dimpling has the potential advantage of restricting the thickness in the test area, and thus minimising the risk of flash-over, but requires a thick test-piece. Pressing the dimple into a specially made test-piece may well produce different results from machining the dimple into an otherwise flat test-piece. The present work has shown that dimpling does indeed restrict the location of the breakdown, but gives a higher result for a given thickness than plane electrodes of the same total area as a consequence of the reduced volume under stress. Furthermore, the direction of application of the voltage appears to influence the result. The origins of this behaviour are not clear, but must be related to the direction of current electron flow when breakdown starts. The results of the present study thus show that dimpling is not necessary, and that tests can be readily performed on plane samples with symmetrical disc electrodes.

Problems and Errors in Testing

Although the test data are presented and analysed without reference to errors of measurement, a number of issues arose during testing which may be relevant to setting experimental procedures for testing. These factors are summarised here.

Electroding Application of coaxial electrodes on opposite sides of planar test-pieces requires care and attention to the mounting and masking arrangements employed. It is most convenient if an arrangement can be made which positively locates the test-piece in a mount with coaxial apertures, as indicated in Fig. 4. After coating one side, the mount is simply turned over to coat the other. However, generally this requires precision in machining the mount to suit a given size of test-piece, and thus leads to lack of flexibility in coping with different test-piece sizes. Clearly, mounts for routine testing of particular sizes of test-piece can be made cost-effectively, but a more complex arrangement may be needed to cope with variable-sized components.

The electrodes should have well-defined edges. Intimate contact between the test-piece and the masking aperture is required to avoid tapering the coating excessively at the edges.

The surfaces must be clean and grease-free before electroding. Thin evaporated coatings can be damaged fairly easily, but thicker coatings are more resilient towards accidental handling damage. The current work used aluminium coatings, and some problems arose with poor evaporation conditions resulting in resistive coatings of inadequate thickness. Aluminium has the advantage of easy evaporation at fairly low temperatures, but other types of coating, such as Au/Pd, may be preferable for continuity and resistance to oxidation.

Calibration of HT Voltage It is highly desirable to be able to calibrate the HT source to provide traceability of measurement. However, the system employed gave no direct access to the HT connections which were sealed inside an interlocked cabinet for safety reasons. It is understood that a LT lead system connects the sealed system to a meter on the control panel, which is somewhat inadequate as a means of reading. As supplied, the instrument relies completely on internal correlation of meter reading and HT voltage.

Consideration was given to three means of direct calibration without compromising safety:

1. The use of a battery-powered, separately calibrated digital voltmeter contained completely within the interlocked cabinet. For safety reasons this could not be

digitally connected to an external recording device other than through an optical link;
2. The use of a calibrated HT resistor chain within the interlocked chamber, the voltage over part of which is measured using a low-tension battery energised voltmeter. The same problems with recording would be encountered;
3. Using a pair of polished spheres with a variable but measured gap between them, the HT breakdown voltage could be compared against well-known tables for breakdown of air gaps. However, the difficulty with this method is the need to control the atmospheric temperature, pressure and moisture content.

In view of the time and effort in setting this equipment up, no actions were taken during the course of the project. The data reported are therefore nominal, and reliant on the pre-set relationship between the HT source and the meter reading.

HT Source Control and Measurement The particular instrument employed in the current work generated a ramped voltage at a nominal pre-set rate based on the set peak and a rise time to that peak. On approaching the set peak, integral/derivative type control slowed down the rate to rise to avoid overshoot. This would be important for a withstand test, but for a breakdown test it is highly desirable to maintain the set rate up to failure. With many of the tests requiring voltages approaching the machine limits, some uncertainty in voltage rise rate conditions was introduced. This has been ignored in the analysis. However, it could mean that the duration under electrical stress was longer in some test-pieces than the nominal values. The parallel situation would be to vary the loading rate or displacement rate in mechanical testing, which is never normally undertaken. Correcting for such effects would be impossible without detailed knowledge of the effects of time under electrical stress on breakdown values.

The use of a data-logger measuring the LT response shown on the meter in parallel with the meter was very useful in this work, but highlighted a further potential problem. As breakdown was approached, the apparent voltage rise rate appeared to jump. The effect was variable, but did not seem to occur to any significant extent on the meter. Consequently it was unclear what the true breakdown voltage was. A number of possible approaches to the logged data were evaluated:

1. To use the peak hold value from the peak hold circuit;
2. To take the jump as an indication of the initiation of breakdown, and to use a value which was extrapolated from below the jump to the centre of the logger time interval when the jump was first detected;
3. To take the value immediately after the jump and before any drift in the plateau representing peak hold.

The principle difficulty was in deciding whether the effects seen were entirely instrumental in origin, resulting from noise or spikes in the supply system or from changes in HT circuit impedance as current began to flow in the test-pieces, or were due to the behaviour of the test-piece itself, in particular the way in which breakdown currents accelerated with time. It was suggested by the instrument manufacturer that the earthing of the instrument needed to be improved, but after installation of a separate ground to outside the laboratory, there was little improvement. After considerable discussion and analysis it was decided to rely exclusively on the peak hold values, so all data are reported in this condition, with the proviso that there may be an error of several hundred volts in the recording of each break down voltage level. It was concluded that the whole system needed redesigning in a manner which was suitable for proper control of voltage rise rate and for accurate recording of the voltage levels at breakdown.

Dimpling Although dimpling was found not to be a necessary part of test-piece preparation for the tests, it actually proved to be quite a difficult operation. In particular, setting the position of the dimpling tool such that is centre was on the line of rotation of the test-piece proved quite difficult, with the result that many samples had a slight pip in the centre of the dimple, making thickness measurement difficult. Further, dimpling tools have to be specially made, and the grit size employed in the present case tended to leave grinding marks on the dimpled surface. If the test-piece was not rotated under the tool, the dimple was not round, but adopted the shape of the tool, and grinding marks were strongly directional. If the test-piece was rotated, the dimple was round and there was less directionality to the grinding marks, but the centre pip tended to occur. If dimpling is essential to reduce the thickness of the test-piece to a level capable of being tested with a given instrument, it is important therefore to pay particular attention to the set-up employed and the dimpling procedure adopted.

Breakdown Mechanisms

It was not the purpose of this work to study the mechanisms of breakdown, but some of the test-pieces were examined under the optical microscope after breakdown. Most examples examined showed a single breakdown position, with damage visible on both surfaces. The size of the damaged region tended to be larger on the grounded side than on the HT side. In some examples, however, there was evidence of two breakdown positions, one usually much more severe than the other. Figures 7 and 8 show two examples from an alumina test series, from a plane test-piece and a dimpled test-piece. In the later case, the larger, presumably primary, failure has occurred in the centre of the dimple as expected, but the secondary site is on the parallel part of the test-piece which is 1 mm thick, thicker than can normally be broken down by the HT limits of the apparatus. The reasons for the double event are completely unclear and can only be speculated about.

2.5 Conclusions

An evaluation has been made of the possibilities for developing electric strength test procedures appropriate for modern high-strength and high permittivity ceramic materials used for passive and active electrical purposes. It has been shown that:

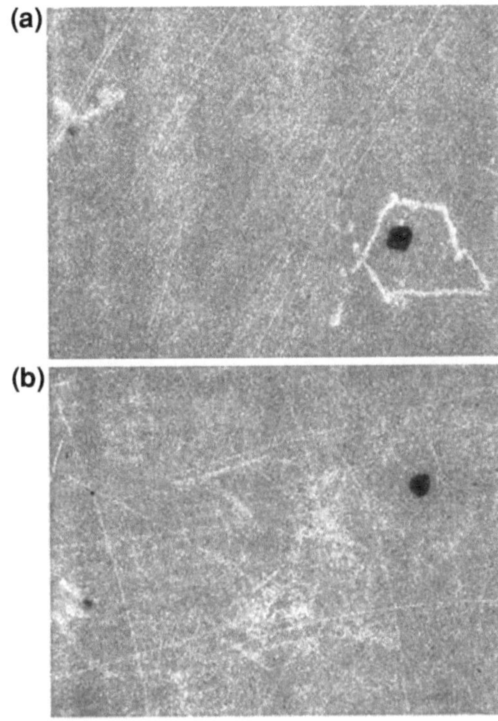

Fig. 7 Optical microscope images of the double breakdown sites in a plane sample from an alumina substrate material as seen from the respective faces

- Thin planar test-pieces ranging in thickness between 1.0 and 0.25 mm and with electric strengths exceeding 100 kV/mm can be electroded and tested in transformer oil without difficulties of edge tracking or flashover;
- Differences in electric strength can be systematically resolved using sufficient test-pieces, and examples have been obtained indicated electrode area effects, thickness effects, and machining/annealing effects;
- Dimpling such test-pieces other than for the purposes of bringing the breakdown voltage within instrumental range has little value, and is unnecessary to prevent edge flashover; producing small dimples in thin test-pieces is not a simple task and requires specialised and precision equipment;
- Test results from a given batch of nominally identical samples have been found to possess a variation in electric strength which is expressible in terms of the two-parameter Weibull distribution analogous to procedures developed to analyse mechanical strength of brittle materials;
- On planar test-pieces, the quality of thin evaporated electrodes must be such that breakdown positions are uniformly distributed over the surface of the electrode, and are not concentrated either at the edge or in the centre; a statistical test for the area distribution has been devised and demonstrated effectively on batches of the same materials with good and bad electroding;

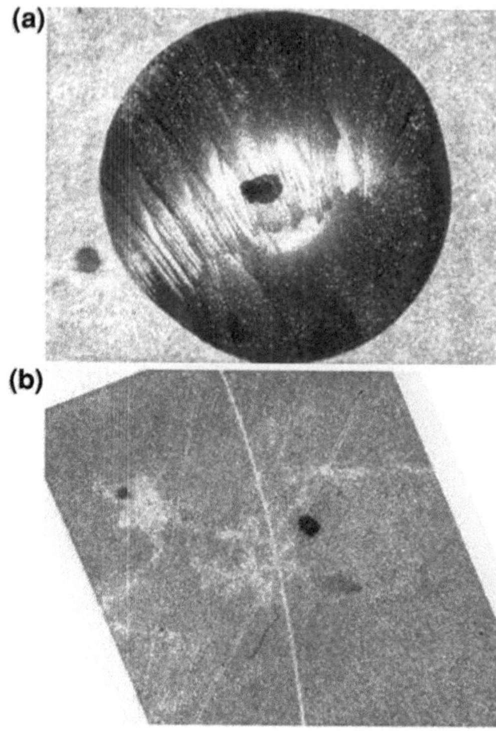

Fig. 8 Optical microscope images of the double breakdown sites in a dimpled sample from alumina substrate material as seen from **a** dimpled side and **b** plane side

- Concerns that breakdowns will occur preferentially around the periphery of the electrodes in planar test-pieces as a result of electric stress concentrations have proved to be unfounded;
- Concerns that have not been satisfactorily dealt with in this project are:
  - Traceable calibration of HT voltage sources contained within safety enclosures;
  - Obtaining consistency in the ramp rate of HT voltage in the instrument employed, in particular whether the response of the LT data-logger added to the test machine was a true reflection of the HT voltage being applied to the test-piece;
  - Explaining the occasional occurrence of double puncturing of the test-piece;
- For testing advanced ceramic products, it is possible to devise a suitable standard with greater flexibility than the existing method given in IEC 60672-2 for traditional ceramic materials.

Acknowledgments The author wishes to thanks collaborators Graham Hill, Roger Morrell, Mark Stewart and Mike Barnett and also to acknowledge the European Commission for funding 'CERAM-ELEC' under the Framework 4 Standards Measurement and Test Programme. Additional support for this Work Package was received from the (then) Engineering Industries Division of Department of Trade and Industry under the Materials Metrology Programme (now National Measurement System, BIS, UK).

Appendix A: Tests on PZT Materials

(See Table 2)

Table 2 Sample breakdown test data

Test no. 5A material: 5	Morgan Matroc TP, PZT-5A (NPL code KXB)
Test-piece dimensions, mm	10 $\Phi \times 1$
Test-piece condition	Supplied electrodes removed
Electrodes	Al, 5 mm Φ
HV ramp rate	0.8 kV/s
Breakdown voltages, kV	11.2, 6.8, 7.3, 12.5, 12.3, 7.7, 13.0, 11.3
Breakdown stress levels, kV/mm	11.1, 6.8, 7.2, 12.2, 12.2, 7.6, 12.8, 11.0
Average breakdown stress level, kV/mm	10.5 ± 2.4
Test no. 5B material: 6	Morgan Matroc TP, PZT-4D (NPL code KXC)
Test-piece dimensions, mm	10 $\Phi \times 1$
Test-piece condition	Supplied electrodes removed
Electrodes	Al, 5 mm Φ
HV ramp rate	0.8 kV/s
Breakdown voltages, kV	11.3, 11.5, 7.5, 14.9, 16.1
Breakdown stress levels, kV/mm	11.1, 11.5, 7.5, 14.7, 16.1
Average breakdown stress level, kV/mm	12.2 ± 4.7
Test no. 5C Material: 6	Morgan Matroc TP, PZT-4D (NPL code KXC)
Test-piece dimensions, mm	10 $\Phi \times 1$
Test-piece condition	Supplied electrodes removed, surfaces ground with 1200 grit SiC paper
Electrodes	Al, 5 mm Φ
HV ramp rate	0.8 kV/s
Breakdown voltages, kV	15.4, 6.6, 7.4
Breakdown stress levels, kV/mm	15.0, 6.6, 7.3
Average breakdown stress level, kV/mm	9.6 ± 4.7
Test no. 8A Material: 5	Morgan Matroc TP, PZT-5A (NPL code KXB)
Test-piece dimensions, mm	10 $\Phi \times 1$
Test-piece condition	Supplied electrodes removed
Electrodes	Al, 5 mm Φ
HV ramp rate	0.8 kV/s
Breakdown voltages, kV	6.8, 8.5, 10.3, 12.2, 14.0, 9.7, 13.9, 12.0, 14.9, 13.8, 10.8, 14.2, 7.6, 8.2, 14.2, 14.0
Breakdown stress levels, kV/mm	6.8, 8.5, 10.1, 11.9, 13.7, 9.6, 13.6, 11.8, 14.5, 13.6, 10.6, 13.8, 7.5, 8.1, 13.8, 13.8
Average breakdown stress level, kV/mm	11.4 ± 2.5
Test no. 8B Material: 6	Morgan Matroc TP, PZT-4D (NPL code KXC)
Test-piece dimensions, mm	10 $\Phi \times 1$
Test-piece condition	Supplied electrodes removed
Electrodes	Al, 5 mm Φ
HV ramp rate	0.8 kV/s

(continued)

Table 2 (continued)

Breakdown voltages, kV	14.1, 13.9, 15.4, 15.1, 15.9, 15.4, 14.7, 7.9, 16.3, 14.3, 9.3, 16.2, 11.3, 10.4, 14.6, 12.5
Breakdown stress levels, kV/mm	13.9, 13.7, 15.2, 15.0, 15.7, 15.0, 14.5, 7.8, 16.1, 14.1, 9.2, 15.9, 11.2, 10.3, 14.5, 12.3
Average breakdown stress level, kV/mm	13.4 ± 2.5

References

1. Scott, J.F.: Nanoferroelectrics: statics and dynamics. J. Phy. Condens Matter. **18**(17), R361–R386 (2006)
2. Barnett, M., Morrell, R., Hill, G.J., Cain, M.G.: "EC SMT Project, CERAMELEC", National Physical Laboratory. Technical Report, CMMT(A)294, NULL (2000)
3. IEC: Ceramic and glass insulating materials—Part 2: methods of test, IEC (1999) (IEC 60672–2 ed. 2.0 b:1999)
4. Standard, A.: Standard test method for dielectric breakdown voltage and dielectric strength of solid electrical insulating materials at commercial power frequencies. Technical Report, West Conshohocken, PA (1994)
5. Zheng, D., Swingler, J., Weaver, P.: Current leakage and transients in ferroelectric ceramics under high humidity conditions. Sens. Actuators A. Phys. **158**(1), 106–111 (2010)
6. Wolak, M.A., Pan, M.-J., Wan, A., Shirk, J.S., Mackey, M., Hiltner, A., Baer, E., Flandin, L.: Dielectric response of structured multilayered polymer films fabricated by forced assembly. Appl. Phys. Lett. **92**(11), 113301 (2008)
7. Boggs, S.: A rational consideration of space charge. IEEE Electr. Insul. Mag. **20**(4), 22–27 (2004)
8. Lebedev, S.M., Gefle, O.S., Pokholkov, Y.P., Gockenbach, E., Borsi, H., Wasserberg, V., Abedi, N., Szczechowski, J.: Influence of high-permittivity barriers on PD activity in three-layer dielectrics. J. Phys. D. Appl. Phys. **37**(22), 3155–3159 (2004)
9. Dawber, M., Rabe, K.M., Scott, J.F.: Physics of thin-film ferroelectric oxides. Rev. Mod. Phys. **77**, 1083–1130 (2005)
10. Bartnikas, R.: Engineering dielectrics. In: Bartnikas, R. (ed.) Electrical Properties of Solid Insulating Materials: Measurement Techniques. ASTM Special Technical Publication 926, Philadelphia (1987)
11. Scott, J.F.: Ferroelectric Memories, Series in Advanced Microelectronics, vol. 3. Springer, Berlin (2000)
12. Milliken, A.D., Bell, A.J., Scott, J.F.: Dependence of breakdown field on dielectric (inter-electrode) thickness in base-metal electroded multilayer capacitors. Appl. Phys. Lett. **90**(11), 112910 (2007)
13. Weibull, W.: A statistical distribution function of wide applicability. J. Appl. Mech. **18**(3), 293–297 (1951)
14. Dissado, L.A.: Theoretical basis for the statistics of dielectric breakdown. J. Phys. D. Appl. Phys. **23**(12), 1582 (1990)
15. Tuncer, E., James, D.R., Sauers, I., Ellis, A.R., Pace, M.O.: On dielectric breakdown statistics. J. Phys. D. Appl. Phys. **39**(19), 4257–4268 (2006)
16. BSI.: Advanced technical ceramics. Mechanical properties of monolithic ceramics at room temperature. Statistical analysis, BSI (2007) (BS EN 843–5:2006)

Standards for Piezoelectric and Ferroelectric Ceramics

Markys G. Cain and Mark Stewart

1 Standards for Piezoelectric and Ferroelectric Materials

This chapter provides detail of standards-related activities for piezoelectric materials—mainly bulk ceramic types. The production of standards, particularly international ones, is a long process. The existence or not of a standard in a particular area is not solely because of the need for such a standard, but is due to the concerted effort of individuals. Documents must be kept current, and may be withdrawn or cancelled if there is nobody prepared to maintain it after publication. There are strong groups for piezoelectric materials standards in Europe (CENELEC) and America (IEEE-UFFC), but some of the most quoted standards have been withdrawn (IEEE 176-1987 [1], IEEE 180-1986, MIL-STD 1376B (SH)). An updated and regularly reviewed set of standards can be found at http://www.piezoinstitute.com.

1.1 Standards Organisations with Piezoelectric Related Standards

- CENELEC (European Committee for Electrotechnical Standardization)—www.cenelec.eu
- IEC (International Electrotechnical Commission)—www.iec.ch
- IEEE-UFFC (Ultrasonics, Ferroelectrics and Frequency Control Society of the Institute of Electrical and Electronics Engineers)—www.ieee-uffc.org
- US Military Standards—www.everyspec.com
- VAMAS (Versailles Project on Advanced Materials and Standards)—www.vamas.org

M. G. Cain (✉) · M. Stewart
National Physical Laboratory, Hampton Road, Teddington, Middlesex TW11 0LW, UK
e-mail: markys.cain@npl.co.uk

M. G. Cain (ed.), *Characterisation of Ferroelectric Bulk Materials and Thin Films*, Springer Series in Measurement Science and Technology 2,
DOI: 10.1007/978-1-4020-9311-1_12,

- EIA (Electronic Industries Alliance)—http://www.eciaonline.org/eiastandards
- JISC (Japanese Industrial Standards Committee)—www.standards.jisc.ac.uk

CENELEC

CENELEC is the European Committee for Electrotechnical Standardization and it has been officially recognised as the European Standards Organisation in its field by the European Commission in Directive 83/189/EEC. A comprehensive series of piezoelectric standards were developed for CENELEC under BTTF 63-2. There were two working groups in this committee and it is WG-2 that was producing standards related to piezoelectric materials, under the stewardship of Wanda Wolny. The first three have been approved, and published, and are available from the various national standards bodies. As with most standards within CENELEC there are English, French and German versions of these standards.

EN 50324-1:2002 Piezoelectric properties of ceramic materials and components

Part 1: Definitions and classifications: This standard relates to piezoelectric transducer ceramics for application both as transmitters and receivers in electroacoustics and ultrasonics over a wide frequency range. They are used for generation and transmission of acoustic signals, for achievement of ultrasonic effects, for transmission of signals in communication electronics, for sensors and actuators, and used for generation of high voltages in ignition devices. Piezoelectric ceramics can be manufactured in a wide variety of shapes and sizes. Commonly used shapes include discs, rectangular plates, bars, tubes, cylinders and hemispheres as well as bending elements (circular and rectangular), sandwiches and monolithic multilayers [2].

EN 50324-2:2002 Piezoelectric properties of ceramic materials and components

Part 2: Methods of measurement and properties—low power: The methods of measurement described in this specification are for use with piezoelectric components produced from the ceramic materials described in EN 50324-1 Definitions and classification. Methods of measurement for specific dielectric, piezoelectric and elastic coefficients are generally applicable to piezoelectric ceramics. The polycrystalline nature of ceramics, statistical fluctuations in composition and the influence of the manufacturing process, result in specified material coefficients being typical mean values. These values are provided for design information only. Piezoelectric transducers can have widely differing shapes and may be employed in a range of vibrational modes. Material parameters however, are measured on simple test-pieces, (discs, rods) using specific geometric and electrical boundary conditions. Consequently, the results of the tests provide basic material parameters only and must be used as a guide to the actual properties of manufactured commercial components [3].

EN 50324-3:2002 Piezoelectric properties of ceramic materials and components

Part 3: Methods of measurement—high power: This standard relates piezoelectric transducer ceramics for power application over a wide frequency range both as electromechanical or mechanoelectrical converters. This standard covers the large signal characterization of piezoelectric ceramics material only, and not the characterization of a complete assembled transducer. The selection of a material for a given power application is difficult and the advice given in Sect. 2 is mainly indicative [4].

prEN 50ZZZ-1 BTTF 63-2 CONV 12 Properties of multilayer actuators

Part 1: Terms and definitions: This European Standard relates to the definitions for Multilayer actuators. Their applications have been widely increased in various field of industry:

- mechanical engineering: tools positioning, clamps, active wedges, damping, active control, generation sonic or ultrasonic vibrations
- microelectronics: positioning of masks, wafers or magnetic heads, non magnetic actuation, circuit breakers
- fluids: proportional valves, pumps, ink jet, droplet generators, injectors
- optics: mirrors or lenses positioning, focusing, laser cavity tuning, alignment or deformation of fiber, scanners, choppers, interferometers, modulators

Multilayer actuators can be manufactured in a wide variety of sizes. The most common shape is the rectangular bar. Ring multilayer actuators exist also. The measurements under prestress apply to the stack actuators. This standard relates to 'd33 actuators, which elongate in the direction of poling' and is limited to the static and quasistatic applications.

prEN 50ZZZ-2 BTTF 63-2 CONV 12 Properties of multilayer actuators

Part 2: Methods of measurement: This standard describes the measurement techniques, which can be applied to measure the properties of Multilayer Actuators, as defined in prEN 50ZZZ-1. Multilayer actuators can be manufactured in a wide variety of sizes. The most common shape is the rectangular bar. Ring multilayer actuators exist also. The measurements under prestress apply to the stack actuators. This standard relates to 'd33 actuators, which elongate in the direction of poling' and is limited to the static and quasistatic applications.

prEN 50PPP (BTTF 63-2(CONV)12) Properties of piezoelectric thick films

IEC

The IEC, the International Electrotechnical Commission is the international standards and conformity assessment body for all fields of electrotechnology. he standards are maintained by a series of technical committees that cover a narrow subject area. Most of the standards developed in the IEC are maintained and produced by TC49, Piezoelectric and dielectric devices for frequency control and selection. At present the only active work item related to piezoelectric ceramics seems to be a glossary, IEC 61994-4-2.

TC-49 Piezoelectric and dielectric devices for frequency control and selection

- IEC 60483 (1976-01) Guide to dynamic measurements of piezoelectric ceramics with high electromechanical coupling
- IEC 60302 (1969-01) Standard definitions and methods of measurement for piezoelectric vibrators operating over the frequency range up to 30 MHz
- IEC 60642 (1979-01) Piezoelectric ceramic resonators and resonator units for frequency control and selection—Chapter I: Standard values and conditions—Chapter II: Measuring and test conditions
- IEC 60642-2 (1994-02) Piezoelectric ceramic resonator units—Part 2: Guide to the use of piezoelectric ceramic resonator units
- IEC 60642-3 (1992-03) Piezoelectric ceramic resonators—Part 3: Standard outlines
- IEC 61253-1 (1993-12) Piezoelectric ceramic resonators—A specification in the IEC quality assessment system for electronic components (IECQ)—Part 1: Generic specification—Qualification approval
- IEC 61253-2 (1993-12) Piezoelectric ceramic resonators—A Specification in the IEC quality assessment system for electronic components (IECQ)—Part 2: Sectional specification—Qualification approval
- IEC 61253-2-1 (1993-12) Piezoelectric ceramic resonators—A specification in the IEC quality assessment system for electronic components (IECQ)—Part 2: Sectional specification—Qualification approval—Sect. 1: Blank detail specification—Assessment level E
- IEC 61261-1 (1994-03) Piezoelectric ceramic filters for use in electronic equipment—A specification in the IEC quality assessment system for electronic components (IECQ)—Part 1: Generic specification—Qualification approval
- IEC 61261-2 (1994-03) Piezoelectric ceramic filters for use in electronic equipment—A specification in the IEC quality assessment system for electronic components (IECQ)—Part 2: Sectional specification—Qualification approval
- IEC 61261-2-1 (1994-03) Piezoelectric ceramic filters for use in electronic equipment—A specification in the IEC quality assessment system for electronic components (IECQ)—Part 2: Sectional specification—Qualification approval—Sect. 1: Blank detail specification—Assessment level E
- IEC 61994-4-2 TS Ed. 1.0 B 1CD Piezoelectric and dielectric devices for frequency control and selection—Glossary—Part 4–2: Piezoelectric materials—Piezoelectric ceramics

TC-87 Ultrasonics

IEC 61088 (1991-09) Characteristics and measurements of ultrasonic piezoceramic transducers. Specifies the essential electroacoustic characteristics of piezoceramic transducers for industrial application of ultrasonic energy. Also specifies the methods of measuring these characteristics. It has the status of a technical report.

SC-47E Subcommittee 47E: Discrete Semiconductor Devices

IEC 60747-14-1 (2000-10) Semiconductor devices—Part 14–1: Semiconductor sensors—General and classification. Describes general items concerning the specifications for sensors which are basically made of semiconductor materials, but also applicable to sensors using dielectric or ferroelectric materials.

IEEE-UFFC

The IEEE-UFFC is the Ultrasonics, Ferroelectrics and Frequency Control Society of the Institute of Electrical and Electronics Engineers. The IEEE published some of the most quoted standards related to piezoelectric and ferroelectric materials, including:

- ANSI/IEEE 176-1987 IEEE Standard on Piezoelectricity [Description].
- 180-1986 IEEE Standard Definitions of Primary Ferroelectric Electric Terms.
- IEEE Std 319-1971 (R1978), IEEE Standard on Magnetostrictive Materials: Piezomagnetic Nomenclature.
- IEEE Std 100-1996, IEEE Standard Dictionary of Electrical and Electronics Terms.

ANSI publications are available from the Sales Department, American National Standards Institute, 1430 Broadway, New York, NY 10018.
IEEE publications are available from the Institute of Electrical and Electronic Engineers, Service Center, 445 Hoes Lane, Piscat- away, NJ 08854.

Both these standards have been withdrawn by the IEEE although they are still available from some sources as a historical document. There are currently two active groups working under the IEEE-UFFC auspices:

- IEEE Subcommittee on Loss in Acoustic Materials.
- Standards on Characterization of Losses in Electromechanical Materials coordinated by Dr. Stewart Sherrit (JPL). The aim is to develop new ways of analysing loss through use of complex impedance and to write new IEEE standard based on the analysis.
- A Draft of the IEEE standard on Ferroelectricity. This is a definition of terms for ferroelctrics and is being developed by Susan Trolier-McKinstry. A draft of this standard has recently been published in the IEEE-UFFC journal [5].

US Military Standards

The US military have specifications and standards relating to the supply of goods for use in defence, including the following standard:

- MIL-STD 1376B (SH) Piezoelectric Ceramic Material and Measurements—Guidelines for Sonar Transducers.

This standard was cancelled in 1999. It is still widely referred to as the source of the various Navy types for piezoelectric materials.

There is also some standards development going on at the UTMR although it is not clear what form these standards will eventually be published as.

- Committee on Standard Protocols for Single Crystal Piezoelectrics. This is coordinated by Dr. Lynn Ewart-Paine (NUWC) on behalf of Dr. Wallace Smith—ONR and is aimed at producing a protocol for the calculation of the dielectric constant and piezoelectric coefficients for piezoelectric single crystals.

VAMAS

VAMAS is the Versailles Project on Advanced Materials and Standards. It supports world trade in products dependant on advanced materials technologies, through International collaborative projects aimed at providing the technical basis for harmonised measurements, testing, specifications, and standards.The pre-standards research activities of VAMAS are organised into technical committees called Technical Working Areas (TWAs) which are approved by the a steering committee and led by International Chairmen. A recent technical working area, TWA 24 Performance Related Properties for Electro-Ceramics is chaired by Dr. Markys Cain (NPL), and there are a number of proposed projects in this area directly related to piezoelectrics. A VAMAS project—An International Intercomparison of Direct Piezoelectric Coefficient using the Berlincourt Method, has been completed in 2008, and its VAMAS report may be downloaded from http://www.vamas.org [6]. The objective was to determine the experimental variability in the measurement of the piezoelectric coefficient of electroceramic materials via the standard method described as the Berlincourt Method.

The VAMAS TWA-24 home page is hosted alternately by NIST and NPL (www.vamas.org).

EIA

The Electronic Industries Alliance (EIA) is a national trade organization that includes the full spectrum of US manufacturers, representing more than 80% of the electronics industry. There are some standards related to multilayer capacitors, including:

- EIA 521 Application Guide for Multilayer Ceramic Capacitors—Electrical.

This document covers capacitor classes I–IV. Definition of important terms are included. Factors influencing performance as temperature, voltage (a.c. and d.c.), temperature-voltage, ageing, and frequency are discussed in detail. Other topics as piezoelectric properties, corona, dielectric absorption, reliability, and applications are described extensively.

JISC (Japanese Industrial Standards Committee)

It is difficult to find authoritative information on Japanese standards related to piezoelectric materials.

The JISC consists of many national committees and plays central role in standardisation activities in Japan. In essence, the role of the JISC can be summarised as:

- Establishment and maintenance of JIS.
- Administration of accreditation and certification.
- Participation and contribution in international standardisation activities.
- Development of measurement standards and technical infrastructure for standardisation.

There are many standards (JIS: Japanese Industrial Standards) that are concerned with quartz resonators, but no national standards for piezoelectric ceramics or single crystals. The Electronic Materials Manufacturers Association of Japan is a not-for-profit organisation that develops and publishes voluntary consensus standards for electronic devices, including piezoelectric ceramic devices. They have published following relevant standards that are only available in Japanese (NB translation may not be exact):

- EMAS-6100 Electronic test methods for the characterisation of piezoelectric ceramic oscillators.
- This is a very comprehensive document covering the determination of piezoelectric properties at low field using the classic resonance technique.
- EMAS-6008 Test method for mechanical performance and chemical resistance.

This standard covers the mechanical inspection, including mechanical three point bend tests, and resistance to chemical attack and performance in various humid environments.

2 Conclusions

This short chapter represents a snapshot of the standards which are most closely related to piezo materials and technologies. Naturally, the development of standards takes some years to take from a draft to a fully accepted published and accepted international standard. The activity is constantly on the move and the reader is guided to the standards pages of the UK's National Physical Laboratory or other National Measurement Institute (NMI) of other countries, listed in Appendix A.

A National Measurement Institutes: Worldwide Listing 2013

The International Bureau of Weights and Measures (BIPM) was set up by the Metre Convention and has its headquarters near Paris, France. It is financed jointly by its Member States and operates under the exclusive supervision of the CIPM (a Mutual Recognition Arrangement (CIPM MRA) for national measurement standards and for calibration and measurement certificates issued by NMIs). Its mandate is to

provide the basis for a single, coherent system of measurements throughout the world, traceable to the International System of Units (SI):
http://www.bipm.org/en/practical_info/useful_links/nmi.html

- www.nist.gov National Institute of Standards and Technology.
- www.npl.co.uk The UK's National Physical Laboratory.
- www.csiro.au The Commonwealth Scientific and Industrial Research Organisation (CSIRO), Australia.
- www.ansi.org American National Standards Institute, United States of America.
- www.health.gov.au Australian Radiation Laboratory, Australia.
- www.cisti.nrc.ca/inms National Research Council Canada, Canada.
- www.inrim.it Istituto Nazionale di Ricerca Metrologica, Italy.
- http://www.cem.es Centro Espaol de Metrologa (CEM), Spain.
- www.metas.ch The Federal Institute of Metrology (METAS), Switzerland.
- http://www.mirs.gov.si/en Metrology Institute of the Republic of Slovenia (MIRS), Slovenia.
- www.mikes.fi Centre for Metrology and Accreditation, Finland.
- http://www.dfm.dtu.dk Danmarks Nationale Metrologiinstitut (DFM), Denmark.
- www.ptb.de/welcome.html Physikalisch Technische Bundesanstalt (PTB), Germany.
- www.bam.de/english.html Federal Institute for Materials Research and Testing (BAM), Germany.
- http://www.nsai.ie National Standards Authority of Ireland, Ireland.
- www.aist.go.jp Agency of Industrial Science and Technology (AIST), Japan.
- https://www.nmij.jp National Metrology Institute of Japan (NMIJ), Japan.
- http://www.aist.go.jp/aist_e/aist_laboratories/6metrology/index.html National Institute of Materials and Chemical Research (NIMC), Japan.
- http://en.nim.ac.cn National Institute of Metrology (NIM), China.
- http://www.kriss.re.kr/eng/global/01_1.html Korea Research Institute of Standards and Science (KRISS), Korea.
- www.sirim.my Standard and Industrial Research Institute of Malaysia (SIRIM), Malaysia.
- http://www.msl.irl.cri.nz Measurement Standards Laboratory, New Zealand.
- http://www.nmc.a-star.edu.sg National Measurement Centre, Singapore.
- www.spring.gov.sg Singapore Institute of Standards and Industrial Research (SISIR), Singapore.
- http://www.dcmtpin1.org/english/index.html South Africa Bureau of Standards (SABS), South Africa.
- http://www.sp.se/en/index/information/rmp/sidor/default.aspx Swedish National Centre of Metrology, Sweden.
- www.sp.se Swedish National Testing and Research Institute, Sweden.
- http://www.itri.org.tw/eng Industrial Technology Research Institute, Taiwan.
- www.oiml.org International Organisation of Legal Metrology (OIML), France.
- http://www.lne.eu/en/metrology/french-metrology.asp (LNE), France.

References

1. IEEE.: IEEE Standard on Piezoelectricity (1978)
2. Cain, M.G.: Piezoelectric properties of ceramic materials and components: part 1: terms and definitions, BSI (2002)
3. Cain, M.G.: Piezoelectric properties of ceramic materials and components: part 2: methods of measurement: low power, BSI (2002)
4. Cain, M.G.: Piezoelectric properties of ceramic materials and components: part 3: methods of measurement: high power, BSI (2002)
5. Meitzler, A., Trolier-McKinstry, S., Ballato, J., Bhalla, A., Bloomfield, P., Cao, W., Cross, L., Dougherty, J., Fousek, J., Guo, R.: Draft 16 of a working document for a proposed standard to be entitled: IEEE standard definitions of terms associated with ferroelectric and related materials, IEEE Trans. Ultrason. Ferroelectr. Freq. Control **50**(12), 1613–1646 (2003)
6. Lodeiro, M.J., Stewart, M., Cain, M.G.: A Round-robin to Measure the Direct Piezoelectric Coefficient Using the Berlincourt Method. VAMAS, NPL (2004)

Index

M. G. Cain (ed.), *Characterisation of Ferroelectric Bulk Materials and Thin Films*, Springer Series in Measurement Science and Technology 2,
DOI: 10.1007/978-1-4020-9311-1, © Queen's Printer and Controller of HMSO 2014

MIX
Papier aus verantwortungsvollen Quellen
Paper from responsible sources
FSC® C105338

If you have any concerns about our products,
you can contact us on
ProductSafety@springernature.com

In case Publisher is established outside the EU,
the EU authorized representative is:
Springer Nature Customer Service Center GmbH
Europaplatz 3, 69115 Heidelberg, Germany

Printed by Libri Plureos GmbH
in Hamburg, Germany